Lecture Notes in Computer Science 15666

Founding Editors

Gerhard Goos
Juris Hartmanis

Editorial Board Members

Elisa Bertino, *Purdue University, West Lafayette, IN, USA*
Wen Gao, *Peking University, Beijing, China*
Bernhard Steffen, *TU Dortmund University, Dortmund, Germany*
Moti Yung, *Columbia University, New York, NY, USA*

The series Lecture Notes in Computer Science (LNCS), including its subseries Lecture Notes in Artificial Intelligence (LNAI) and Lecture Notes in Bioinformatics (LNBI), has established itself as a medium for the publication of new developments in computer science and information technology research, teaching, and education.

LNCS enjoys close cooperation with the computer science R & D community, the series counts many renowned academics among its volume editors and paper authors, and collaborates with prestigious societies. Its mission is to serve this international community by providing an invaluable service, mainly focused on the publication of conference and workshop proceedings and postproceedings. LNCS commenced publication in 1973.

Giltae Song
Editor

Comparative Genomics

22nd RECOMB International Workshop, RECOMB-CG 2025
Seoul, South Korea, April 24–25, 2025
Proceedings

 Springer

Editor
Giltae Song
Pusan National University
Busan, Korea (Republic of)

ISSN 0302-9743 ISSN 1611-3349 (electronic)
Lecture Notes in Computer Science
ISBN 978-3-031-94927-2 ISBN 978-3-031-94928-9 (eBook)
https://doi.org/10.1007/978-3-031-94928-9

© The Editor(s) (if applicable) and The Author(s), under exclusive license
to Springer Nature Switzerland AG 2026

This work is subject to copyright. All rights are solely and exclusively licensed by the Publisher, whether the whole or part of the material is concerned, specifically the rights of translation, reprinting, reuse of illustrations, recitation, broadcasting, reproduction on microfilms or in any other physical way, and transmission or information storage and retrieval, electronic adaptation, computer software, or by similar or dissimilar methodology now known or hereafter developed.
The use of general descriptive names, registered names, trademarks, service marks, etc. in this publication does not imply, even in the absence of a specific statement, that such names are exempt from the relevant protective laws and regulations and therefore free for general use.
The publisher, the authors and the editors are safe to assume that the advice and information in this book are believed to be true and accurate at the date of publication. Neither the publisher nor the authors or the editors give a warranty, expressed or implied, with respect to the material contained herein or for any errors or omissions that may have been made. The publisher remains neutral with regard to jurisdictional claims in published maps and institutional affiliations.

This Springer imprint is published by the registered company Springer Nature Switzerland AG
The registered company address is: Gewerbestrasse 11, 6330 Cham, Switzerland

If disposing of this product, please recycle the paper.

Preface

This volume presents the proceedings of the 22nd International Conference on Comparative Genomics, RECOMB-CG 2025, held at Yonsei University in Seoul, South Korea, on April 24–25, 2025.

All contributions in this book are original research papers. We received 28 submissions from authors affiliated with institutions in 14 countries: Canada, the USA, India, Bangladesh, Japan, Germany, Mexico, France, Israel, Singapore, South Korea, the Netherlands, Pakistan, and Italy. While these affiliations reflect a broad international presence, we believe the authors themselves represent an even more diverse geographic background.

The program committee (PC) carried out an outstanding job in the review process. Each submission was evaluated through approximately three independent reviews. Ultimately, the PC faced the difficult task of selecting 21 papers for oral presentation at the conference. Due to the inherent constraints of a conference setting, several high-quality submissions could not be included in this volume.

The accepted papers span a range of subfields within comparative genomics, including: (1) Cancer genomics, (2) Homology and reconciliation, (3) Phylogenetics, (4) Sequence analysis, (5) Epidemiology, and (6) Genome evolution.

In addition to the research presentations, the conference featured three distinguished invited speakers: Alexander Urban, Stanford University, USA; Choongwon Jeong, Seoul National University, South Korea; and Eun-kyeong Jo, Chungnam National University; South Korea.

We extend our sincere thanks to the program committee members for their dedication and careful evaluation, and to the publication team for their efforts in assembling this volume.

April 2025   Giltae Song

Organization

Program Committee Chair

Giltae Song — Pusan National University, South Korea

Organizing Committee Chair

Jaebum Kim — Konkuk University, South Korea

Steering Committee

Marília Braga	Bielefeld University, Germany
Dannie Durand	Carnegie Mellon University, USA
Jens Lagergren	KTH Royal Institute of Technology, Sweden
Aoife McLysaght	Trinity College Dublin, Ireland
Luay Nakhleh	Rice University, USA
David Sankoff	University of Ottawa, Canada

Program Committee

Lars Arvestad	Stockholm University, Sweden
Mukul S. Bansal	University of Connecticut, USA
Anne Bergeron	Université du Québec à Montréal, Canada
Sèverine Bérard	Université de Montpellier, France
Paola Bonizzoni	Università di Milano-Bicocca, Italy
Marília Braga	Bielefeld University, Germany
Broňa Brejová	Comenius University, Slovakia
Cedric Chauve	Simon Fraser University, Canada
Fabio Pardi	CNRS, France
Miklós Csűrös	University of Montreal, Canada
Daniel Doerr	Heinrich Heine University Düsseldorf, Germany
Nadia El-Mabrouk	University of Montreal, Canada
Oliver Eulenstein	Iowa State University, USA
Guillaume Fertin	University of Nantes, France
Martin Frith	University of Tokyo, Japan

Pawel Gorecki	University of Warsaw, Poland
Wataru Iwasaki	University of Tokyo, Japan
Katharina Jahn	Freie Universität Berlin, Germany
Lingling Jin	University of Saskatchewan, Canada
Flora Jay	University of Paris-Saclay, CNRS, France
Jaebum Kim	Konkuk University, South Korea
Manuel Lafond	Université de Sherbrooke, Canada
Hayan Lee	Temple University, USA
Kevin Liu	Michigan State University, USA
Istvan Miklos	Alfréd Rényi Institute of Mathematics, Hungary
Aïda Ouangraoua	Université de Sherbrooke, Canada
Teresa Przytycka	National Center of Biotechnology Information, USA
Aakrosh Ratan	University of Virginia, USA
Maribel Hernandez Rosales	CINVESTAV Irapuato, Mexico
Michael Sammeth	Coburg University, Germany
Mingfu Shao	Pennsylvania State University, USA
Sagi Snir	University of Haifa, Israel
Giltae Song	Pusan National University, South Korea
Yanni Sun	City University of Hong Kong, China
Wing-Kin Sung	Chinese University of Hong Kong, China
Krister Swenson	CNRS, Université de Montpellier, France
Olivier Tremblay-Savard	University of Manitoba, Canada
Tamir Tuller	Tel Aviv University, Israel
Celine Scornavacca	CNRS, Université de Montpellier, France
Jean-Stéphane Varré	Université de Lille, France
Fábio Henrique Viduani Martinez	Federal University of Mato Grosso do Sul, Brazil
Yong Wang	Academy of Mathematics and Systems Science, China
Yufeng Wu	University of Connecticut, USA
Tomas Vinar	Comenius University, Slovakia
Louxin Zhang	National University of Singapore, Singapore
Jie Zheng	ShanghaiTech University, China

Contents

Cancer Genomics

Novel Driver Mutations in GCB Lymphoma Patients that Affect
Transcription Factors Binding 3
   Ofek Shami-Schnitzer and Tamir Tuller

Inferring Phylogenetic Trees of Cancer Evolution from Longitudinal
Single-Cell Copy Number Profiles 9
   Yushu Liu and Luay Nakhleh

Homology and Reconciliation

Tree Decomposition for Reconstructing Ancestral RNA Sequences
of Multiple Families 27
   Songdi Hu, Vladimir Reinharz, and Olivier Tremblay-Savard

Whole-Genome Duplication Detection with Phylogenomics
Reconciliation: A Scalable Approach 51
   Reza Kalhor, Manuel Lafond, and Celine Scornavacca

A Sankoff-Rousseau-Like Algorithm for Minimizing Lateral Gene
Transfers and Losses on Single Origin Characters 69
   Alitzel López Sánchez, Guillaume E. Scholz, Peter F. Stadler,
   and Manuel Lafond

tMHG-Finder: Tree-Guided Maximal Homologous Group Finder
for Bacterial Genomes 87
   Yongze Yin, Bryce Kille, Huw A. Ogilvie, Todd J. Treangen,
   and Luay Nakhleh

Phylogenetics

Phylogenetic Network Diversity Parameterized by Reticulation Number
and Beyond 107
   Leo van Iersel, Mark Jones, Jannik Schestag, Celine Scornavacca,
   and Mathias Weller

On the Robustness to Gene Tree Rooting (or Lack Thereof) of Triplet-Based
Species Tree Estimation Methods 131
   *Tanjeem Azwad Zaman, Rabib Jahin Ibn Momin,
and Md. Shamsuzzoha Bayzid*

Exact Counts of Binary Phylogenetic Networks with Two and Three
Reticulation Events (Extended Abstract) 137
   Hao Yu and Louxin Zhang

Ancestral Pangenomes and Their Phylogenetic Reconstruction 141
   Xintong Zhou and David Sankoff

QT-WEAVER: Correcting Quartet Distribution Improves Phylogenomic
Analyses Despite Gene Tree Estimation Error 150
   Navid Bin Hasan, Sohaib, and Md. Shamsuzzoha Bayzid

Fast Calculation of Cherry Distance on Level-1 Orchard Networks:
Optimization, Heuristic and Implementation 157
   Kaari Landry and Olivier Tremblay-Savard

Sequence Analysis

Detecting and Mapping Local Model Violations During Biomolecular
Sequence Analysis: a *RE*sampling and *V*isual *EvAL*uation Approach 181
   Meijun Gao and Kevin J. Liu

A Simple Way to Find Related Sequences with Position-Specific
Probabilities 197
   Martin C. Frith

Position Specific Scoring Is All You Need? Revisiting Protein Sequence
Classification Tasks 202
   *Sarwan Ali, Taslim Murad, Prakash Chourasia, Haris Mansoor,
Imdad Ullah Khan, Pin-Yu Chen, and Murray Patterson*

Epidemiology

Residual Immunity and Seasonality of an Epidemic 219
   Siyu Chen and David Sankoff

Adapting the Cov2clusters Tool for Clustering MPOXV Whole Genome
Sequences 231
   *Eric CH Chen, Tara Newman, John Tyson, Anthea Lam, Michael Chan,
Agatha Jassem, Natalie Prystajecky, Shannon Russell, and James Zlosnik*

Genome Evolution

Probability-Based Sequence Comparison Finds the Oldest Ever Nuclear
Mitochondrial DNA Segments in Mammalian Genomes 247
   Muyao Huang and Martin C. Frith

Unraveling Insect Immunity: A Cross-Order Comparative Genomic
Analysis of Key Immune Proteins 251
   Triveni Shelke, Vanika Gupta, and Ishaan Gupta

Analyzing Sequence Similarity Distributions in Salmonidae: A Branching
Process Approach 268
   Yue Zhang

Evolutionary Reconstruction of Hormone-bHLH Regulatory Networks
in Solanaceae: Phylogenomics Insights from PSTVd-Tomato Interactions 284
   Katia Aviña-Padilla, Octavio Zambada-Moreno,
   Manuel A. Barrios-Izás, Michelle Bustamante-Castillo,
   and Maribel Hernández-Rosales

Author Index 293

Cancer Genomics

Novel Driver Mutations in GCB Lymphoma Patients that Affect Transcription Factors Binding

Ofek Shami-Schnitzer[1] and Tamir Tuller[2(✉)]

[1] The George S. Wise Faculty of Life Sciences, Tel Aviv University, 69978 Tel-Aviv, Israel
[2] Department of Biomedical Engineering, The Engineering Faculty, Tel Aviv University, 69978 Tel-Aviv, Israel
tamirtul@tauex.tau.ac.il

Abstract. Mutations in the DNA can affect cancer development and progression not only by changing the amino acid chain but also by affecting different regulatory elements such as transcription factors. This study presents a pipeline to identify mutation blocks—highly mutated genomic regions affecting transcription factors binding. Analysis of GCB lymphoma data revealed 56 mutation blocks affecting expression in key oncogenes such as BCL2, MYC, SGK1, PIM1 and linked to transcription factors such as MSC, TCFL5, HOXB7, FOXP3, ZBTB6. These mutation blocks suggest a selection for mutations that alter gene regulation, contributing to lymphoma development.

Keywords: GCB lymphoma · DLBCL · Transcription factors

1 Introduction

DNA mutations drive cancer not only by altering protein sequences but also by influencing different regulatory mechanisms [1–9].

Transcription factors (TFs) regulate gene expression by binding to specific DNA sequences and can play a key role in cancer by changing the expression of the gene due to mutations [10–14].

BCL2 gene, is a key apoptosis regulator that its overexpression is linked to diffuse large B-cell lymphoma (DLBCL) [15, 16]. The overexpression is often a result of BCL2 (14; 18) (q32; q21) translocation, which increases mutation rates, including synonymous mutations [17, 18]. Germinal Center B Cell-Like (GCB) is a DLBCL subgroup defined by gene expression profiling [19].

A prior study found two synonymous BCL2 mutations affecting MSC repressor binding in GCB lymphoma [8]. Instead of focusing on specific mutations, we identified "mutation blocks"—small genomic regions with high mutation density. Since TFs bind short sequences, we hypothesize that mutations within these regions may similarly impact cancer by altering TFs binding. We analyzed large-scale genomic data, including GCB lymphoma patients, to identify mutation blocks with regulatory potential.

2 Results

2.1 A Pipeline for Finding Mutations Affecting TFs Binding Sites

A multistage data pipeline was developed to identify mutation blocks—genomic regions with high mutation rates, mutations in these regions correlate with changes in gene expression and there is a TF related mechanism that may explain the change. Visualization of the pipeline is available in Fig. 1.

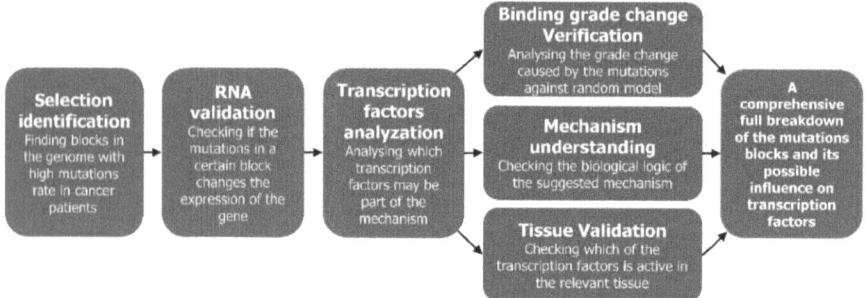

Fig. 1. – a visualization of the pipeline for finding mutations effecting TFs binding sites. Each block in the visualization represents the different stages of the pipeline.

The pipeline output consists of trios linking: (1) **Mutation blocks** with high mutation rates in a cancer type, (2) **Genes** showing altered expression where these blocks are located, and (3) **Transcription factors** affected by these mutations, active in the relevant tissue, and with a plausible regulatory mechanism.

2.2 Systems Biology Study of Mutations Blocks

The pipeline identified 56 mutation blocks in oncogenes with high mutation rates. Patients with these mutations showed significant gene expression changes, with a plausible TF related mechanism for each block. We hypothesize that these mutations alter TFs binding affinity, affecting gene regulation.

Certain genes and TFs appeared multiple times in the results, suggesting a broader role in GCB. Their expression was altered in multiple genomic regions. Figure 2 presents a heatmap of the 56 mutation blocks, their associated genes, and the TFs they may affect. Notably, while each mutation block impacts a single gene, it can influence the binding of multiple TFs.

Several mutations blocks were found in the **genes** BCL2 (14 mutations blocks), MYC (five blocks), SGK1 (five blocks) and PIM1 (three blocks). All of these genes were upregulated by these mutations blocks and all genes were previously shown to have connections to DLBCL [18, 20–23].

Several mutations blocks were found in the following TFs: MSC (six mutations blocks), TCFL5 (four mutations blocks), HOXB7 (three mutations blocks), FOXP3 (three mutations blocks) and ZBTB6 (three mutations blocks). All mutations may lead

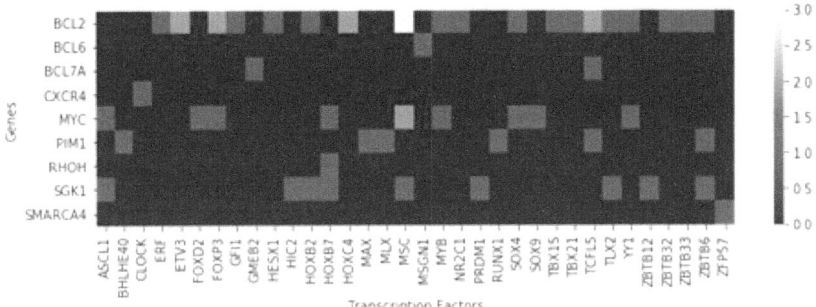

Fig. 2. – A heatmap representing the possible connection between genes and TFs across GCB lymphoma patients. The color intensity indicates the frequency (number of times) where each TF was found to be potentially connected to a mutations block in a specific gene. For example, two mutation blocks in the BCL2 gene were found to be potentially connected to TCFL5, which was also identified as potentially connected to mutation blocks in the BCL7A and PIM1 genes.

to overexpression of the target genes of these TFs due to reduced affinity of repressors or increase binding of an activator. Some of these TFs are known for their connection to cancer and DLBCL specifically[24–27].

Example Analysis: Mutation Block 63319682 – MSC Binding & BCL2 Expression
This mutation block (chr18: 63319682-63319692) in the 5' UTR of BCL2 includes 14 mutations. Patients with these mutations showed significantly higher BCL2 expression ($P = 0.0013$, Fig. 3A), which may contribute to cancer progression.

Analysis of TFs relevant to lymphocytes identified MSC as a key TF. Mutations in this block likely disrupt MSC binding, a known BCL2 repressor, leading to BCL2 overexpression (Fig. 3B). Additional MSC-related mutation block analyses are in Supplementary Data Sect. 2 along with a csv table summarizing all 56 mutations blocks.

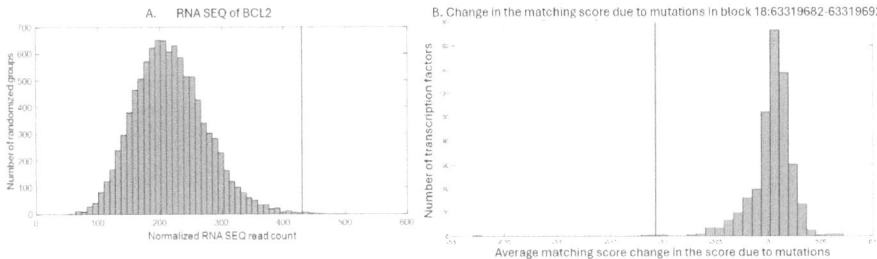

Fig. 3. (A) RNA sequencing data analysis: 10,000 random patient groups were tested for mean BCL2 expression. The red line marks expression in patients with mutations in block 63319682-63319692. (B) Histogram of TF binding changes due to mutations. Most TFs show minimal effects, while MSC (red line) exhibits significant binding disruption. (Color figure online)

3 Discussion

This study highlights the critical role of silent mutations in GCB lymphoma, particularly in TFs binding sites. We identify 56 mutation blocks affecting oncogenes (BCL2, MYC, SGK1, PIM1) and TFs (MSC, TCFL5, HOXB7, FOXP3, ZBTB6), suggesting selective pressure for regulatory disruption. These TFs may serve as biomarkers or therapeutic targets. While correlations between mutation blocks, TFs binding, and gene expression are strong, experimental validation is needed.

Our findings support TFs binding disruption as a key driver in GCB lymphoma, offering a framework for studying regulatory mutations in other cancers.

4 Methods

A pipeline was developed to identify cancer driver mutations affecting TFs binding. It analyzes patient mutations data in stages (see Fig. 1):

Selection Identification: Mutation data from 241 GCB lymphoma patients was acquired [28] and standardized to GRCh38 [29]. Mutation blocks were identified using a 10-nucleotide sliding window, selecting the top 5% most mutated regions (per chromosome) and merging overlapping blocks.

RNA Validation: Using highly mutated blocks and the COSMIC database [28], we identified cancer-related genes within these regions. RNA-seq data from GCB patients was used to compare gene expression between mutated and randomly selected patient groups (10,000 iterations, p-value calculated).

Transcription Factor Analysis: This stage has three sub-steps which together identify relevant TFs significantly impacted by mutations:

- **Binding Change Verification:** TFs binding scores (JASPAR PSSM [29]) were calculated for 746 TFs in mutation blocks comparing scores before and after mutations and identifying significantly affected TFs (z-score > 3 or < -3).
- **Tissue Validation:** GTEx database [30] confirmed which TF is expressed in lymphocytes.
- **Mechanism Validation:** TFs roles (activator/repressor) were assessed via gene ontology [31, 32] to ensure logical correlation with observed gene expression and binding changes.

Systems Biology Analysis: A macro-level analysis aggregated results identified recurring TFs and genes affected across the dataset, revealing broader regulatory trends.

Acknowledgments. The study was partially supported by the koret-berkeley-tau initiative.

References

1. Futreal, P.A., Coin, L., Marshall, M., Down, T., Hubbard, T., Wooster, R., et al.: A census of human cancer genes. Nat. Rev. Cancer **4**, 177 (2004)
2. Diederichs, S., Bartsch, L., Berkmann, J.C., Fröse, K., Heitmann, J., Hoppe, C., et al.: The dark matter of the cancer genome: aberrations in regulatory elements, untranslated regions, splice sites, non-coding RNA and synonymous mutations. EMBO Mol. Med. **8**, 442 (2016). https://doi.org/10.15252/emmm.201506055
3. Gutman, T., Goren, G., Efroni, O., Tuller, T.: Estimating the predictive power of silent mutations on cancer classification and prognosis. npj Genomic Med. **6**, 67 (2021). https://doi.org/10.1038/s41525-021-00229-1
4. Zhang, X., Meyerson, M.: Illuminating the noncoding genome in cancer. Nature Cancer. **1**, 864–872 (2020). https://doi.org/10.1038/s43018-020-00114-3
5. Cartegni, L., Chew, S.L., Krainer, A.R.: Listening to silence and understanding nonsense: exonic mutations that affect splicing. Nat. Rev. Genet. **3**, 285–298 (2002). https://doi.org/10.1038/nrg775
6. Lynn, N., Tuller, T.: Detecting and understanding meaningful cancerous mutations based on computational models of mRNA splicing. npj Syst. Biol. Appl. **10**, 25 (2024). https://doi.org/10.1038/s41540-024-00351-7
7. Gartner, J.J., Parker, S.C.J., Prickett, T.D., Dutton-Regester, K., Stitzel, M.L., Lin, J.C., et al.: Whole-genome sequencing identifies a recurrent functional synonymous mutation in melanoma. Proc. Natl. Acad. Sci. U.S.A. **110**, 13481 (2013). https://doi.org/10.1073/pnas.1304227110
8. Shami-Schnitzer, O., Zafir, Z., Tuller, T.: Novel driver synonymous mutations in the coding regions of GCB lymphoma patients improve the transcription levels of BCL2. In: Bebis, G., Alekseyev, M., Cho, H., Gevertz, J., Rodriguez Martinez, M. (eds.) ISMCO 2020. LNCS, vol. 12508, pp. 108–118. Springer, Cham (2020). https://doi.org/10.1007/978-3-030-64511-3_11
9. Gutman, T., Tuller, T.: Computational analysis of MDR1 variants predicts effect on cancer cells via their effect on mRNA folding. PLoS Comput. Biol. **20**, e1012685 (2024). https://doi.org/10.1371/journal.pcbi.1012685
10. Nebert, D.W.: Transcription factors and cancer: an overview. Toxicology **181–182**, 131–141 (2002). https://doi.org/10.1016/S0300-483X(02)00269-X
11. Bushweller, J.H.: Targeting transcription factors in cancer — from undruggable to reality. Nat. Rev. Cancer **19**, 611–624 (2019). https://doi.org/10.1038/s41568-019-0196-7
12. Morova, T., McNeill, D.R., Lallous, N., Gönen, M., Dalal, K., Wilson, D.M., et al.: Androgen receptor-binding sites are highly mutated in prostate cancer. Nat. Commun. **11**, 832 (2020). https://doi.org/10.1038/s41467-020-14644-y
13. Katainen, R., Dave, K., Pitkänen, E., Palin, K., Kivioja, T., Välimäki, N., et al.: CTCF/cohesin-binding sites are frequently mutated in cancer. Nat. Genet. **47**, 818–821 (2015). https://doi.org/10.1038/ng.3335
14. Kaiser, V.B., Taylor, M.S., Semple, C.A.: Mutational biases drive elevated rates of substitution at regulatory sites across cancer types. PLoS Genet. **12**, e1006207 (2016). https://doi.org/10.1371/journal.pgen.1006207
15. Cory, S., Adams, J.M.: The Bcl2 family: regulators of the cellular life-or-death switch. Nat. Rev. Cancer **2**, 647 (2002)
16. Vaux, D.L., Cory, S., Adams, J.M.: Bcl-2 gene promotes haemopoietic cell survival and cooperates with c-myc to immortalize pre-B cells. Nature **335**, 440 (1988)
17. Lohr, J.G., Stojanov, P., Lawrence, M.S., Auclair, D., Chapuy, B., Sougnez, C., et al.: Discovery and prioritization of somatic mutations in diffuse large B-cell lymphoma (DLBCL)

by whole-exome sequencing. Proc. Natl. Acad. Sci. U.S.A. **109**, 3879 (2012). https://doi.org/10.1073/pnas.1121343109
18. Monni, O., Franssila, K., Joensuu, H., Knuutila, S.: BCL2 overexpression in diffuse large B-cell lymphoma. Leuk. Lymphoma **34**, 45–52 (1999). https://doi.org/10.3109/10428199909083379
19. Blenk, S., Engelmann, J., Weniger, M., Schultz, J., Dittrich, M., Rosenwald, A., et al.: Germinal center B cell-like (GCB) and activated B cell-like (ABC) type of diffuse large B cell lymphoma (DLBCL): analysis of molecular predictors, signatures, cell cycle state and patient survival. Cancer Inform. **3**, 117693510700300000 (2007). https://doi.org/10.1177/117693510700300004
20. Ott, G., Rosenwald, A., Campo, E.: Understanding MYC-driven aggressive B-cell lymphomas: pathogenesis and classification. Blood **122**, 3884–3891 (2013). https://doi.org/10.1182/blood-2013-05-498329
21. Slack, G.W., Gascoyne, R.D.: MYC and aggressive B-cell lymphomas. Adv. Anat. Pathol. **18** (2011)
22. Gao, J., Sidiropoulou, E., Walker, I., Krupka, J.A., Mizielinski, K., Usheva, Z., et al.: SGK1 mutations in DLBCL generate hyperstable protein neoisoforms that promote AKT independence. Blood **138**, 959–964 (2021). https://doi.org/10.1182/blood.2020010432
23. Brault, L., Menter, T., Obermann, E.C., Knapp, S., Thommen, S., Schwaller, J., et al.: PIM kinases are progression markers and emerging therapeutic targets in diffuse large B-cell lymphoma. Br. J. Cancer **107**, 491–500 (2012). https://doi.org/10.1038/bjc.2012.272
24. Ushmorov, A., Leithäuser, F., Ritz, O., Barth, T.F.E., Möller, P., Wirth, T.: ABF-1 is frequently silenced by promoter methylation in follicular lymphoma, diffuse large B-cell lymphoma and Burkitt's lymphoma. Leukemia **22**, 1942 (2008)
25. Errico, M.C., Jin, K., Sukumar, S., Carè, A.: The widening sphere of influence of HOXB7 in solid tumors. Can. Res. **76**, 2857–2862 (2016). https://doi.org/10.1158/0008-5472.CAN-15-3444
26. Martin, F., Ladoire, S., Mignot, G., Apetoh, L., Ghiringhelli, F.: Human FOXP3 and cancer. Oncogene **29**, 4121–4129 (2010). https://doi.org/10.1038/onc.2010.174
27. Zhao, Y., Cui, W., Feng, Z., Xue, J., Gulinaer, A., Zhang, W.: Expression of Foxp3 and interleukin-7 receptor and clinicopathological characteristics of patients with diffuse large B-cell lymphoma. Oncol. Lett. **19**, 2755–2764 (2020). https://doi.org/10.3892/ol.2020.11374
28. Tate, J.G., et al.: COSMIC: the catalogue of somatic mutations in cancer. Nucleic Acids Res., gky1015–gky1015 (2018). https://doi.org/10.1093/nar/gky1015
29. Mathelier, A., Zhao, X., Zhang, A.W., Parcy, F., Worsley-Hunt, R., Arenillas, D.J., et al.: JASPAR 2014: an extensively expanded and updated open-access database of transcription factor binding profiles. Nucleic Acids Res. **42**, D142–D147 (2013). https://doi.org/10.1093/nar/gkt997
30. Lonsdale, J., Thomas, J., Salvatore, M., Phillips, R., Lo, E., Shad, S., et al.: The genotype-tissue expression (GTEx) project. Nat. Genet. **45**, 580–585 (2013). https://doi.org/10.1038/ng.2653
31. Ashburner, M., Ball, C.A., Blake, J.A., Botstein, D., Butler, H., Cherry, J.M., et al.: Gene ontology: tool for the unification of biology. The gene ontology consortium. Nat. Genet. **25**, 25–29 (2000). https://doi.org/10.1038/75556
32. Aleksander, S.A., et al.: The gene ontology knowledgebase in 2023. Genetics **224** (2023). https://doi.org/10.1093/genetics/iyad031

Inferring Phylogenetic Trees of Cancer Evolution from Longitudinal Single-Cell Copy Number Profiles

Yushu Liu and Luay Nakhleh[✉]

Department of Computer Science, Rice University, Houston, TX, USA
{yushu.liu,nakhleh}@rice.edu

Abstract. Understanding evolutionary dynamics is critical for unraveling the complex progression of diseases such as cancer. Cancer evolution is inherently a temporal process driven by the accumulation of mutations and clonal expansions over time. Traditional phylogenetic methods often rely solely on static, cross-sectional data, limiting their ability to infer the timing of key evolutionary events. To address this challenge, we developed NestedBD-Long, a novel method that integrates temporal data from longitudinal sampling into phylogenetic analyses using the birth-death evolutionary model on copy numbers. This approach allows for the direct mapping of real-world time onto inferred evolutionary trees, providing a clearer and more accurate representation of cancer's evolutionary trajectory. Evaluations demonstrate that NestedBD-Long outperforms traditional approaches, with accuracy improving as the number of temporal sampling points increases. This advancement provides a powerful framework for studying tumor progression, treatment resistance, and metastatic spread by capturing the dynamics between evolutionary events and real-world timelines. NestedBD-Long is available at https://github.com/Androstane/NestedBD.

Keywords: Copy number aberrations · single-cell DNA sequencing data · phylogenetic inference · longitudinal analysis

1 Introduction

Copy number alterations (CNAs), involving amplifications or deletions of DNA segments, are key genomic events driving cancer progression [6]. These alterations can disrupt gene dosage, leading to either the overexpression or loss of critical genes, ultimately contributing to uncontrolled cell growth and tumor evolution [3,19]. The study of CNAs is crucial for understanding the mechanisms underlying tumorigenesis and identifying potential therapeutic targets in cancer [8].

Single cell DNA sequencing (scDNAseq) has revolutionized cancer genomics by enabling the study of genomic heterogeneity at the resolution of individual cells [11,27]. This technology provides a detailed view of clonal architecture, capturing the diversity of copy number alterations (CNAs) within tumors and facilitating the reconstruction of their evolutionary trajectories. scDNAseq has been applied to various cancer types, yielding valuable insights into tumor progression and therapy resistance [9,17].

In recent years, researchers have increasingly employed longitudinal single-cell sequencing—where sequencing is performed at multiple time points–to monitor temporal changes in genomic features throughout disease progression. This approach offers a dynamic perspective on tumor evolution, particularly under the selective pressures of treatment [23,30]. By analyzing genomic alterations before, during, and after therapeutic interventions, researchers can track the emergence of resistant subclones, identify critical evolutionary bottlenecks, and refine treatment strategies. Longitudinal sequencing enables the identification of key evolutionary events, such as the rise of therapy-resistant subclones, thus providing opportunities for more precise therapeutic adjustments.

Despite its promise, there remain limited computational methods that fully leverage longitudinal sequencing data in phylogenetic inference. For example, while tools such as LACE and scLongTree [1,15,20] can infer phylogenetic trees from longitudinal single-nucleotide variant (SNV) data, they assume uniform time intervals between consecutive sampling points–failing to account for cases where one interval is significantly longer than another. Moreover, to our knowledge, no existing methodologies are explicitly designed for the inference of longitudinal single-cell CNA data.

In this paper, we address this gap by introducing a novel approach, NestedBD-Long, for inferring phylogenetic trees from longitudinal single-cell CNA data, explicitly incorporating real-world temporal information (e.g., days or months) via a birth-death evolutionary model introduced in [18]. NestedBD-Long is based on NestedBD, utilizing the same evolutionary model and leveraging BEAST2 [2] for Bayesian inference, but extends its capability to handle longitudinal data. This feature enables a time-resolved analysis of tumor evolution, distinguishing our method from existing tools such as MEDICC2 [14] and Lazac [22], which can process CNA data but do not integrate explicit temporal sampling. Figure 1 provides an overview of our proposed method.

We evaluated NestedBD-Long on both simulated and real biological datasets. The simulations were designed to replicate the complexities of tumor evolution across multiple time points, capturing key genomic alterations and clonal dynamics. Real datasets provided a benchmark for assessing the practical applicability of our approach in real-world cancer studies. Our results demonstrate that explicitly incorporating temporal information substantially improves phylogenetic accuracy, particularly as the number of temporal sampling points increases. Moreover, NestedBD-Long consistently outperforms traditional approaches by more accurately reconstructing evolutionary trajectories and providing deeper insights into tumor progression. This capability to map inferred evolutionary time onto real-world temporal scales represents a significant advancement in cancer phylogenetics, facilitating more precise analyses of tumor evolution, treatment response, and resistance mechanisms.

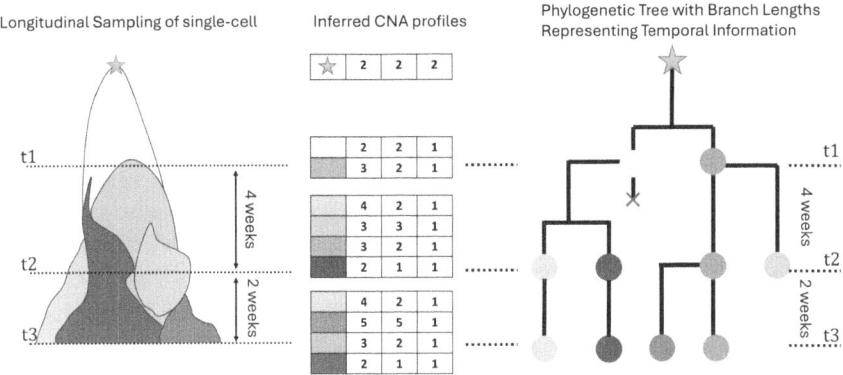

Fig. 1. Overview of NestedBD-Long. From binned copy number profiles collected at multiple time points, the method infers a single-cell phylogenetic tree with branch lengths mapped to real-world time and branch-specific mutation rates.

NestedBD-Long is available at https://github.com/Androstane/NestedBD.

2 Methods

2.1 Temporal Integration of Longitudinal CNA Data

To infer phylogenetic trees from longitudinal CNA data, we employed a birth-death skyline model [10] that incorporates sampled ancestors. While this model was employed in the context of epidemiology and fossil calibration, we adapt it here given the analogy between ancestral samples and fossil data. This model accounts for serially sampled cells from different time points and allows sampled lineages to persist and contribute to subsequent clonal expansions. The model is parameterized by four rates:

- the *birth rate* (λ), representing the probability of speciation;
- the *death rate* (μ), representing the probability of extinction;
- the *sampling rate* (ψ), representing the probability of observing a lineage through sampling; and,
- the *post-sampling removal probability* (r), representing the probability that a sampled lineage is removed from further evolution. Each sampled lineage is either removed with probability r or remains in the population, contributing to future diversification as a sampled ancestor.

This results in three classes of nodes in the reconstructed phylogeny T:

- bifurcation nodes, representing cell divisions,
- sampled tip nodes, corresponding to sampled lineages at the tips; and,
- sampled internal nodes, denoting lineages that were sampled but continued to evolve.

Given the model specification, the probability of observing a phylogenetic tree T can be computed as:

$$P(T \mid \lambda, \mu, \psi, r) = P_{\text{lineages}}(T \mid \lambda, \mu) \cdot P_{\text{sampling}}(T \mid \psi) \cdot P_{\text{removal}}(T \mid r). \quad (1)$$

This equation decomposes the likelihood into three independent terms. The first term, $P_{\text{lineages}}(T \mid \lambda, \mu)$, describes the probability of generating the observed tree topology given the birth and death rates. It accounts for the balance between clonal expansion and extinction, which determines the overall structure of the tree. A higher birth rate relative to the death rate results in larger trees with more bifurcations, while a high extinction rate leads to smaller trees with frequent lineage loss. The second term, $P_{\text{sampling}}(T \mid \psi)$, represents the likelihood of observing a lineage through sampling at different time points. Since tumor samples are collected longitudinally, this term conditions the inference on the observed data, incorporating the probability of detecting a given lineage at a particular time. The third term, $P_{\text{removal}}(T \mid r)$, accounts for whether sampled lineages persist in the tumor or are removed from further evolution. When $r = 1$, all sampled cells are assumed to be terminal, meaning that once a cell is sampled, it does not give rise to any new descendants in the reconstructed tree. When $r < 1$, some sampled cells persist and may be resampled at a later time point, leading to the presence of sampled ancestor nodes in the tree.

To infer phylogenies under this model, we employ a Bayesian inference framework implemented in BEAST2 [2], incorporating the evolutionary model from NestedBD [18]. Specifically, we estimate the posterior distribution of phylogenetic trees using a combination of the birth-death evolutionary model of copy number profiles and the birth-death skyline model leveraging Markov Chain Monte Carlo (MCMC) for parameter inference. The posterior distribution is given by:

$$P(T, R, d \mid D) \propto P(D \mid T, R, d) \cdot P(T \mid \lambda, \mu, \psi, r) \cdot f(R) \cdot f(d), \quad (2)$$

where D is the estimated copy number profile, d is the distance between the common ancestor of all cells and its diploid ancestor, and R is the collection of parameters that define the clock model on T. We specified the sampled ancestor tree model using the Sampled Ancestor package in BEAST2 [10]. The diversification rate $(r_d = \lambda - \mu)$ was assigned a Uniform(0, 1000000) prior, while the extinction fraction $(r_e = \mu/\lambda)$ and the sampling rate (ψ) were assigned Uniform(0, 1) priors. We assumed the post-sampling removal probability $r = 1$, consistent with the lineage removal process described in Sect. 'Longitudinal Sampling' and the biological constraint that a cell, once removed during sequencing, can no longer undergo further evolution. Additional parameter specifications follow those in NestedBD [18].

Given the likelihood, we employ standard tree moves in [2] to explore the tree space. The tree moves used in this study with their weights also follow specification in [18].

While the model allows for rate variation over time, we assume constant rates to mitigate potential parameter identifiability issues. For simulated datasets,

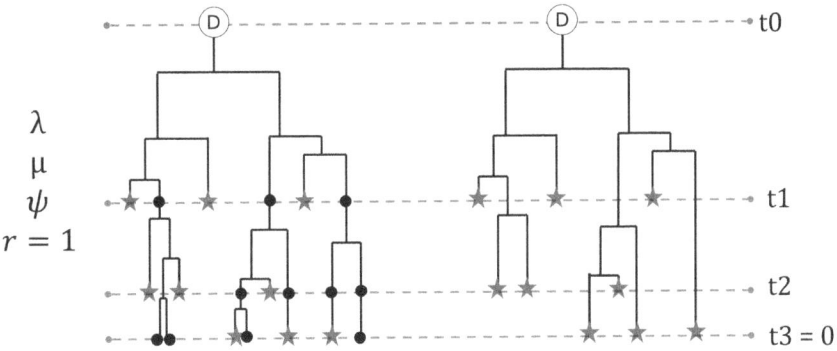

Fig. 2. Full tree versus reconstructed tree. The left panel is a full tree generated by the sampled ancestor birth-death process with a post-sampling removal probability of $r = 1$, meaning all sampled lineages are terminal. The tree follows a birth-death process, where each lineage evolves with a birth rate λ and a death rate μ at each interval. At each sampling time point t_1 to t_3, all existing lineages are independently sampled with probability ψ. Red stars represent sampled nodes that do not contribute to future evolution (or sampled tips at t_3), while black dots represent unsampled lineages that continued to evolve(or unsampled tips at t_3). The right panel shows the corresponding reconstructed tree, which includes only observed sampled nodes and their ancestral relationships. The node labeled D at t_0 represents the healthy diploid cell, serving as the ancestral state before clonal evolution.

node ages were defined by preset sampling times, while for biological datasets, tree heights were rescaled so that the oldest sampled cell was assigned an age of 1. A figure illustrating a sampled ancestor tree versus a reconstructed tree under the specified parameters is shown in Fig. 2.

2.2 Simulation

To evaluate the robustness of NestedBD-Long under diverse conditions, we varied simulation parameters, including temporal sampling intervals, evolutionary scenarios, and noise levels in the simulated CNA profiles. For each parameter combination, we generated 10 independent replicates to ensure statistical reliability. The performance of the methods is then assessed by aggregating results across these replicates, allowing for a comprehensive comparison of accuracy and consistency across different evolutionary scenarios.

Simulating Phylogenetic Trees. As the first step, we generated an ultrametric tree with 1000 extant leaves under a birth-death model with equal birth and death rates using the simulator described in [25]. The root node was set as a diploid state without any CNAs, while each leaf node represented a single-cell sample from the patient. Internal nodes in the simulated trees represented ancestral cells that existed in the past but were not directly sampled. To maintain consistency across simulations, the branch lengths of the generated trees were scaled such that all trees had a total height of 1, ensuring that the expected number of evolutionary events remained uniform regardless of the number of sampled cells. To model potential sampling time points, we introduced additional internal nodes along each lineage at specific heights of $\frac{1}{3}$, $\frac{1}{2}$ and $\frac{3}{4}$ of the tree's height, to represent potential sampling events.

Simulating CNAs. Given an input tree with 1000 leaves, we used the simulator described in [28] to model CNA events along the branches. The simulations were performed with $x = 22$ chromosomes, a mean of $\lambda = 2$ CNAs per edge, and a bin size of $b = 500$ kbp. The number of clones representing distinct subclonal lineages was set to $c \in \{0, 4, 10\}$. Clonality was introduced by selecting ancestral nodes to define diverging subclonal populations, with selection based on subtree size or branch length. Boundary error rates, representing misidentifications at CNA breakpoints, were defined as $r_b \in \{0.08, 0.14\}$ for low-noise and high-noise datasets, respectively. Additionally, a fixed jitter error rate of $r_j = 0.15$ was applied to introduce random fluctuations in copy number states, reflecting read count nonuniformity in scDNA-seq data.

Longitudinal Sampling. Cell sampling was performed from the simulated tree in Sect. 'Simulating Phylogenetic Trees' with 1000 leaves and three potential sampling times set at heights of $\frac{1}{3}, \frac{1}{2}, \frac{3}{4}$ of the tree's height. We varied the number of sampling points, $n = 1, 2, 3, 4$, ensuring that for each n, the current time t_0 (corresponding to the leaves) was always included, along with $n - 1$ additional time points randomly selected from the predefined heights. Sampling was conducted sequentially, from the earliest to the latest time points, by randomly selecting 30 nodes. To mirror biological sequencing, once a cell is selected, its entire lineage, including all descendant cells, is removed from the tree. This process was repeated at each time point until 30 cells were obtained for all designated sampling times. If an insufficient number of cells remained at a later time point due to earlier sampling, the procedure was restarted until a successful sampling process could be completed.

2.3 Evaluating Inference Accuracy

We assessed the topological accuracy of each method by calculating the normalized Robinson-Foulds (RF) distance [21] between the inferred and true topologies. For methods that provide a direct point estimate of the tree topology, we used the inferred tree as is. In cases where a method employs Bayesian inference and offers a posterior distribution rather than a point estimate, we summarized this distribution by taking 2,000 samples from the MCMC chain to compute the inferred topology and branch lengths. We then determined the maximum clade credibility tree (MCC), defined as the tree with the highest product of posterior clade probabilities, and summarized branch lengths using the median node heights across the samples.

3 Results

To our knowledge, no existing method explicitly infers phylogenetic trees from longitudinal single-cell CNA data. While LACE [1] and scLongtree [15] incorporate longitudinal information, they use SNVs rather than CNAs as input and are therefore not applicable to this study (though generating CNA and SNV data from the same sample and comparing the methods is a direction for future research). Consequently, we compare NestedBD-Long to MEDICC2 [13], Lazac [22] and NestedBD [18], which infer phylogenetic trees from single-cell CNA data but do not explicitly model longitudinal sampling. We evaluate the performance of these methods on simulated datasets by assessing the accuracy of inferred trees under varying sampling time points, clonality levels, and CNA profiling error rates. Additionally, we apply these methods to biological single-cell copy number profiles from TNBC patients to compare their inferred phylogenies and examine how well they capture temporal structure.

3.1 Performance on Simulated Data

Accuracy with Varying Sampling Time Points. We evaluated the performance of each method on simulated copy number profiles by computing the normalized RF distance between the inferred and ground truth trees. This analysis assesses the impact of increasing the number of sampling time points on topological reconstruction accuracy. The results are summarized in Fig. 3.

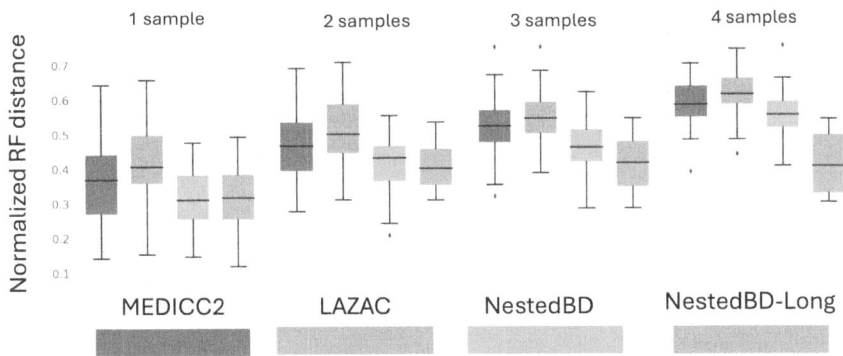

Fig. 3. Accuracy of the inferred trees on the simulated data by number of sampling time points. The box plots summarize the distributions of normalized RF distances between the ground truth and inferred trees for each method using 10 replicates.

The results indicate that the Bayesian methods NestedBD and NestedBD-Long outperform the distance-based methods Lazac and MEDICC2. Notably,

NestedBD-Long maintains consistent performance even as the number of longitudinal sampling points increases, underscoring the benefits of integrating temporal information into tree inference. Specifically, NestedBD-Long maintains a narrow range of normalized RF distances across various sampling points, whereas other methods experience wider variability and increased RF distances as the number of sampling points grows. The gap between methods opens up as more sampling points are introduced, highlighting the challenges non-Bayesian methods face in handling the added complexity. In summary, these findings demonstrate that leveraging temporal information through NestedBD-Long can improve the accuracy of phylogenetic analyses in longitudinal studies.

Impact of Clonality. We examined how clonality influences the performance of each method. The results are summarized in Fig. 4. Notably, no clear trend emerges between clonality and tree accuracy, likely because none of the models explicitly incorporates clonality into their inference process. As a result, the performance of all methods remains relatively consistent across different clonality levels. NestedBD-Long exhibits particularly low variability, demonstrating robustness in handling both clonal and non-clonal data. In contrast, MEDICC2 and Lazac display higher variability, especially at increased clonality levels, suggesting less stable performance. These findings highlight that while clonality does not directly affect tree accuracy, maintaining stable performance across varying clonality levels is crucial for reliable phylogenetic inference, particularly when clonality is not explicitly modeled.

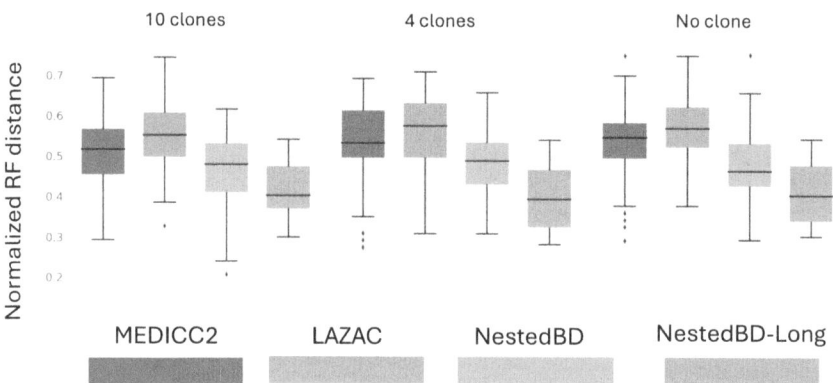

Fig. 4. Accuracy of the inferred trees on the simulated data by Clonality. The box plots summarize the distributions of normalized RF distances between the ground truth tree and the inferred trees for each method using 10 replicates.

Impact of Error in CNA Profiles. The effect of varying error levels in CNA profiles on tree inference accuracy is summarized in Fig. 5. As expected, all methods exhibit increased normalized RF distance with higher error levels, indicating a decline in reconstruction accuracy as data errors increase. NestedBD-Long and NestedBD demonstrate greater robustness, maintaining consistently lower RF distances across different error conditions. In contrast, Lazac shows significant variability, particularly struggling in high-error conditions, suggesting reduced reliability in noisy data. These results indicate that Bayesian methods like NestedBD-Long and NestedBD offer improved stability and resilience to errors, making them more suitable for phylogenetic inference in error-prone single-cell datasets.

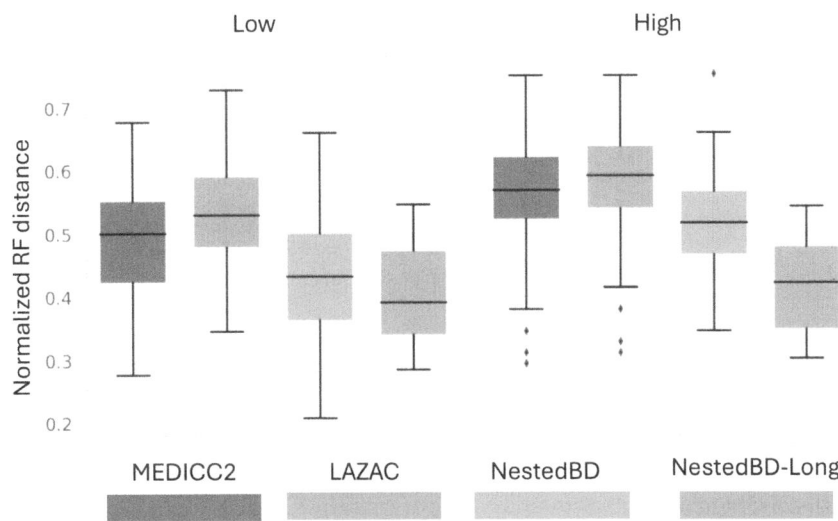

Fig. 5. Accuracy of the inferred trees on the simulated data by level of error in CNA profile. The box plots summarize the distributions of normalized RF distances between the ground truth tree and the inferred trees for each method using 10 replicates. 'High' and 'Low' correspond to the error levels in the simulated CNA profiles (see Sect. Simulating CNAs).

3.2 Performance on Biological Data

We applied all four phylogenetic inference methods to single-cell copy number profile datasets from two TNBC patients, KTN132 and KTN152, obtained from [16]. These datasets contain longitudinal samples from two and three time points, respectively. For each dataset, NestedBD and NestedBD-Long were run for 80 million iterations, with the first 20% of samples discarded as burn-in. The posterior distribution was summarized using the MCC tree, with median node heights. To assess how phylogenetic structure aligns with sampling time points, we computed the pairwise distances between nodes sampled at the same time point and across different time points, normalizing by the largest distance for each method. The results are summarized in Fig. 6.

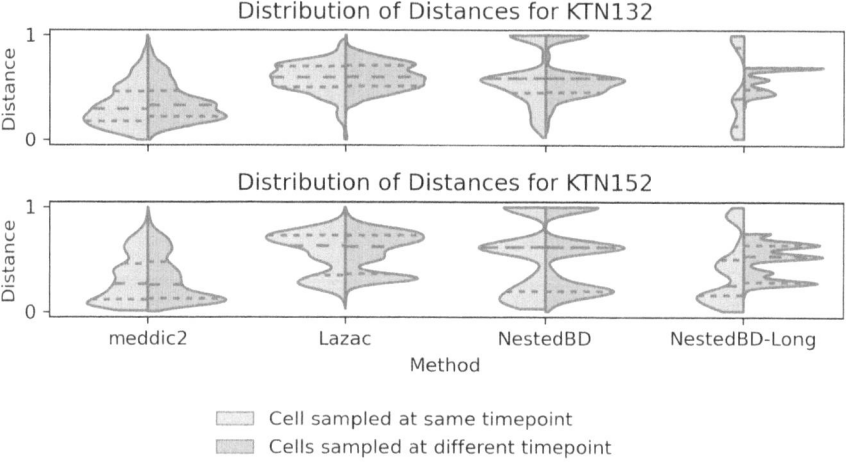

Fig. 6. **Normalized Pairwise Distance Distributions by Sampling Timepoints.** This violin plot illustrates the distribution of normalized pairwise distances between nodes sampled at the same and different time points. The analysis is conducted using four methods—MEDICC2, Lazac, NestedBD, and NestedBD-Long—on single-cell copy number profiles from TNBC patients KTN132 and KTN152 from [16].

For methods that do not explicitly incorporate temporal information, including MEDICC2, Lazac and NestedBD, the distribution of phylogenetic distances between samples from the same and different time points are largely overlapping. It is worth noting that NestedBD inferred a tree with pairwise distances grouped into two or three peaks, which may reflect differences in how it handles rate variation with the birth-death evolutionary model compared to the distance-based method. In contrast, NestedBD-Long, which explicitly models temporal constraints, produces markedly different distributions for pairwise distances between cells sampled at the same time point and those sampled at different time points. This suggests that integrating sampling time allows NestedBD-Long to better capture temporal structure in evolutionary relationships. These results highlight the potential impact of incorporating temporal information when analyzing longitudinal single-cell data.

For each inferred tree, we annotated branches defining major cell clades with cancer-related oncogenes and tumor suppressor genes (TSGs) affected by CNAs, based on data from [26]. The phylogenies inferred using NestedBD-Long for KTN132 and KTN152 are shown in Fig. 7 and Fig. 8.

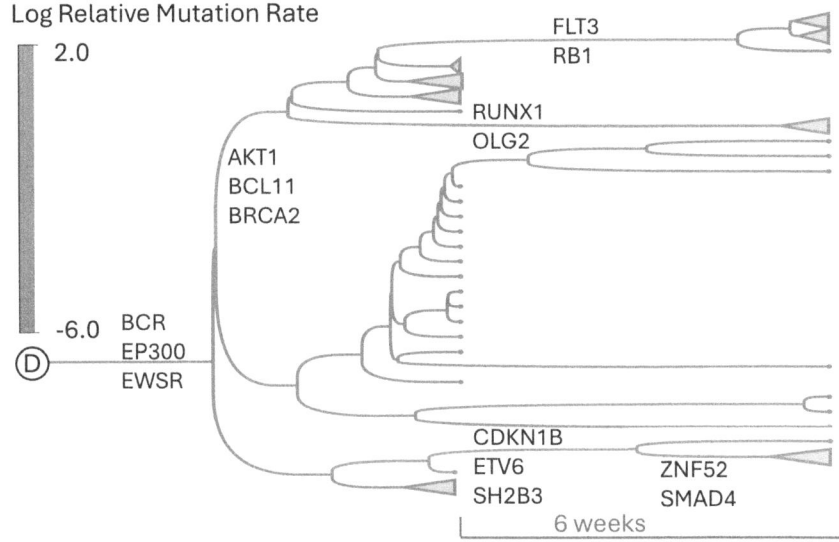

Fig. 7. Inferred tree using NestedBD-Long for TNBC patient KTN132. Phylogenetic tree inferred using NestedBD-Long for patient KTN132. Branches defining major cell clades are annotated with TNBC-related oncogenes and tumor suppressor genes (TSGs) impacted by CNAs.

It is worth noting in both figures, branches leading to major clones between sampling points consistently show high mutation rates, indicating rapid clonal evolution and genomic alterations within short intervals. Without longitudinal sampling, scaling mutation rates per unit time would be challenging, risking misinterpretation of clonal expansion dynamics. By integrating multiple time points, NestedBD-Long enables precise temporal reconstruction, distinguishing fast-growing clones from slower-diverging lineages. This is crucial for understanding the timing of key oncogenic events and their role in tumor progression.

For patient KTN132, the inferred tree highlights a subclonal lineage with the RB1 mutation. The high mutation rate on this branch suggests that RB1 loss, known to promote metastasis [12], may have facilitated rapid clonal expansion. We also observed another subclone with the SMAD4 mutation emerging after the first treatment. Its role in drug resistance [29] aligns with the fact that the mutation occurred later in time, likely as a response to treatment. A similar trend is observed in patient KTN152, where branches with key mutations and rapid clonal expansion exhibit increasing mutation rates. Notably, a subclone carrying a TCF3 mutation, implicated in the control of breast cancer growth and initiation [24], shows accelerated mutation accumulation, resembling the pattern seen

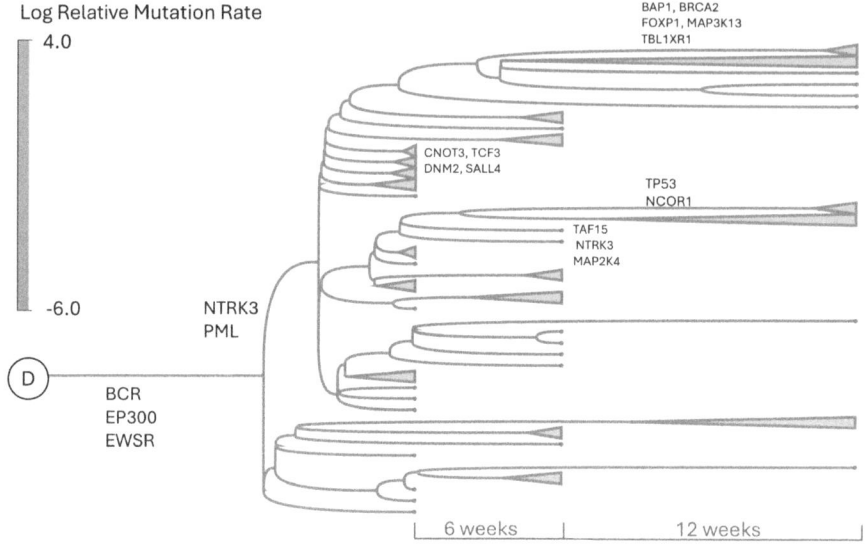

Fig. 8. **Inferred tree using NestedBD-Long for TNBC patient KTN152.** Phylogenetic tree inferred using NestedBD-Long for patient KTN152. Branches defining major cell clades are annotated with TNBC-related oncogenes and tumor suppressor genes (TSGs) impacted by CNAs.

in KTN132. Additionally, mutations associated with drug resistance including SALL4, FOXP1, and DNM2, emerge at later time points, suggesting a potential response to treatment-induced selective pressures [4,5,7]. This temporal pattern suggests distinct selective pressures acting at different stages of tumor evolution, and reinforces incorporating time information in phylogenetic inference is essential for distinguishing early mutations that drive metastasis from later adaptations to therapeutic pressure.

4 Discussion

Longitudinal single-cell sequencing has become an essential tool for tracking tumor evolution and treatment response by capturing dynamic changes in clonal architecture over time. However, existing phylogenetic methods often overlook the importance of explicitly incorporating sampling time, potentially limiting their accuracy in reconstructing evolutionary trajectories. In this study, we introduced NestedBD-Long, which applies the birth-death skyline model to CNA data, integrating temporal sampling to improve evolutionary inference and address limitations in existing methods.

Our evaluations on simulated datasets demonstrate that as the number of temporal sampling points increases, the gap between methods opens up, with NestedBD-Long consistently achieving the highest accuracy. Unlike methods that do not incorporate temporal data, which show increased variability and declining performance, NestedBD-Long maintains stable and precise phylogenetic reconstructions. These results highlight the advantage of explicitly modeling temporal information, particularly in complex evolutionary settings where accurate lineage tracing is critical. On biological datasets, we observed that NestedBD-Long produces distinct distributions of pairwise distances between cells sampled at the same and different time points, demonstrating its ability to capture temporal structure in evolutionary relationships. In contrast, methods that do not account for sampling time exhibit overlapping distance distributions, suggesting a lack of temporal resolution. Furthermore, by following longitudinal constraints, NestedBD-Long can provide more biologically meaningful insights into clonal evolution.

While NestedBD-Long improves temporal resolution in tumor phylogenies, future work could further refine its inference by incorporating additional evolutionary factors such as selection pressures and subclonal interactions. Additionally, extending this framework to infer phylogenies from SNV data or enabling joint inference from both SNV and CNA data could provide a more comprehensive view of tumor evolution. Overall, these findings emphasize the importance of integrating sampling time into phylogenetic reconstruction. By explicitly modeling longitudinal data, NestedBD-Long enhances the accuracy and stability of phylogenetic inference, making it a valuable tool for studying cancer evolution and treatment response.

Acknowledgements. This research was supported in part by the National Science Foundation, grants IIS-1812822 and IIS-2106837 (L.N.).

Disclosure of Interests. The authors have no competing interests to declare that are relevant to the content of this article.

References

1. Ascolani, G., et al.: Lace 2.0: an interactive r tool for the inference and visualization of longitudinal cancer evolution. BMC Bioinform. **24**(1), 99 (2023)
2. Bouckaert, R., et al.: Beast 2.5: an advanced software platform for Bayesian evolutionary analysis. PLoS Comput. Biol. **15**(4) (2019). https://doi.org/10.1371/journal.pcbi.1006650, https://www.scopus.com/inward/record.uri?eid=2-s2.0-85065051451&doi=10.1371%2fjournal.pcbi.1006650&partnerID=40&md5=f0893e29c605d259e243757ee60124c6, cited By 588
3. Burrell, R.A., McGranahan, N., Bartek, J., Swanton, C.: The causes and consequences of genetic heterogeneity in cancer evolution. Nature **501**(7467), 338–345 (2013)
4. Chen, Y.Y., et al.: Knockdown of sall4 inhibits the proliferation and reverses the resistance of MCF-7/ADR cells to doxorubicin hydrochloride. BMC Mol. Biol. **17**, 1–11 (2016)
5. Chernikova, S.B., et al.: Dynamin impacts homology-directed repair and breast cancer response to chemotherapy. J. Clin. Investig. **128**(12), 5307–5321 (2018)

6. Ciriello, G., Miller, M.L., Aksoy, B.A., Senbabaoglu, Y., Schultz, N., Sander, C.: Emerging landscape of oncogenic signatures across human cancers. Nat. Genet. **45**(10), 1127–1133 (2013)
7. De Silva, P., et al.: Foxp1 negatively regulates tumor infiltrating lymphocyte migration in human breast cancer. EBioMedicine **39**, 226–238 (2019)
8. Fisher, R., Pusztai, L., Swanton, C.: Cancer heterogeneity: implications for targeted therapeutics. Br. J. Cancer **108**(3), 479–485 (2013)
9. Funnell, T., et al.: Single-cell genomic variation induced by mutational processes in cancer. Nature **612**(7938), 106–115 (2022). https://doi.org/10.1038/s41586-022-05249-0
10. Gavryushkina, A., Welch, D., Stadler, T., Drummond, A.J.: Bayesian inference of sampled ancestor trees for epidemiology and fossil calibration. PLoS Comput. Biol. **10**(12), e1003919 (2014)
11. Gawad, C., Koh, W., Quake, S.R.: Single-cell genome sequencing: current state of the science. Nat. Rev. Genet. **17**(3), 175–188 (2016)
12. Jones, R.A., et al.: Rb1 deficiency in triple-negative breast cancer induces mitochondrial protein translation. J. Clin. Investig. **126**(10), 3739–3757 (2016)
13. Kaufmann, T.L., et al.: Medicc2: whole-genome doubling aware copy-number phylogenies for cancer evolution. Genome Biology **23**(1) (2022). https://doi.org/10.1186/s13059-022-02794-9
14. Kaufmann, T.L., et al.: Medicc2: whole-genome doubling aware copy-number phylogenies for cancer evolution. Genome Biol. **23**(1), 241 (2022)
15. Khan, R., Mallory, X.: sclongtree: an accurate computational tool to infer the longitudinal tree for scdnaseq data. bioRxiv pp. 2023–11 (2023)
16. Kim, C., et al.: Chemoresistance evolution in triple-negative breast cancer delineated by single-cell sequencing chemoresistance evolution in triple-negative breast cancer delineated by single-cell sequencing. Cell **173**(4) (2018). https://doi.org/10.1016/j.cell.2018.03.041
17. Leighton, J., Hu, M., Sei, E., Meric-Bernstam, F., Navin, N.E.: Reconstructing mutational lineages in breast cancer by multi-patient-targeted single-cell DNA sequencing. Cell Genomics **3**(1) (2023)
18. Liu, Y., Edrisi, M., Ogilvie, H.A., Nakhleh, L.: NestedBD: Bayesian inference of phylogenetic trees from single-cell DNA copy number profile data under a birth-death model. Algorithms Mol. Biol. **19**, 18 (2024)
19. McGranahan, N., Swanton, C.: Biological and therapeutic impact of intratumor heterogeneity in cancer evolution. Cancer Cell **27**(1), 15–26 (2015)
20. Ramazzotti, D., et al.: Lace: Inference of cancer evolution models from longitudinal single-cell sequencing data. J. Comput. Sci. **58**, 101523 (2022)
21. Robinson, D., Foulds, L.: Comparison of phylogenetic trees. Math. Biosci. **53**(1–2), 131–147 (1981). https://doi.org/10.1016/0025-5564(81)90043-2
22. Schmidt, H., Sashittal, P., Raphael, B.J.: A zero-agnostic model for copy number evolution in cancer. bioRxiv (2023). https://doi.org/10.1101/2023.04.10.536302, https://www.biorxiv.org/content/early/2023/04/12/2023.04.10.536302
23. Sharma, A., et al.: Longitudinal single-cell RNA sequencing of patient-derived primary cells reveals drug-induced infidelity in stem cell hierarchy. Nat. Commun. **9**(1), 4931 (2018)
24. Slyper, M., et al.: Control of breast cancer growth and initiation by the stem cell-associated transcription factor TCF3. Can. Res. **72**(21), 5613–5624 (2012)
25. Sukumaran, J., Holder, M.T.: DendroPy: a Python library for phylogenetic computing. Bioinformatics **26**(12), 1569–1571 (2010). https://doi.org/10.1093/bioinformatics/btq228, https://doi.org/10.1093/bioinformatics/btq228

26. Tate, J.G., et al.: COSMIC: the catalogue of somatic mutations in cancer. Nucleic Acids Res. **47**(D1), D941–D947 (2018). https://doi.org/10.1093/nar/gky1015
27. Wang, Y., et al.: Clonal evolution in breast cancer revealed by single nucleus genome sequencing. Nature **512**(7513), 155–160 (2014)
28. Weiner, S., Bansal, M.S.: Cnasim: improved simulation of single-cell copy number profiles and DNA-Seq data from tumors. Bioinformatics **39**(7), btad434 (2023). https://doi.org/10.1093/bioinformatics/btad434
29. Xu, W., et al.: Association of smad4 loss with drug resistance in clinical cancer patients: a systematic meta-analysis. PLoS ONE **16**(5), e0250634 (2021)
30. Zhang, K., et al.: Longitudinal single-cell RNA-Seq analysis reveals stress-promoted chemoresistance in metastatic ovarian cancer. Science Advances **8**(8), eabm1831 (2022)

Homology and Reconciliation

Tree Decomposition for Reconstructing Ancestral RNA Sequences of Multiple Families

Songdi Hu[1], Vladimir Reinharz[2], and Olivier Tremblay-Savard[1]($\boxtimes$)

[1] University of Manitoba, Winnipeg, MB R3T 2N2, Canada
olivier.tremblay-savard@umanitoba.ca
[2] Université du Québec à Montréal, Montréal, QC H2X 3Y7, Canada

Abstract. Ancestral sequence reconstruction aims to infer the content of certain biological sequences of interest for ancestral species by comparing extant sequences. Since the search space is quite large, a lot of research has been devoted to the design of efficient and accurate methods to solve different variations of this problem. However, ancestral sequence reconstruction becomes even more complex when the goal is to reconstruct the ancestors of sequences that are not well conserved in extant species. This is the case with non-coding RNA (ncRNA) sequences, for which the structure (formed by base pairing) is more conserved than the actual sequences. One recent approach to tackle the ancestral reconstruction of ncRNA sequences involved considering the sequences of two related ncRNA families simultaneously [26]. Although this helped avoid biases in the reconstruction, some cost calculations had to be simplified for efficiency. In this work, the goal is to improve the cost calculation of that approach by using a more advanced structural model and tree decomposition to partition the cost calculation into subproblems. Our results demonstrate an important gain in accuracy and a significant reduction in the number of optimal sequences inferred. Our software is available on GitHub.

Keywords: Bioinformatics databases · Biology and genetics · Algorithms · Trees · Graph algorithms

1 Introduction

Ancestral reconstruction is an important problem in comparative genomics that aims to reconstruct genomes or sequences as they might have existed millions of years ago. This involves inferring everything from large ancestral chromosome sequences to ancestral gene orders or gene sequences [1,14,25]. Traditionally, the inference of ancestral sequences relies on having a phylogeny—typically a binary tree depicting the evolutionary relationships among the species under study. Starting from the tree's leaves, which are labelled with the current sequences of each species, ancestral reconstruction methodologies are applied to infer the sequences at the internal nodes, which represent ancestors. This reconstruction is done in a manner that minimizes a specific cost, which is defined by a given evolutionary model to penalize various mutations that alter the sequences along the tree branches. In comparative genomics, this challenge is known as the *small parsimony problem*, and numerous strategies have been devised to tackle it [3,6,8,19,21,26].

However, traditional techniques, such as the Fitch [6] or Sankoff [3,21] parsimony algorithms, or the PAML maximum likelihood approach [28], are generally applicable only to sequences exhibiting a significant level of conservation. Particularly, when reconstructing ancestral non-coding RNA (ncRNA) gene sequences—genes that are transcribed into RNA but not translated into a protein—these methods face limitations. This is because ncRNA gene sequences are typically not well-conserved; rather, it is the structure of the base pairs, which imparts functionality to the ncRNA sequences, that remains stable [11]. Consequently, more innovative approaches that consider both the sequential and structural data are required.

In [2], a method was presented to reconstruct ancestral RNA secondary structures and sequences from an alignment. This approach was based on a stochastic context-free grammar (SCFG) to model RNA evolution. Although this approach was a promising first step into this direction, its major limitation was the memory and runtime requirements, which constrained its use to smaller instances.

More recently, [26] proposed an approach that targets the problem from a different perspective. The authors argued that the reconstruction of ancestral RNA sequences of a single ncRNA family with a single secondary structure using a covariation model can introduce bias into the resulting sequences [4,13,29]. In other words, the problem with this approach is that the inferred ancestors tend to be more fit to the structure than the extant sequences, which does not make sense in terms of the evolutionary process and natural selection. Thus, the authors of [26] proposed the idea of considering simultaneously the structures of two related ncRNA families in the inference process. These two families need to be part of the same Rfam clan, meaning that there is a valid assumption that they have evolved from a common ancestral ncRNA that could fold into the two structures. This specific ancestor is what the method aims to infer.

Other approaches that consider sequence and secondary structure are focusing on solving a different problem: the design of multi-stable RNA sequences. Such an approach was presented in [7] with the end goal of developing RNA-based switches that can adopt different structures and perform various functions. In this work, the authors first introduced how the number of sequences compatible with the target structures can be predicted based on the paired positions of the given structure(s). These sequences are optimized to favour desired energy landscapes, such as having one or more deep energy minima corresponding to the target structures. To prevent misfolding, the approach is designed in a way that these local minima, calculated based on the target structures, are separated by a high energy barrier. Additionally, their approach is capable of designing RNAs that can change their preferred structures depending on the temperature, with a desired energy barrier. In [10], an approach was developed to efficiently sample RNA sequences that are capable of adopting multiple predefined structures with specific energy and GC content. The proposed algorithmic framework combined a fixed-parameter tractable (FPT) sampling algorithm with multi-dimensional Boltzmann sampling over distributions controlled by expressive RNA energy models to generate RNA sequences with the desired structural attributes. Variants exploring alignments and reconstruction of secondary structures themselves have been explored in [17].

This paper focuses on the challenge of inferring ancestral ncRNA sequences using the achARNement methodology (hereafter called achARNement1) of [26] as a foundation. As mentioned above, this method utilizes both sequential and structural information from two homologous families of ncRNAs to perform the inference. However, to avoid the extensive computational load caused by potential interconnections among the base pairs of different structures, the original algorithm is restricted to considering only one level of such dependencies. More specifically, for a position that is paired in both structures, the achARNement1 method considers the position paired in the current structure, together with the average cost of any possible nucleotide paired with it in the other structure (see [26] for more details). In contrast, in this study, we designed and implemented a different strategy that employs tree decomposition and tree optimization techniques to manage the high time complexity induced by the inter-dependencies of the structures. We will refer to our new approach as achARNement2 in the following sections. As a result, the refined algorithm can account for more dependencies during the reconstruction process while maintaining a reasonable running time. A comparison is made between the results from achARNement2 and three other existing algorithms, including achARNement1. Our proposed algorithm has demonstrated improved performance over all other methodologies regarding both accuracy and the breadth of the solution space.

2 Problem Statement

In this study, we aimed to develop a model that accounts for both paired and unpaired positions, substitutions, and base-stacking interactions to optimize ancestral sequence accuracy and structural fidelity. More specifically, for the given input data, the goal is to infer the ancestral sequences at each internal node of the species tree (including the root) by considering the sequence and structure conservation of all input ncRNAs (one from each family). Consecutive positions in a sequence (representing base stacking) is another constraint we will consider in addition to base pairs, as they can affect the stability of the structures. With these new factors, the model will become more general and complex, which motivates the use of techniques such as tree decomposition to make the cost calculations more efficient.

Formally, we describe our problem as follows:

- **Input:** (a) Two ncRNA families that are in the same Rfam clan [5] as well as their secondary structures. (b) Multiple species having one copy of each

ncRNA from each family and their corresponding sequences. (c) A species tree of the species considered.
- **Variables:** Let X be the set of all sites in an RNA sequence, $x_i \in X$ be the single sites; $D = \{A, C, G, U\}$, the alphabet or domain, denotes the possible nucleotides to assign to a single site; d_i represents the domain of the site at position i.
- **Factor(s)/constraint(s):** In our problem, the factor for each site/variable x_i is to find a/some nucleotide(s) a that minimizes the cost of having $x_i = \{a|a \in d_i\}$. The overall cost consists of two components:
  (i) **Substitution cost**, based on the nucleotide(s) from its children for the same site/position.
  (ii) **Structure cost**, estimated based on the given secondary structures.

  More specifically, the structure cost depends on the nucleotide selections of adjacent sites and the positions they are paired with in the current and the other structure. The full objective function is described below in Eq. 1.
- **Output:** A set of most parsimonious ancestral sequences (and their cost) for each internal node of the species tree.

3 Methodology

3.1 Using Tree Decomposition

The achARNement1 method followed a dynamic programming design that helped to bound the time-complexity and increased the accuracy of the reconstruction. However, to consider all possible structural dependencies, the running time would not be viable without reducing a considerable amount of calculations. More precisely, the time-complexity would be exponential when, in the worst case, the dependency graph (representing interconnected base pairs in-between structures; see the top-left box in Fig. 1) contains all positions of both sequences in one path (*i.e.* each position in one structure is dependent on a different corresponding position in another structure).

Our solution to the problem is to utilize the power of a dynamic program combined with a tree decomposition. However, since treewidth can have a great impact on the running time, we applied the *Tree diet* method [18] for optimizing (*i.e.* limiting the treewidth of) our tree decomposition. It is a method aimed to remove a minimal set of edges such that a given tree decomposition can be slimmed down to a prescribed treewidth.

Our approach to ancestral sequence reconstruction consists of two parts: a) a preprocessing step that prepares the necessary data for our main algorithm; b) our main algorithm that performs the reconstruction based on the inputs and produces ancestral sequences.

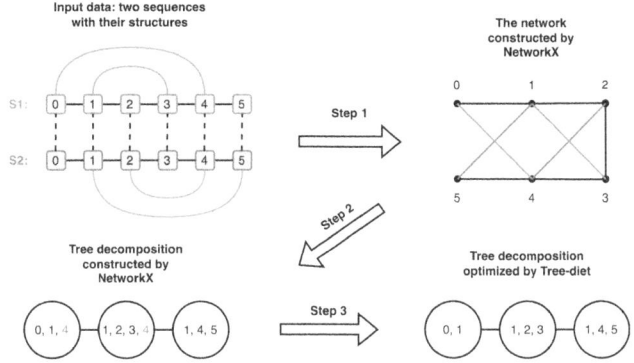

Fig. 1. The figure presents an example of the three steps for preprocessing. In the last step, the vertices marked in red are picked and removed by Tree diet in order to achieve the desired treewidth (tw = 2 in this example). (Color figure online)

3.2 Preprocessing

The preprocessing (see Fig. 1) works directly on the input data and prepares it for the main algorithm. It involves the following steps:

(i) Build a network based on the two given secondary structures (base pairings).
(ii) Build a tree decomposition that maps the vertices and edges of the network.
(iii) Optimize the tree decomposition to a desired treewidth by removing a minimal set of edges with "Tree diet".

The edges of the network in step (i) are based on the stacking relations between adjacent positions in the sequence, and the base pairing relations from two given structures. The vertices are simply representing the positions/sites.

We used the NetworkX package [9] to build the network and construct a tree decomposition. To apply the Tree diet, the tree decomposition needs to be stored in a compatible format. We built a pipeline (which is included in our tool package) for the conversion. To be able to differentiate the edges for weighing the structure costs, the pipeline colours the edges of basepairs based on the structure they belong to (e.g. edges from structure one can be coloured blue, and edges from structure two can be coloured red, as in Fig. 1).

Tree diet takes three parameters as input: the network on which the tree decomposition is based, the bags stored as a linked list (the first bag linked to an empty one as the root), and a desired treewidth. There is also an option to keep specified edges (called "backbones") in the network during the optimization (edge removal). The resulting tree decomposition from step (iii) is ready for use by our main algorithm.

Our proposed algorithm, which we have named achARNement2, uses a new cost calculation approach, but like its predecessor [26], it consists of three main steps, as presented below.

3.3 Bottom-Up Step

A postorder traversal is performed on the species tree as shown in Algorithm 1 of the Appendix, which is initially called on the root of the species tree. The goal is to fill up a 3-dimensional cost matrix for each node of the input phylogeny, representing each possible nucleotide assignment (a tuple of nucleotides) in each bag for each structure family. The bags obtained from the tree decomposition are available as a global variable in the form of a list, ordered from the leaves to the root. Each bag contains a tuple of integers, corresponding to positions in the RNA sequence. For example, for a bag containing the tuple (1, 5, 7), a possible nucleotide assignment could be (A, A, A), representing the nucleotides assigned to positions 1, 5 and 7 respectively. When Algorithm 1 reaches the leaves (see lines 1 to 7), it simply assigns a cost of 0 for the nucleotides that are present in the leaf sequences, and ∞ for all the other possible nucleotides. Otherwise, when we are dealing with an internal node (lines 8 to 13), we make a recursive call to the left and right children (postorder) to get the cost matrices for them. Then the cost matrix can be computed for the internal node for each structure family and each bag (one at a time).

The cost matrix calculation is handled by Algorithm 2, which finds the minimal cost for each possible nucleotide assignment in each family for each bag. This cost includes a substitution cost (from the children assignments) and a structure cost (minimum free energy). More specifically, to calculate the cost of each possible nucleotide assignment for a specific bag, we need to consider all possible nucleotide assignments in the left and right child (lines 1–2) to find the ones yielding the minimum cost. The formula used by achARNement2 to calculate the cost of a particular nucleotide assignment a_p at the tuple of positions p, based on a given nucleotide assignment from a child node ch_p for a structure family sf is presented in Eq. 1 below:

$$c(ch_p) + s(ch_p, a_p) + W * (G * mfe(a_p, sf) + (1 - G) * mfe(a_p, \overline{sf})) \qquad (1)$$

In that formula, the term $c(ch_p)$ represents the optimal cost in the child node that has been previously determined for assigning that specific set of nucleotides at the positions designated by p. The function $s(ch_p, a_p)$ quantifies the substitution cost of substituting every nucleotide from one nucleotide assignment to the other, and it is simply the sum of each individual substitution cost at a specific site. The substitution costs selected for this study are presented in Table 1, where the costs for transitions are lower than the transversion costs, which is representative of the fact that transitions occur more frequently during evolution. The term $mfe(a_p, sf)$ denotes the Minimum Free Energy, which is calculated using the ViennaRNA package (more details below). The MFE is calculated for the nucleotide assignment a_p in both the current structure (represented by sf) and the other structure (represented by $\overline{sf}$). The term G represents a percentage which is used to define the contribution of each MFE (from each structure family) in the total structure cost. This percentage changes following a linear gradient depending on the depth in the phylogeny, where it will be 100% at the leaves and 50% at the root. This is to reflect the trend of divergence in evolution; sequences at the leaves are fully specialized into one specific structure, whereas sequences at the root were able to fold into two different configurations equally. Finally, a structure weight W is assigned to the whole structure cost before adding it to the substitution cost. This constant is used to normalize the MFE values so that they do not dominate the whole cost formula, because they can be within 3 orders of magnitude of the substitution costs. A weight of 0.001 was found to provide a good balance between substitution and structure costs after initial testing on smaller datasets, so this was selected as the default value. Note that the user is free to adjust this constant, if they want to give more or less weight to the structure cost.

To be more specific on the MFE calculation, the energy of each bag assignment is estimated using the "energy_of_structure()" method provided by the ViennaRNA package version 2.6 [15]. This method needs to be called on a full target structure and RNA sequence. Since we have access to the two full target structures, but not to the full RNA sequences when working on a specific bag, we first filled out the RNA sequence with wildcard symbols ($), which are ignored by the MFE calculation method. Then, we replaced some of the wildcard symbols with the given nucleotide assignment at the positions of the bag. This allowed us to get only the energy contribution of the nucleotide assignment for that particular bag.

3.4 Middle Step

The minimal cost matrices (one for each family), which were obtained during the bottom-up phase for both families, are linked at the root of the phylogenetic tree. Specifically, under the premise of an ancestral duplication event, it is necessary to establish a combined cost matrix at the root—integrating the cost matrices of both families—prior to determining the nucleotide selections (enumerating optimal sequences). Consequently, a merging procedure is implemented to generate this integrated cost matrix at the root. This is done using the approach of Algorithm 2 but by considering the two cost matrices (one for each structure family) as the left and right child of the root node.

3.5 Top-Down Step

The goal of the top-down step is to enumerate all the optimal sequences by making selections that follow the principle of parsimony. However, since the "options" are stored as assignments of nucleotides for each bag from the tree decomposition with their total costs instead of independent position costs, the way of enumerating and constructing the complete optimal sequences is more involved. By definition of the tree decomposition, for every edge (u, v) in the network of the two secondary structures, there exists at least one bag that contains both u and v. This means that, as long as the network is a connected graph, any vertex from the network must be included in at least two bags from the corresponding tree decomposition. In other words, every position in the original sequence will be covered by at least two bags from the tree decomposition. Making selections simply following the principle of parsimony for each bag independently will cause a series of nucleotide selections for the bags that might be conflicting on the "overlapped" positions (different selections of nucleotide on the same position). As a result, some, if not most, of the selections would not be transformable into complete sequences.

Instead, to efficiently enumerate the optimal sequences from the potential nucleotide assignments for the bags, we describe a "greedy" method that returns complete sequences by selecting nucleotides for each bag following the reverse order given by the tree decomposition. When multiple assignments (of nucleotides) are estimated to give the same minimal cost, a recursive call will be invoked for each of them. A sequence template (of equal length to the total size of the input sequences) is passed between recursive calls with fixed nucleotides at previously assigned sites and placeholder characters for positions waiting to be assigned. Starting from the root of the tree decomposition, the algorithm progresses by replacing the placeholders on the template with nucleotides that correspond to the minimal cost in the cost matrix for each bag. Consequently, following the assignment in one bag, the sequence template guides the nucleotide selections for subsequent bags. Consistent with the principles of tree decomposition, adjacent bags share at least one vertex (position), ensuring that the search space of possible nucleotide assignments for all unassigned adjacent bags is reduced following each assignment. For instance, in our case, the total number of possible assignments for a bag is $|D|^k$ where $k \leq tw + 1$ (size of the largest bag), D being the alphabet of an RNA sequence. If a previous bag assigns a nucleotide, the domain of the current bag will be $1 \cdot |D|^{k-1}$.

The whole process of the top-down step starts with a call to Algorithm 3 on the root of the phylogeny. At the end of Algorithm 3, once all the optimal sequences have been found and saved for the root node, the "selectNucleosChildren" method (see Algorithm 4) is called to repeat the same process for each descendant node in the tree. The only difference is that for any internal node that is not the root, the optimal sequences of the parent influence the choice of optimal sequences for its children. Indeed, we need to identify the optimal sequences in the children that made the optimal cost possible for the parent sequences. Both of these algorithms make use of the "selectNucleos4Bag' method (see Algorithm 5) to assign nucleotides for each bag following the order given by the tree decomposition. With a sequence template that gets gradually filled throughout the process, after the assignment of the last bag, complete sequences with their costs are returned. These optimal sequences (one set for each structure family) are stored in each node of the phylogeny.

3.6 Software Implementation

We implemented a software based on the algorithm described above in Python, which is available in a software package named achARNement2 on GitHub. Besides our new algorithm, `achARNement2` also includes the implementation of

Fitch, Sankoff, as well as achARNement1. To prepare the optimized tree decomposition, we built a pipeline using the popular toolkit *NetworkX* for building the tree decomposition and *Tree diet* for tree optimization. We used tools from the *ViennaRNA* package for the MFE (structure cost) estimation. See Sect. 5.2 of the Appendix for a detailed analysis of the worst-case runtime complexity.

3.7 Data Generator

For the simulated dataset, we used the data generator provided with achARNement1, which is described briefly below. We also reused the three randomly designed structures of length 100 used in the study of [26] (with two similar structures and one that differs greatly from the other two). Using these structures, we can make three structure pairs (for family 0, 1, and 2, we have family pairs 01, 02, and 12) for testing.

For each of these pairs of structures, we generated 100 bi-stable sequences with *Frankenstein* [16], a genetic algorithm approach for solving the inverse folding problem. These resulting sequences are then used as "seeds" to populate the root ancestor of each of the 100 generated phylogenies for each structure pair. The generated phylogenies were all complete binary trees of depth 6 (with 64 leaves). For each generated phylogeny, the sequences in the internal nodes, as well as the leaves, are generated in a top-down manner. We used five different mutation rates (1%, 5%, 10%, 15%, and 20%) in our tests. The mutation rates are applied to both parent sequences (one from each family) between each level in the tree (e.g. 5% difference between any internal node and its direct children) in a process that is repeated a thousand times to generate a pool of possible descendant sequences. From this pool of sequences, the actual child sequences (one for each family) are sampled according to their MFE for each structure. In other words, mutated sequences from the generated pool that can fold better into the target structure have a higher chance of being selected. In the end, a total of 1500 such trees (100 seeds, three structure pairs, five mutation rates) are generated for the evaluation of achARNement2.

4 Results and Discussion

4.1 Experiments on Simulated Data

For comparison, we analyzed the simulated dataset with four algorithms: Fitch, Sankoff, achARNement1, and achARNement2. Using multiple mutation rates as the independent variable in our tests allowed us to evaluate how the performance of these different methods is affected by having an increasing sequence divergence at the leaves, which can severely affect the amount of signal that can be obtained from them. All the tests were conducted on Ubuntu 22.04.5 using identical hardware: a 20-core Raptor Lake CPU running at 4.5 GHz, with 64 GiB of memory, ensuring a direct comparison of the runtime performance of the four algorithms. For all the experiments, we bound the treewidth to 2 (*i.e.* largest bag of size 3) using Tree diet, as preliminary tests showed that this value resulted in an ideal compromise of accuracy vs runtime. Moreover, all the data points presented below represent an average of 100 replicates (with standard deviation shown in each plot).

Results

The results in terms of error percentages for all the reconstructed sequences at all internal nodes of the phylogenies are presented in Fig. 2. In the figure, achARNement2 shows an improvement in terms of accuracy by exhibiting a lower error percentage over the other three algorithms, especially for family pair 01. We believe that it is due to the high similarity between structure 0 and 1. Also, from the figure, the error percentage of achARNement2 is considerably lower for structured (paired) regions compared to unstructured regions (about 15% max for structured and 22% for unstructured). This meets our expectations based on the fact that our model takes advantage of more structural information compared to the other three algorithms. Another interesting detail is that, in the unstructured positions, achARNement2 outperformed the other algorithms even more. This could be related to the extra base stacking information considered by our approach. Fitch, Sankoff, and achARNement1 tend to have larger solution spaces for optimal sequences than achARNement2. As a result, when the mutation rate exceeds 10%, Fitch and Sankoff could not finish generating the optimal sequences. achARNement1, however, finished most of the instances of a 15% mutation rate (except for four). For achARNement2, we used the extra structure information as a constraint, which helped eliminate an enormous number of potential sequences (as shown in Fig. 3 below), and made it possible to significantly reduce the solution space and produce results even for a mutation rate as high as 20%.

Besides the error percentage, achARNement2 showed a significant improvement with regards to the total number of reconstructed sequences at the root (see Fig. 3). As mentioned above, by utilizing more structural information, achARNement2 eliminates a massive amount of potential sequences. In fact, the average number of inferred sequences by our new algorithm is several orders of magnitude lower than previous approaches for most mutation rates. It is important to mention that severe non-uniqueness of optimal solutions is a well-documented challenge of ancestral reconstruction methods [22,23,30,31]. Being able to significantly reduce the number of optimal solutions inferred with achARNement2 is therefore a crucial achievement and improves its usability in downstream analyses, where sequence ancestors could be selected for further experimentation (e.g. expression, binding, folding, etc.).

Figure 4 illustrates the average runtime across different mutation rates for each algorithm. According to the results, achARNement2 exhibits slower speed compared to the other three algorithms, particularly at lower mutation rates. However, its runtime remains highly stable, even at the highest tested mutation rate (20%). This is mostly due to the fact that it can keep the number of optimal sequences under control even under high mutation rates. In contrast, the other three algorithms demonstrate considerable sensitivity to mutation rates, which

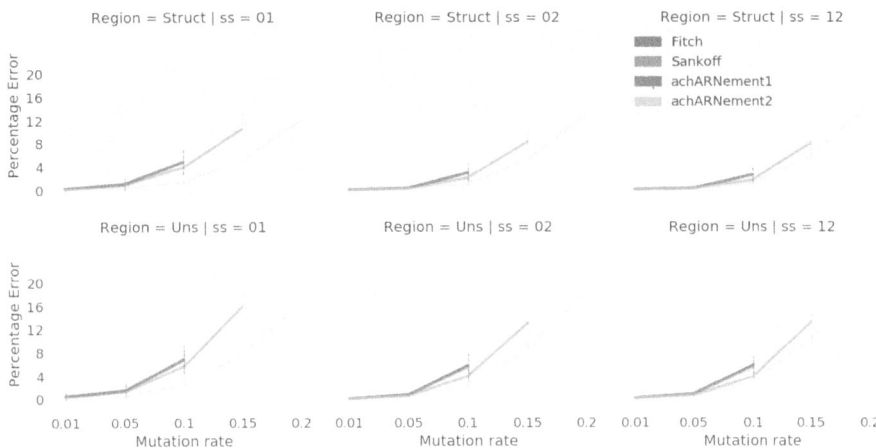

Fig. 2. The average error percentage of all optimal sequences for both families in a tree. Each column represents a pair of secondary structures. Figures in the top row show error percentages for the structured regions of the sequences. Figures in the bottom row illustrate error percentages for the unstructured regions.

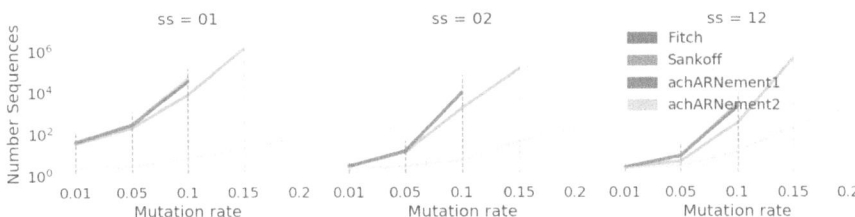

Fig. 3. Average number of optimal sequences for the root. Five mutation rates were tested from 1% to 20%. Each column shows the number of sequences each algorithm inferred for a given family pair (01, 02, and 12).

prevents them from successfully completing instances with high mutation rates in a reasonable amount of time.

Upon conducting an in-depth analysis, we made pairwise comparisons between the ancestral sequences produced at the root of the phylogeny by the four distinct algorithms for each mutation rate. Our objective was to determine if the different approaches are exploring a new area of the search space or reducing the solution set of other methods. It is noteworthy that no method generated the exact same set of output sequences (see Fig. 5), and only a few sets are entirely subsumed by another. The analysis indicates a correlation between lower mutation rates and higher inclusion levels. This is likely because, at lower mutation rates, the extant sequences at the leaves bear greater structural and sequential resemblance to their progenitor, so most methods will have small and similar solution spaces. Observing the inclusion percentage at a mutation rate of

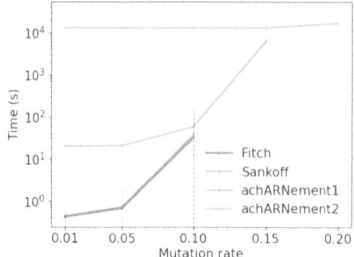

Fig. 4. The comparison of running times between the four algorithms. The x-axis shows the five mutation rates that we consider (1%, 5%, 10%, 15%, and 20%). The y-axis indicates the running time in seconds (logscale).

0.1, a significant majority of the optimal sequences generated by achARNement2 are encompassed within those produced by either Sankoff or Fitch (over 95%). On the other hand, the intersection of the results from achARNement2 and achARNement1, relative to the total output from achARNement2, stands at approximately 63% for the same mutation rate. Notably, the divergence in output sequences intensifies at a mutation rate of 0.15, reducing the intersection to just under 10%. Despite Fitch and Sankoff generating a comparable number of sequences at a mutation rate of 0.1, their outputs are considerably divergent (approximately 75%). Compared to achARNement1, achARNement2 shows a substantially higher inclusion rate against outputs from Fitch and Sankoff, even surpassing the inclusion rate observed between Fitch and Sankoff themselves. It is highly plausible that the majority of optimal sequences reconstructed by achARNement2 are contained within the intersection of sequences from Fitch and Sankoff.

We also evaluated the MFE, for both structures, of each ancestral sequence inferred by the different methods. To represent the ability of the sequences to fold into both structures, we calculated the harmonic mean of both MFE values. In the last three columns of Fig. 5, we report the minimum, average and maximum harmonic means of MFE for inferred sequences exclusive to the base algorithm (exc. base), common to both algorithms (common) and exclusive to the other algorithm (exc. other). Note that a lower value (greater negative value) represents greater stability. The table shows that the average harmonic mean of sequences produced by Fitch, Sankoff, and achARNement1 only (not present in the intersections with achARNement2) are worse than both the averages of sequences exclusive to achARNement2 and the ones contained in the intersection. The only exception is for the comparison with Sankoff for the 1% mutation rate, for which achARNement2 produced only one sequence that was exclusive and had a higher harmonic mean of MFE. These results show the capability of achARNement2 of reconstructing most of the sequences with the ability to fold well into the two structures while getting rid of the worst sequences inferred (exclusively) by other methods.

mut	algo_base	algo_other	# of seqs (base)	# of seqs (other)	common	Inclu/seq	HM (exc.base)	HM (common)	HM (exc.other)
0.01	A2	A1	333	1942	318	95.50%	-19.15, -13.19, -9.77	-25.25, -16.07, -6.57	-20.97, -11.29, -5.54
0.05	A2	A1	423	11089	363	85.82%	-23.32, -15.25, -8.37	-25.25, -16.4, -6.79	-25.65, -12.91, -4.11
0.1	A2	A1	1315	496510	833	63.35%	-23.39, -16.4, -4.95	-25.42, -16.04, -7.07	-24.92, -11.79, -0.48
0.15	A2	A1	14078	10627280	1354	9.62%	-24.01, -14.57, -3.88	-23.17, -16.86, -8.48	-24.41, -11.69, 60.59
0.01	A2	Sankoff	333	2395	332	99.70%	-9.77, -9.77, -9.77	-25.25, -15.96, -6.57	-21.05, -11.05, -5.54
0.05	A2	Sankoff	423	16530	422	99.76%	-15.96, -15.96, -15.96	-25.25, -16.24, -6.79	-25.65, -12.5, -4.11
0.1	A2	Sankoff	1315	1732020	1272	96.73%	-20.67, -13.37, -8.95	-25.42, -16.26, -4.95	-24.92, -11.35, 0.65
0.01	A2	Fitch	333	2256	333	100.00%	None, None, None	-25.25, -15.94, -6.57	-21.05, -10.99, -5.54
0.05	A2	Fitch	423	15152	423	100.00%	None, None, None	-25.25, -16.24, -6.79	-25.65, -12.72, -4.11
0.1	A2	Fitch	1315	1570976	1262	95.97%	-20.67, -13.88, -10.38	-25.42, -16.26, -4.95	-24.92, -11.34, 0.31
0.01	A1	Sankoff	1942	2395	1942	100.00%	None, None, None	-25.25, -12.07, -5.54	-21.05, -10.25, -5.81
0.05	A1	Sankoff	11089	16530	11089	100.00%	None, None, None	-25.65, -13.02, -4.11	-23.32, -11.73, -4.77
0.1	A1	Sankoff	496510	1732020	359810	72.47%	-21.38, -11.89, -2.61	-25.42, -11.76, -0.48	-24.56, -11.25, 0.65
0.01	A1	Fitch	1942	2256	1852	95.37%	-16.91, -12.61, -6.92	-25.25, -12.05, -5.54	-21.05, -10.2, -6.06
0.05	A1	Fitch	11089	15152	10619	95.76%	-22.36, -12.6, -6.28	-25.65, -13.04, -4.11	-23.32, -12.29, -4.77
0.1	A1	Fitch	496510	1570976	330404	66.55%	-24.12, -11.79, -2.61	-25.42, -11.79, -0.48	-24.56, -11.23, 0.31
0.01	Fitch	Sankoff	2256	2395	2251	99.78%	-10.11, -9.83, -9.5	-25.25, -11.72, -5.54	-16.91, -11.84, -5.81
0.05	Fitch	Sankoff	15152	16530	15059	99.39%	-18.97, -12.11, -4.77	-25.65, -12.82, -4.11	-22.36, -10.32, -4.77
0.1	Fitch	Sankoff	1570976	1732020	1179192	75.06%	-21.85, -10.29, 0.0	-25.42, -11.7, 0.31	-24.12, -10.63, 0.65

Fig. 5. The figure illustrates the inclusion relationship and harmonic mean of the minimum free energy (MFE) values for sequences generated by the four algorithms using a simulated dataset. The first column shows the mutation rate of the instance. The second column shows the name of the "base" algorithm (A1 refers to `achARNement1`, A2 for `achARNement2`). The third column shows the name of the other algorithm to which we compare the "base". "Inclu/seq" is the average inclusion percentage of the sequences from the base algorithm. The final trio of columns illustrates the harmonic means of minimum free energy (MFE) in both structures as estimated from sequences exclusive to the base algorithm, sequences common to both algorithms, and sequences exclusive to the other algorithm. Each of these columns contain the minimum, average, and maximum values of harmonic mean.

4.2 Experiments on Biological Data

Biological Datasets

We also assessed the four methods on the Glm and FinP-traJ clans (which were used in [26]) from Rfam. After we retrieved the sequences and structures of the two families from Rfam, we intersected the genome IDs of these sequences to get the set of common genomes (all bacterial genomes). A bacterial genome tree based on these organisms was taken from BVBRC [20]. To prepare the structures and sequences for the reconstruction approach, we aligned them on their seed alignments using CARNA [24]. Finally, all gapped positions in the resulting alignment were removed since only point mutations are allowed in our model.

Results for the Glm Clan

The Glm clan includes two bacterial small non-coding RNAs, GlmY and GlmZ, which are similar in structure yet perform different functions. These RNAs function hierarchically where GlmY acts upstream and indirectly influences glmS mRNA translation by affecting GlmZ, which directly interacts with glmS mRNA to enhance its translation [27]. For our test, we reused the 74 bacterial genomes, their alignments, as well as the corresponding phylogeny analyzed in [26].

For the root node, Fitch and Sankoff produced the same 786432 sequences. For achARNement1, it produced 196608 ancestral sequences. In contrast, achARNement2 produced only 256 ancestral sequences, which is several magnitudes fewer than the other three algorithms.

In terms of runtime, achARNement2 took roughly 2 h to complete the analysis of the GLM clan. In comparison, achARNement1 Fitch and Sankoff took respectively 28 s, 12 s and 12 s to complete.

Results for the FinP-TraJ Clan

The FinP-traJ clan in the Rfam database consists of RNA families that include FinP RNA and traJ mRNA, which are involved in a regulatory system affecting bacterial conjugation, specifically in the F-plasmid of *Escherichia coli* and related bacteria [12]. This system plays a crucial role in controlling bacterial fertility through interaction between these two RNA components. Like the Glm clan, we used the same dataset that was used in the study of achARNement1, consisting of 54 genomes and the corresponding phylogeny. Compared to the two families in the Glm clan, the ones from FinP-traJ are more varied in structure.

For the three algorithms that we included for comparison, the results are as follows: Fitch and Sankoff inferred the same 12582912 sequences at the root, while the achARNement1 inferred fewer sequences with a total of 3145728. As for achARNement2, only 128 sequences were inferred, which is significantly lower than previous approaches.

In terms of runtime, achARNement2 took roughly 33 min to complete the analysis of the FinP-traj clan. In comparison, achARNement1 Fitch and Sankoff took respectively 28 s, 151 s and 167 s to complete.

Comparison of Reconstructed Ancestors

Figure 6 depicts the results derived from biological datasets regarding the inclusion of ancestral sequences produced by different methods. It is observed that the average harmonic mean of MFE across both structures for sequences reconstructed by achARNement2 is significantly lower than that of achARNement1. This suggests that achARNement2 is more effective in reconstructing sequences that align well with both structures, thereby indicating its superior fitness. Similarly to the analysis performed on simulated data, the intersection of output sequences from the various algorithms is examined for the GLM and FinP-traJ clans. Notably, all sequences reconstructed by achARNement2 at the roots are subsets of the optimal sequences determined by achARNement1, Sankoff, and Fitch. This outcome demonstrates that achARNement2 is capable of eliminating a greater number of candidate sequences, benefiting from the newly implemented method for estimating structure costs.

5 Conclusion

Our improved approach (achARNement2) resulted in a significantly lower error rate on simulated data and a greatly reduced solution space both on simulated

clan	algo_base	algo_other	# of seqs (base)	# of seqs (other)	common	HM (common)	HM (exc.other)
GLM	A2	A1	256	196608	256	-10.02, -10.02, -10.02	-10.03, -8.67, -7.77
GLM	A2	Sankoff	256	786432	256	-10.02, -10.02, -10.02	-10.03, -8.31, -7.06
GLM	A2	Fitch	256	786432	256	-10.02, -10.02, -10.02	-10.03, -8.31, -7.06
GLM	A1	Sankoff	196608	786432	196608	-10.03, -8.67, -7.77	-9.73, -8.19, -7.06
GLM	A1	Fitch	196608	786432	196608	-10.03, -8.67, -7.77	-9.73, -8.19, -7.06
GLM	Sankoff	Fitch	786432	786432	786432	-10.03, -8.31, -7.06	None, None, None
FinP	A2	A1	128	3145728	128	-12.65, -12.64, -12.62	-13.01, -10.49, -7.45
FinP	A2	Sankoff	128	12582912	128	-12.65, -12.64, -12.62	-13.01, -10.06, -6.31
FinP	A2	Fitch	128	12582912	128	-12.65, -12.64, -12.62	-13.01, -10.06, -6.31
FinP	A1	Sankoff	3145728	12582912	3145728	-13.01, -10.49, -7.45	-12.43, -9.92, -6.31
FinP	A1	Fitch	3145728	12582912	3145728	-13.01, -10.49, -7.45	-12.43, -9.92, -6.31
FinP	Sankoff	Fitch	12582912	12582912	12582912	-13.01, -10.06, -6.31	None, None, None

Fig. 6. The figure presents the outcomes derived from the four algorithms using biological data pertaining to the GLM and FinP-traJ clan. The initial column lists the clan's name. The subsequent two columns denote the algorithms termed "base" and "other." The inclusion percentage is always 100%, so there are no sequences exclusive to the base algorithm. The last two columns show the harmonic means of minimum free energy (MFE) in both structures calculated for sequences common to both algorithms and sequences exclusive to the other algorithm (the three values are minimum, average, and maximum, respectively).

and real data. These improvements are achieved by allowing the algorithm to consider more structural information compared to achARNement1. More specifically, we developed a method inspired by the tree decomposition and dynamic programming that solves the problem of "inter-dependency" (complexity induced by dependencies among secondary structures) that achARNement1 avoided for better time efficiency. A greedy approach is also presented to enumerate the resulting optimal sequences from potential bag assignments.

The application of the new approach to structure cost estimation demonstrated its effectiveness on both simulated and biological datasets. For both simulated and real datasets, it showed a considerable improvement in reducing the solution space of the reconstruction, producing a set that is at times several orders of magnitude smaller than that of the other three algorithms (Fitch, Sankoff, and achARNement1). On simulated data, our results show that the proposed algorithm consistently outperformed the other three existing algorithms in terms of error percentage, especially for higher mutation rates. An even more interesting observation is that a majority of the sequences inferred by achARNement2 can be found in the results of the other three algorithms, suggesting that the inferred ancestral sequences are as reliable as those of the other three algorithms. For the biological data analysis, the harmonic mean of MFE of the output sequences showed that in general, the sequences inferred by achARNement2 are more capable of folding into both structures than solutions produced by Fitch, Sankoff, and achARNement1.

On the flip side, the runtimes of achARNement2 obtained for all the tests were longer than the ones from the other three methods used for comparison. However, this was compensated by significantly lower error rates and smaller sizes of the reconstructed ancestor sets, which are of great importance for the applicability of the method.

While the results of this study are promising, further development is needed to enhance the robustness and applicability of the approach. One potential direction is to incorporate insertions and deletions (indels) in the sequence reconstruction process. Indels are prevalent in genomic sequence alignments, and their inclusion in our ancestral reconstruction approach could provide valuable insights into sequence evolution. Accounting for indels would also eliminate the requirement for input sequences/structures to be of equal length, which would greatly simplify the preprocessing tasks. Moreover, this adaptation would enable the model to handle more structurally diverse families, broadening its applicability to a wider array of Rfam clans. As for scalability, it would be very interesting to adapt the model to more than two structure families (e.g. three structures). Potentially, this could lead to an even better solution reduction and a smaller error percentage. Speed improvements could be obtained by simplifying the energy computation, focusing more on local stack energy instead of full energy. Finally, `achARNement2` could be used in the context of RNA sequence design. By selecting the desired structures and populating the leaves of the input phylogeny with samples of sequences folding into the two structures, the ancestors produced at the root would be good candidates for multi-functional RNAs.

Acknowledgements. This work was funded by the NSERC Discovery (grant number RGPIN-2016-06051) and FRQNT-NSERC NOVA (grant number 586843-23) programs.

Disclosure of Interests. Authors have no competing interests to declare.

Appendix

5.1 Nucleotide substitution matrix

Table 1 shows the simple transition/transversion matrix used for substitution costs, which was also used in [26].

Table 1. Nucleotide substitution matrix.

	A	C	G	U
A	0	2	1	2
C	2	0	2	1
G	1	2	0	2
U	2	1	2	0

5.2 Pseudocode and runtime complexity

In this section, we present the pseudocode of achARNement2, divided into 5 algorithms which are presented in the subsections below. Worst-case runtime complexities are described for each algorithm.

Bottom-Up Step

Algorithms 1 and 2 constitute the bottom-up step of achARNement2. Algorithm 1 is called on the root of the input phylogeny and uses recursion to complete a post-order traversal of it. Its aim is to calculate a cost matrix for every node of the phylogeny, which stores for each of the two families and each bag, the optimal cost of each possible nucleotide assignment based on the left and right child nodes (this cost is calculated using Algorithm 2). Leaf nodes are base cases, where the cost matrix is initialized to ∞ except for the nucleotides that represent the input sequence.

Algorithm 1: calculateCosts(n)

This method represents the bottom-up step, where we do a post-order traversal of the tree in order to calculate the most parsimonious costs for each possible nucleotide at every site.

---

input : A node **n** of the species tree with extant sequences at the leaves for the structure families studied.

output: A cost matrix **n.costMatrix** for every internal node of the tree for every structure family

1 **if** *n is a leaf* **then**
2   seq = a string that stores the nucleotides of the leaf sequence;
3   All cells in n.costMatrix are first initialized to ∞;
4   **for** *every structure family, fam* **do**
5   **for** *every bag in the tree decomposition* **do**
6   nucAssign = tuple([seq[pos] for pos in bag]);
7   n.costMatrix[fam][bag][nucAssign] = 0;
8 **else**
9   calculateCosts(n.leftChild);
10   calculateCosts(n.rightChild);
11   **for** *every structure family, fam* **do**
12   **for** *every bag in the tree decomposition* **do**
13   computeBagCost(n, fam, bag);

Algorithm 2 runs in $O(4^{(2k)})$ time, where k is the size of the largest bag (treewidth + 1), as the cost between all possible assignments on the parent node need to be calculated considering all possible assignments on the child nodes. Note that calculating Eq. 1 requires calling the "energy_of_structure()" method from the ViennaRNA package. We consider this to be taking constant time (since the size of the structures is the same for each call).

Consequently, Algorithm 1 (*i.e.* the whole bottom-up step) runs in $O(NB \cdot 4^{(2k)})$ time, where N is the total number of nodes in the phylogeny and B is the number of bags in the tree decomposition. Note that the number of bags is directly proportional to the length of the input sequences and inversely proportional to the treewidth.

The middle step only involves running Algorithm 2 one additional time, as described in the main text.

Algorithm 2: computeBagCost(n, fam, bag)
This method calculates the cost of all possible nucleotide assignments and saves them into the cost matrix.

input : A node **n**, the structure family **fam** and a **bag**.
output: Filled out **n.costMatrix** for the family and bag considered.

1 **for** *every possible assignment for the given bag, nucAssign* **do**
2   **for** *every possible assignment for the same bag on its child nodes, nucAssignChild* **do**
3   costLeft = Equation 1 ; // with $ch_p \leftarrow$ nucAssignChild, $a_p \leftarrow$ nucAssign and sf $\leftarrow$ fam on the left child node
4   costRight = Equation 1 ; // with $ch_p \leftarrow$ nucAssignChild, $a_p \leftarrow$ nucAssign and sf $\leftarrow$ fam on the right child node
5   Save the minimum values for costLeft and costRight in *minLeft* and *minRight*;
6   totalCost = *minLeft* + *minRight*;
7   *n.costMatrix[fam][bag][nucAssign]* = totalCost;

Top-Down Step

Algorithm 3 kickstarts the whole top-down procedure by first initializing the optimal sequences for the ancestor at the root of the phylogeny, making use of Algorithm 5 to accomplish that. It then calls Algorithm 4 on the root of the phylogeny, which has the role of identifying optimal sequences for its children.

Algorithm 3: selectNucleos(n)
The method shows how we enumerate the optimal sequences based on the cost matrices.
input : The root of the phylogeny, invoked using **n**.
output: Saves a list of optimal sequences for every structure family, for all internal nodes (including the root) of the tree.

1 **for** *for every structure family, fam* **do**
2   seqTemplate = a list of placeholder characters for the sequence length;
3   optimalSeqs = a pair initialized to (∞, [empty list]);
4   n.selectNucleos4Bag(fam, firstBag, seqTemplate, accCost = 0, optimalSeqs);
5   sequenceList = optimalSeqs[1];
6   Save sequenceList for fam in the root node;
7 **selectNucleosChildren**(n);

Algorithm 4 uses optimal sequences already generated for the current node (n) to produce the optimal sequences related to these choices on the left and right child nodes, making use of Algorithm 5. It will then call itself recursively on both children until leaves are reached.

Algorithm 5 aims to find all the optimal assignments of nucleotides. It considers all possible assignments of the positions in the bag, except when a position has already been fixed in a previous bag ($O(4^k)$ in the worst case, where k is the size of the largest bag). It is a recursive method that will be called initially on the first bag of the tree decomposition and then on the following bags for each optimal sequence template produced ($O(4^k)$ if all sequences are optimal, which is unlikely in practice). Consequently, in the worst case, it runs in time $O(4^{Bk})$, where B is the number of bags. However, achARNement2, with its improved cost calculation approach, produces a small number of optimal sequences, so the average case is much faster in practice. To better represent the effect of the number of optimal sequences obtained on the overall runtime, we define variable P, which is $O(4^{Bk})$ in the worst case. So, the worst case runtime complexity of Algorithm 5 can also be represented by $O(P)$.

Building upon the runtime complexity of Algorithm 5, Algorithm 4 traverses all N nodes of the phylogeny and for each structure family (2, a constant factor ignored below), considers each optimal sequence of the parent (of which there are P, as described above) and uses Algorithm 5 on each of these. This results in a runtime complexity of $O(NP^2)$.

Algorithm 3 (which corresponds to the whole top-down step) only needs to do one call to Algorithm 5 before running Algorithm 4, which keep its worst-case runtime complexity at $O(NP^2)$. Once again, note that the number of optimal sequences P is very low in practice, even when simulating high mutation rates. This explains why the speed of achARNement2 is not significantly affected by increasing the mutation rate (as seen in Fig. 4).

Algorithm 4: selectNucleosChildren(n)
It is part of the top-down step of the algorithms, invoked for each node to select nucleotides.

input : A node **n** from the species tree, for which we have a list of optimal sequences.
output: Saves a list of optimal sequences for every structure family of the child nodes.

1 **if** *n.leftChild and n.rightChild are leaves* **then**
2   return ;   // No more selection needed, task done.
3 **for** *every structure family, fam* **do**
4   **for** *every optimal sequence for fam, parentSeq* **do**
5   assignedBags = all assignments of nucleotides corresponding to bags from parentSeq;
6   seqTemplate = a list of placeholder characters for the seq. length;
7   optimalSeqsLeft = a pair initialized to (∞, [empty list]);
8   **n.selectNucleos4Bag**(fam, firstBag, seqTemplate, accCost = 0, optimalSeqsLeft, assignedBags, n.leftChild);
9   seqTemplate = a list of placeholder characters for the seq. length;
10   optimalSeqsRight = a pair initialized to (∞, [empty list]);
11   **n.selectNucleos4Bag**(fam, firstBag, seqTemplate, accCost = 0, optimalSeqsRight, assignedBags, n.rightChild);
12   **if** *the left child is not a leaf* **then**
13   Append optimalSeqsLeft[1] to the left child;
14   **if** *the right child is not a leaf* **then**
15   Append optimalSeqsRight[1] to the right child;
16 **if** *the left child is not a leaf* **then**
17   selectNucleosChildren(n.leftChild);
18 **if** *the right child is not a leaf* **then**
19   selectNucleosChildren(n.rightChild);

Algorithm 5: selectNucleos4Bag(fam, bag, seqTemplate, acc-Cost, optSeqs, assignedBags[Optional], childNode[Optional])

This method, called from the node **n**, recursively selects nucleotides for each bag.

input : A structure family, **fam**; a **bag**; a **seqTemplate**; **accCost** which accumulates the cost for each bag assignment; **optSeqs**, a tuple (x,y) where x is the cost, y is a list of optimal sequences; **assignedBags**, a list of bag assignments derived from the optimal sequences of the direct ancestor; **childNode**, either left or right child node of **n**.

output: An updated tuple **optSeqs**, representing the optimals sequence(s) constructed and their cost.

1. Search *positions* of bag with fixed nucleotides in the *seqTemplate*;
2. assignOptions = All assignments of the bag with fixed nucleotide(s) at *positions* of the seqTemplate.;
3. currentDict = empty dictionary ;
4. optAssignList = empty list ;
5. minCost = ∞;
6. **for** *every nucAssign in assignOptions* **do**
7.   **if** *assignedBags is None* **then**
8.   costNucleos = n.costMatrix[fam][bag][nucAssign];
9.   **else**
10.   costNucleos = Equation 1 ;   // with $ch_p \leftarrow$ nucAssign, $a_p \leftarrow$ assignedBags[bag] and sf $\leftarrow$ fam on childNode
11.   **if** *costNucleos < minCost* **then**
12.   optAssignList = [nucAssign];
13.   minCost = costNucleos;
14.   **else if** *costNucleos == minCost* **then**
15.   optAssignList += nucAssign;
16.   newCost = accCost + minCost;

// Continued on the next page.

```
17  seqTemplateList = a list of new templates created by filling the assignments
    from the optAssignList into the seqTemplate received as a parameter;
18  if not the last bag then
19    for template in seqTemplateList do
20      candidateSeqs = a pair initialized to (∞, [empty list]);
21      if assignedBags is None then
22        n.selectNucleos4Bag(fam, nextBag, template, newCost,
                              candidateSeqs);
23      else
24        n.selectNucleos4Bag(fam, nextBag, template, newCost,
                              candidateSeqs, assignedBags, childNode);
25      if candidateSeqs[0] < optSeqs[0] then
26        optSeqs = candidateSeqs;
27      else if candidateSeqs[0] == optSeqs[0] then
28        optSeqs[1] += candidateSeqs[1];
29  else
30    optSeqs[0] = newCost;
31    optSeqs[1] = seqTemplates;
```

Combined Complexity

After combining the different steps of the approach, the final complexity of achARNement2 is $O(NB \cdot 4^{(2k)} + NP^2)$.

References

1. Bourque, G., Pevzner, P.A.: Genome-scale evolution: reconstructing gene orders in the ancestral species. Genome Res. **12**(1), 26–36 (2002)
2. Bradley, R.K., Holmes, I.: Evolutionary Triplet Models of Structured RNA. PLoS Comput. Biol. **5**(8), e1000483 (2009)
3. Clemente, J.C., Ikeo, K., Valiente, G., Gojobori, T.: Optimized ancestral state reconstruction using Sankoff parsimony. BMC Bioinform. **10**(1), 51 (2009)
4. Eddy, S.R., Durbin, R.: RNA sequence analysis using covariance models. Nucleic Acids Res. **22**(11), 2079–2088 (1994)
5. EMBL-EBI: Glossary: Clan. https://docs.rfam.org/en/latest/glossary.html
6. Fitch, W.M.: Toward defining the course of evolution: minimum change for a specific tree topology. Systematic Biol. **20**(4), 406–416 (1971)
7. Flamm, C., Hofacker, I.L., Maurer-Stroh, S., Stadler, P.F., Zehl, M.: Design of multistable RNA molecules. RNA **7**(2), 254–265 (2001)
8. Gupta, A., Maňuch, J., Stacho, L., Zhu, C.: Small phylogeny problem: character evolution trees. In: Sahinalp, S.C., Muthukrishnan, S., Dogrusoz, U. (eds.) CPM 2004. LNCS, vol. 3109, pp. 230–243. Springer, Heidelberg (2004). https://doi.org/10.1007/978-3-540-27801-6_17
9. Hagberg, A., Conway, D.: NetworkX: Network analysis with Python (2020). https://networkx.org/

10. Hammer, S., Wang, W., Will, S., Ponty, Y.: Fixed-parameter tractable sampling for RNA design with multiple target structures. BMC Bioinform. **20**(1), 1–13 (2019)
11. Hoeppner, M.P., Gardner, P.P., Poole, A.M.: Comparative analysis of RNA families reveals distinct repertoires for each domain of life. PLoS Comput. Biol. **8**(11), e1002752 (2012)
12. Jerome, L.J., van Biesen, T., Frost, L.S.: Degradation of FinP antisense RNA from F-like plasmids: the RNA-binding protein, FinO, protects FinP from ribonuclease E. J. Mol. Biol. **285**(4), 1457–1473 (1999)
13. Knudsen, B., Hein, J.: RNA secondary structure prediction using stochastic context-free grammars and evolutionary history. Bioinformatics **15**(6), 446–454 (1999)
14. Liberles, D.A.: Ancestral Sequence Reconstruction. Oxford University Press on Demand (2007)
15. Lorenz, R., et al.: ViennaRNA package 2.0. Algorithms Mol. Biol. **6**, 1–14 (2011)
16. Lyngsø, R.B., Anderson, J.W., Sizikova, E., Badugu, A., Hyland, T., Hein, J.: Frnakenstein: multiple target inverse RNA folding. BMC Bioinform. **13**, 1–12 (2012)
17. Marchand, B., Anselmetti, Y., Lafond, M., Ouangraoua, A.: Median and small parsimony problems on RNA trees. Bioinformatics **40**(Supplement_1), i237–i246 (2024)
18. Marchand, B., Ponty, Y., Bulteau, L.: Tree diet: reducing the treewidth to unlock FPT algorithms in RNA bioinformatics. Algorithms Mol. Biol. **17**(1), 1–17 (2022). article no. 8
19. Miklós, I., Kiss, S.Z., Tannier, E.: Counting and sampling SCJ small parsimony solutions. Theoret. Comput. Sci. **552**, 83–98 (2014)
20. Olson, R.D., et al.: Introducing the bacterial and viral bioinformatics resource center (BV-BRC): a resource combining PATRIC, IRD and ViPR. Nucleic Acids Res. **51**(D1), D678–D689 (2023)
21. Sankoff, D.: Minimal mutation trees of sequences. SIAM J. Appl. Math. **28**(1), 35–42 (1975)
22. Sankoff, D., Blanchette, M.: The median problem for breakpoints in comparative genomics. In: Jiang, T., Lee, D.T. (eds.) COCOON 1997. LNCS, vol. 1276, pp. 251–263. Springer, Heidelberg (1997). https://doi.org/10.1007/BFb0045092
23. Sankoff, D., Blanchette, M.: Multiple genome rearrangement and breakpoint phylogeny. J. Comput. Biol. **5**(3), 555–570 (1998)
24. Sorescu, D.A., Möhl, M., Mann, M., Backofen, R., Will, S.: CARNA–alignment of RNA structure ensembles. Nucleic Acids Res. **40**(W1), W49–W53 (2012)
25. Tesler, G.: Efficient algorithms for multichromosomal genome rearrangements. J. Comput. Syst. Sci. **65**(3), 587–609 (2002)
26. Tremblay-Savard, O., Reinharz, V., Waldispühl, J.: Reconstruction of ancestral RNA sequences under multiple structural constraints. BMC Genomics **17**, 175–186 (2016)
27. Urban, J.H., Vogel, J.: Two seemingly homologous noncoding RNAs act hierarchically to activate glmS mRNA translation. PLoS Biol. **6**(3), e64 (2008)
28. Yang, Z.: PAML 4: phylogenetic analysis by maximum likelihood. Mol. Biol. Evol. **24**(8), 1586–1591 (2007)
29. Yao, Z., Weinberg, Z., Ruzzo, W.L.: CMfinder–a covariance model based RNA motif finding algorithm. Bioinformatics **22**(4), 445–452 (2006)

30. Zheng, C., Chen, E., Albert, V.A., Lyons, E., Sankoff, D.: Ancient eudicot hexaploidy meets ancestral Eurosid gene order. BMC Genomics **14**, 1–13 (2013)
31. Zheng, C., Zhu, Q., Adam, Z., Sankoff, D.: Guided genome halving: hardness, heuristics and the history of the Hemiascomycetes. Bioinformatics **24**(13), i96–i104 (2008)

Whole-Genome Duplication Detection with Phylogenomics Reconciliation: A Scalable Approach

Reza Kalhor[1](✉), Manuel Lafond[1](✉), and Celine Scornavacca[2](✉)

[1] Department of Computer Science, Université de Sherbrooke, Sherbrooke, Canada
{Manuel.Lafond,Reza.Kalhor}@USherbrooke.ca
[2] Institut des Sciences de l'Evolution de Montpellier (Université de Montpellier, CNRS, IRD, EPHE), Montpellier, France
celine.scornavacca@umontpellier.fr

Abstract. Gene duplication plays a crucial role in species adaptation and the emergence of functions, making the inference of past duplications key to understanding evolution. Whole-genome duplications (WGD), which copy all the gene families in a single event, had a significant impact on the evolution of plants, yeast, and even vertebrates. Genome-scale data is often used to infer WGDs, for instance syntenic blocks and gene counts, but ancient WGDs are notoriously difficult to retrace, as their signals get lost in rearrangements and losses. Reconciliation of species and gene phylogenies may find ancient duplications, although current tools often assume independence between gene families and can miss WGDs, in which all genes are interdependent. Phylogenomics reconciliation addresses this by reconciling multiple gene families, but current models restrict the space of possible reconciliations, ignore gene losses due to fractionation, or rely on conserved syntenies across multiple species, limiting the number of genes that can be considered simultaneously.

In this work, we focus on a reconciliation model that is synteny-free, that considers losses, and that allows flexible remapping of gene duplications. Reconciliation under this model was shown NP-hard and current algorithms are far from achieving the scalability needed for phylogenomics analysis. We introduce novel algorithmic techniques that can handle large datasets containing tens of thousands of gene trees, which was not possible before. Our experiments on simulations and real data demonstrate that the traditional lca-mapping can make incorrect predictions in cases of WGD followed by fractionation, to which our approach is more robust. Tests on real data also show that our approach can recover WGDs that can be missed by other phylogenomics reconciliation methods.

Keywords: Phylogenomics · Reconciliation · Whole genome duplications · Algorithms

1 Introduction

Gene duplication is a fundamental mechanism for the development of novel functions and species adaptation [38]. Inferring past duplication events is therefore essential in understanding evolution, with applications ranging from reconstructing ancestral genomes [14] to inferring functionally similar genes [22,30,45]. *Reconciliation* is commonly used to infer gene-level events by explaining the discrepancies between a gene tree and a species tree [17]. In this paper, we consider duplications and losses as the main gene-level processes at work, for which there exist several reconciliation tools [6,13,15,24]. Most compare individual gene trees against a species tree, assuming that gene families can be reconciled independently, although it is well-known that macro-evolutionary events can affect multiple genes. A striking example includes *whole genome duplication* (WGD), which doubles all of the gene content [29,38]. WGDs play a pivotal role in genetic diversity, having impacted eukaryotes evolution [31,39]. In plants, WGDs are believed to be largely responsible for speciation events [10], and approximately one-third of modern vascular plant species underwent WGD [48]. A WGD took place roughly 100–150 Mya ago during the evolution of Saccharomyces yeasts [16,47] and milestones in fungal diversity were linked to genome doubling [1].

While genome-scale data such as conserved syntenic blocks or gene copy number counts can provide information on recent events, ancient WGDs involving distant species have time to scramble the genomes and leave little trace of conservation [43,44]. This is exarcerbated by *fractionation*, which tends to delete one of the gene copies after the event [7,20,33,43]. Other approaches based on K_s rates suffer from similar issues [46]. This motivates approaches based on phylogenies and reconciliation. Since reconciling gene families separately may miss WGD events, *phylogenomics reconciliation* incorporates multiple gene families simultaneously for the prediction of ancestral events. Guigó et al. [21] were among the first to extend reconciliation to multiple gene trees. They used heuristics until Page and Cotton [40] formalized it into the *Episodes Clustering* (EC) problem, which asks to map the duplication events into a minimum number of species. This was later extended to a variant called *Minimum Episodes* (ME) [5], where the number of duplication rounds within each species is also counted. There are very efficient algorithms for these problems [5,8,32,36,42] and related variants (see e.g. [18], and [42] for a survey), but these models have inherent drawbacks. First, for these methods, the speciation events predicted by the *lca-mapping* – which maps genes to their latest possible species– usually cannot be contested. This in turn fixes the set of duplications *a priori*, before searching for segmental events, which significantly reduces the space of considered reconciliations. Second, these models do not consider gene losses, even though they are informative as they often eliminate redundancy after large-scale duplications.

Recently, Delabre et al. [4,11] introduced a multi-gene reconciliation model that does handle losses. They aim to reconstruct the evolution of syntenic blocks, which can infer segmental duplication events. However, the scale of such events is limited by the size of detectable syntenies among dozens of species, which is usually small. The model also requires all the gene families within a block to evolve in compatible gene trees, which is seldom true in practice.

In this work, we focus on a reconciliation approach that avoids the aforementioned drawbacks: it infers segmental duplications in a model that is *synteny-free*, that *considers losses*, and that is *not restricted by the lca-mapping*. The model was introduced in [12], where it was shown that finding a most parsimonious reconciliation is NP-hard (see [12, Theorem 3]). A $O(\lceil \delta/\lambda \rceil^{d+1} \cdot n)$ time algorithm is known, where δ is the cost of a segmental duplication, λ is the cost of a single loss, d is the minimum number of segmental duplications, and n is the total number of gene and species tree nodes in the instance.

Unfortunately, the implementation of this algorithm can only handle small datasets, as it could not terminate even on modest instances with around 20 species and 100 gene trees. Moreover, the base $\lceil \delta/\lambda \rceil$ of the exponential complexity only allows small duplication/loss ratios, which has the same undesired effect of limiting the range of possible species of ancestral genes as previous work. Due to these limitations, the potential of the novel model has not been fully explored. We thus turn to heuristics to explore this question, and introduce an approach that starts with a given initial reconciliation, then explores the solution space by performing gene remapping moves until convergence.

Even with this simple strategy, the gargantuan size of the solution space comes with algorithmic challenges. In datasets of moderate size, there can be hundreds of thousands of possible moves, which need to be re-evaluated at each iteration. Minimizing the computation time per iteration is therefore crucial, leading to the combinatorial problem of evaluating the change in cost of each move. We propose a dynamic programming algorithm that computes the cost impact of each such move in $O(1)$ amortized time, allowing phylogenomics reconciliation that scales to thousands of gene trees.

Our implementation `FastMultRec` was evaluated on real and simulated data. To our knowledge, the added value of phylogenomics reconciliation against the traditional lca-mapping has not been established clearly in the literature, and sophisticated reconciliation models might not be needed. Our experiments say otherwise though. Indeed, we show on simulations that the lca-mapping may indicate incorrect WGD locations, mostly due to fractionation and complex loss patterns that follows WGD, to which our approach exhibits more robustness. This trend is confirmed when analyzing three real datasets exhibiting one or more WGD: our approach correctly finds all known WGDs with little false positive, which is not the case for the lca method. We also compare our approach with `MetaEC` [18], the most recent implementation for the EC problem. We confirm that restricting the search space is indeed an important limitation, as the explored scenarios often provide the same information as the lca method.

2 Methods

We use the standard notions of trees, forests, height, and distance. The root of a tree T is denoted $r(T)$; its set of leaves is $L(T)$ and its height is $h(T)$. The *distance* between two nodes u, v of T is $dist_T(u,v)$, the number of edges on the $u - v$ path. The set of trees in a forest F is denoted $t(F)$. The set of nodes of F is $V(F)$, and the leaves of F is $L(F) = \bigcup_{T \in t(F)} L(T)$, and the *height* of F is $h(F) = \max_{T \in t(F)} h(T)$. The subgraph of F induced by X is denoted $F[X]$. For $u, v \in V(F)$ in the same tree, we write $u \preceq_F v$ if u is a descendant of v, and $u \prec_F v$ when $u \neq v$. Two nodes u, v of F are *incomparable* if neither is an ancestor of the other. If $X \subseteq L(F)$ is a set of leaves from the same tree of $t(F)$, we write $\text{lca}_F(X)$ for the lowest common ancestor of X. We may drop the subscript F in the notation if unambiguous.

Reconciliations. A *gene tree* is a tree in which the leaves contain extant genes, and a *gene forest* is a forest G of gene trees. A *species tree* is a tree S in which $L(S)$ is a set of extant species. A *leaf species map* is a function $\sigma : L(G) \to L(S)$ that assigns each extant gene to its extant species.

Given a gene forest G, a species tree S, and a leaf species map σ, a *reconciliation* is a map $m : V(G) \to V(S)$ that assigns each gene forest node to a species node. Moreover, m must satisfy the following conditions:
1. *preserve leaf maps*: for each $u \in L(G)$, $m(u) = \sigma(u)$;
2. *time consistency*: for each non-root node $u \in V(G)$ with parent p_u, we have $m(u) \preceq_S m(p_u)$.

The *lca-mapping* between G and S is the unique reconciliation μ that assigns each node to the lowest possible species according to time-consistency (Fig. 1 top row). For each internal node u, $\mu(u) = \text{lca}_S(\{\sigma(w) : w \in L(G) \text{ and } w \preceq u\})$.

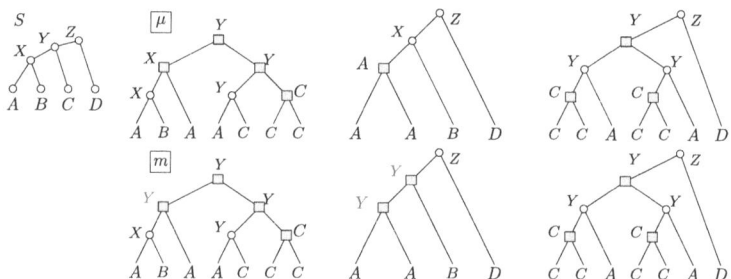

Fig. 1. Two possible reconciliations for a species tree S and a forest of three gene trees. Each leaf is labeled by σ, the species that contains it, and internal nodes are labeled according to μ (top) or m (bottom). Blue squares are Dup nodes, white circles are $Spec$ nodes, and light gray edges are losses. The reconciliation m needs three segmental duplications (gray rectangles), whereas μ needs five. (Color figure online)

Speciation, Duplication, and Loss. We use an established framework [19,49] to infer whether each internal node of G underwent a speciation ($Spec$) or duplication (Dup) event under a reconciliation m. In essence, we infer a speciation at u if its children branched into two separate species, and a duplication otherwise. The *event labeling under* m is the function $ev_m : V(G) \setminus L(G) \to \{Spec, Dup\}$ that, for each $u \in V(G)$ with children u_1, u_2, puts:

$$ev_m(u) = \begin{cases} Spec & \text{if } m(u) \text{ has children } x_1, x_2 \text{ s.t. } m(u_1) \preceq_S x_1, m(u_2) \preceq_S x_2 \\ Dup & \text{otherwise.} \end{cases}$$

A node $u \in V(G)$ with $m(u) = s$ and $ev_m(u) = Dup$ is called an s-*duplication* (with respect to m). As an example, in middle gene tree of Fig. 1, the node mapped to X according to μ is a *Spec* (top), but the same node mapped to Y according to m is a Dup (bottom).

Gene losses are counted on the branches of G by listing the species that should be present but are not. For an edge $uv \in E(G)$, the number of losses $l_m(uv)$ on uv is $dist_S(m(u), m(v)) - 1$ if $ev_m(u) = Spec$, and $dist_S(m(u), m(v))$ if $ev_m(u) = Dup$. The total number of losses under m is $l_m = \sum_{uv \in E(G)} l_m(uv)$.

Segmental Duplications. For a species $s \in V(S)$, the *duplication forest* of s (under m), denoted $F_m(s)$, is the subgraph of G induced by s-duplications. Formally: $F_m(s) = G[\{v \in V(G) : m(v) = s, ev_m(v) = Dup\}]$. A *segmental duplication* is a single duplication event that may contain one or *or more* genes. Such genes must coexist in the same species and must be incomparable in G and $F_m(s)$, and so it is not hard to see that the smallest possible number of segmental duplications in species s is the height of $F_m(s)$, which we recall is $h(F_m(s))$. The total number of segmental duplications in m is thus $d_m = \sum_{s \in V(S)} h(F_m(s))$. For example in Fig. 1, $d_\mu = 5$ (two duplications in Y, one in X, A, C), whereas $d_m = 3$ (two duplications in Y, one in C). We can finally state our problem.

The MOST PARSIMONIOUS SEGMENTAL RECONCILIATION problem

Input: a gene forest G, a species tree S, a leaf species map σ, and a duplication cost δ and loss cost λ.

Find: a reconciliation m between G and S of minimum cost $\delta \cdot d_m + \lambda \cdot l_m$.

Efficient exploration of the solution space

We propose the following strategy to explore the space of reconciliations:

1) start with an initial reconciliation m (typically the lca-mapping μ);
2) for each gene node $u \in V(G)$, and for each species s, compute the change in cost if we remap u to s;
3) choose one of the moves from the previous step, and apply it;
4) repeat step 2 and 3 until no more moves need to be considered.

The types of "remapping moves" that we consider at step 2 follows.

Up-Moves. Let $u \in V(G)$ and $s \succ m(u)$. Remapping u to s may create time-inconsistencies with its parent p_u, for instance if we remap u to a species that pre-existed $m(p_u)$. Our moves can thus also remap ancestors to preserve time-consistency. See Fig. 2 left box for an example, where remapping u to Z enforces remapping all the ancestors in Y. The *up-remapping of u to s* is another mapping $m[u \to s]$ in which:

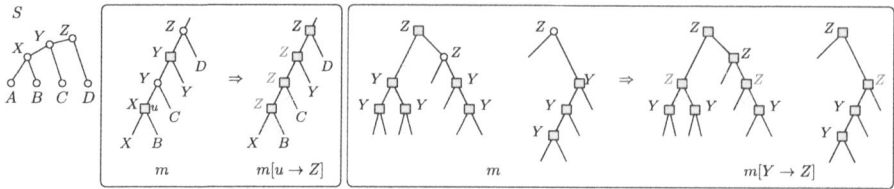

Fig. 2. An illustration of up-moves and bulk up-moves. Left: the species tree S. Middle: a portion of a gene tree undergoing an up-move that remaps u from X to Z. For time-consistency, the parent and grand-parent of u must also be remapped to Z, creating a chain of duplications in Z and additional losses. Right: a bulk up-move that takes the highest segmental duplication in Y spanning two gene trees, which are the roots of the trees in $F_m(Y)$, and remaps them to Z.

(1) $m[u \to s](w) = s$ for every ancestor $w \succeq u$ such that $m(w) \prec s$;
(2) $m[u \to s](w) = m(w)$ for every other node of G.

Note that (1) includes the remapping of u. The astute reader may ask whether up-moves that remap parents are necessary, as it would be simpler to only consider moves that do not remap the parent. Let us call the latter a *simple* up-move. The middle gene tree in Fig. 1 shows the advantage of the more complex type of up-moves. In that tree, we would like the A-duplication to be remapped to Y to save one duplication, but with simple moves we would need to remap the parent from X to Y first. However, that moves does not decrease the number of segmental duplications, but has more losses, so it is strictly worse. The more complex remap from A to Y lets us find an improved reconciliation directly, without having to go through a worse reconciliation first.

Bulk Up-Moves. We may need to remap several genes upwards to save a segmental duplication. One way is to take the genes in the same segmental duplication in some species s, and move that duplication to an ancestor of s. For $s \in V(S)$ and $t \succ s$, the *bulk up-remapping from s to t* is another mapping $m[s \to t]$ in which, for every root u of the forest $F_m(s)$, we apply the up-move that remaps u to t (in any order). Such bulk up-moves always reduce the height of $F_m(s)$, although they can increase the duplication height of other species. This is illustrated in Fig. 2, where the roots of the Y duplication forest are all remapped to Z, thereby reducing the number of segmental duplications in Y, but increasing those in Z in this case.

Down-Moves. A down-move remaps $u \in V(G)$ to a species s below $m(u)$. Since $\mu(u)$ is the lowest possible, such an s must satisfy $\mu(u) \preceq s \prec m(u)$, and we may need to remap descendants of u to s to avoid time-inconsistencies. The *down-remapping of u to s* is another mapping $m[u \to s]$ in which:
(1) $m[u \to s](w) = s$ for every descendant $w \preceq u$ such that $m(w) \succ s$;
(2) $m[u \to s](w) = m(w)$ for every other node of G.

Note that if a down-move can be applied on some $u \in V(G)$, then $m(u) \succ \mu(u)$, that is, before the remap u is mapped above its lca. In this case, it is known that u must be a *Dup* node (see e.g. [12, Lemma 2]). If the down-move is $m[u \to \mu(u)]$, then u is remapped to $\mu(u)$, and it is *possible* that u changes from a *Dup* node to a *Spec*, and even its parent p_u can become a *Spec*. This contrasts with up-moves: because these remap a node u above $\mu(u)$, the up-remapped node will always be a *Dup*, and up-moves can only turn *Spec* into *Dup* nodes. Because of these differences, the two types of moves require different algorithms.

Bulk Down-Moves. As for up-moves, several genes may be remapped "down-wards" at once to save a duplication (or more). Here, we take the deepest segmental duplication in s, and move it to a lower species.

For $s \in V(S)$ and $t \prec s$, the *bulk remapping from s to t* is another mapping $m[s \to t]$ that takes each deepest leaf u of $F_m(s)$, that is, every u of maximum depth $h(F_m(s))$, and applies the down-remapping of u to t. If such a down-move is not possible because $\mu(u) \preceq t \prec s$ does not hold, then $m[s \to t]$ is *invalid*.

Strategies and Bottlenecks. We consider two strategies to implement Step 3) above, which chooses a move. The *greedy* strategy simply chooses the move that yields the lowest reconciliation cost. The *stochastic* strategy assigns each possible reconciliation m' a weight $w_{m'} = e^{c/(kT)}$, where c is the cost of m', k is the Boltzmann constant and T is a user-specified *temperature* parameter, and then chooses a move randomly according to their weights. We stop when convergence is reached or when a specified number of iterations is achieved.

Step 2), which computes the change in cost of possible moves, is an important bottleneck. Its complexity controls the number of iterations we can perform in reasonable time. A naive implementation takes time $O(|V(G)|^2 h(S))$ to compute every entry, per iteration. Since we need to scale to gene forests with thousands of trees and tens of thousands of nodes, a quadratic dependency on $|V(G)|$ is not acceptable. In the following, we aim for an amortized time of $O(1)$ per entry.

Computing the Change in Cost of Up-Moves. We use dynamic programming to compute up-moves quickly. The full algorithm is somewhat involved, so here we present the main idea and refer to the appendix for details. Consider an up-move $m[u \to s]$ that remaps u to $s \succ m(u)$. For any $t \in V(S)$, we denote $\Delta(u, s, t) = h(F_{m[u \to s]}(t)) - h(F_m(t))$, which is the change in duplication height in species t after remapping u to s. Then denote $\Delta(u, s) = \sum_{t \in V(S)} \Delta(u, s, t)$, the *total* change in duplication heights after remapping u to s.

If we calculate those in pre-order on G (i.e., parents are handled before children), then we can re-use the information from the parents and only need to calculate a constant number of entries.

Lemma 1. *Let $u \in V(G)$ and let $s \in V(S)$ be an ancestor of $m(u)$. If u is a root, or if u has parent p_u satisfying $s \prec m(p_u)$, then $\Delta(u, s) = \Delta(u, s, s) + \Delta(u, s, m(u))$. Otherwise, u has a parent p_u satisfying $s \succeq m(p_u)$, in which case $\Delta(u, s) = \Delta(p_u, s) - \Delta(p_u, s, s) - \Delta(p_u, s, m(u)) + \Delta(u, s, s) + \Delta(u, s, m(u))$.*

Proof. (sketch). If u is a root, notice that only u gets remapped, and only the heights in $m(u)$ and s can change, so only $\Delta(u, s, s)$ and $\Delta(u, s, m(u))$ are relevant. If $s \prec m(p_u)$, again only u gets remapped, and the type of p_u (*Spec* or

Dup) does not get affected, and the same argument applies. In the last case where $s \succeq m(p_u)$, the up-move only affects duplications in $s, m(u)$, or t with $m(u) \prec t \prec s$. For those, the set of t-duplications that get removed is the same whether we remap u to s or p_u to s, because such a duplication must be a strict ancestor of u. Hence the difference in duplication heights for those t's is already counted in $\Delta(p_u, s)$. It is possible to have differences in s or $m(u)$ though, so we take $\Delta(p_u, s)$, remove $\Delta(p_u, s, s)$ and $\Delta(p_u, s, m(u))$ which could be different, and replace them with $\Delta(u, s, s)$ and $\Delta(u, s, m(u))$. □

As a consequence of Lemma 1, using a top-down approach, one can obtain $\Delta(u, s)$ by reusing $\Delta(p_u, s)$ and computing only four additional values, at most. This is far from being over though, as these are not trivial to obtain in $O(1)$ time each. They require information from other trees in the input forest, which in turn require maintaining auxiliary data structures efficiently—but ultimately it can be done. In a similar fashion, we can also compute the change in number of losses after remapping u to s, which we denote as $\Lambda(u, s)$. This can also be achieved by reusing $\Lambda(p_u, s)$, and then we only need to look around u locally to infer additional losses. Those techniques are all detailed in the appendix, as well as the analogous bottom-up algorithms for down-moves.

Theorem 1. *Given a gene forest G, a species tree S, and a reconciliation m, the values of $\Delta(u, s)$ and $\Lambda(u, s)$ for up-moves and down moves can be computed in amortized time $O(1)$ per entry, i.e., constant time per entry on average.*

Estimating Bulk Moves. It is quite challenging to compute the exact change in cost of every possible bulk move efficiently. We thus focus on moves that are guaranteed to be advantageous, and use upper bounds on their change in cost. Recall that in a bulk up-move $m[s \to t]$, we remap all the roots of the s-duplication subtrees of $F_m(s)$ to t. With the data structures necessary to achieve Theorem 1, these are easy to obtain. The move reduces the height in s-duplications by 1, but might increase the height in other species. If such an increase happens, no reduction in segmental duplications occurs and because up-moves create losses, the move will not be advantageous and we ignore it. We can detect this situation efficiently. The number of losses of a bulk up-move can be estimated as the sum of losses created by moving each gene individually, stored when computing the Λ values. For bulk down-moves, we use a similar idea—when considering a move $m[s \to t]$ with $t \prec s$, we take all the deepest nodes in the forest $F_m(s)$, and try to remap those to t. If this is invalid, or if such a remapping would increase the number of segmentnal duplications, we ignore the move. We can show that the list of candidate bulk moves can be built in total time $O(|V(G)|h(S))$, which does not add complexity to Theorem 1.

3 Results

The source code of our new method `FastMultRec` is available at https://github.com/r3zakalhor/FastMultRec. We implemented the greedy and stochastic version, but we report the results from the greedy version, since, in our simulations, the stochastic version took much longer to terminate and did not produce meaningfully better results[1]. We evaluated our algorithms on (i) simulated trees with known WGD locations on the species tree; and (ii) three real datasets with well-documented WGDs. Our approach was compared with the lca-mapping, and MetaEC [18], which solves the Episode Clustering problem. Our main goal is to validate that phylogenomics reconciliation models provide added value against the lca-mapping.

We do not compare our results with the exact algorithm developed in [12]. The reason for this is that this implementation could not terminate in reasonable time on either our simulated datasets or real data—except for very small δ/λ rates such as $\delta = 3, \lambda = 2$. Even with this setting, the software takes more than 8 h to terminate on a single dataset, and gives the same result as the lca mapping (mainly because such a low δ/λ leaves no room to remap a gene).

Simulation of Species and Gene Trees. We used SimPhy [34] to create species trees and gene trees. Since SimPhy cannot replicate the effects of a WGD, we borrowed ideas from [18] and planted WGDs by simulating a species tree, copying one of its subtrees, and letting Simphy evolve on the modified species tree. One key aspect to consider after a WGD is *fractionation*: since duplicated genes become redundant, there is often a gradual loss of one copy to restore genomic balance and reduce metabolic costs [7,33,43]. Without such loss patterns, the lca-mapping usually performs well, so we focus on the quality of reconciliation when fractionation occurs (which is also more biological relevant). SimPhy does no consider this, so we enforced it in a post-processing step: in each gene tree G, each leaf v was deleted with probability $(g-1)/g$, where g is the number of leaves of G in the same species as v. We used several duplication and loss rates: fixed rates (F) between 1e-6 and 1e-18; lognormally distributed (LN) with location in [-25,-15] and scale in $\{0.1, 0.3, 1\}$. For each parameters, we generated 100 simulations (called runs), each with 100 gene trees and a species tree having between 30 and 80 leaves, chosen randomly.

The results below apply to WGDs inferred on the true simulated gene trees. However, to assess the quality of predictions in the presence of reconstruction errors or incomplete lineage sorting, we performed experiments in which we applied nearest neighbor interchanges (NNI) on the gene trees before reconciling them. For space reasons, these results can be found in the appendix.

Precision and Recall. We assessed our ability in recovering WGD events. We posited that a species is considered to have undergone a WGD if it exhibits a duplication in more than 80% of the gene trees. Given the known planted WGDs, for each approach, we calculated the number of true positives/negatives and false positives/negatives, from which we obtained precision and recall values.

No WGD. As a sanity check, we reconciled the original SimPhy gene trees on which no WGD was planted. We report that none of the lca-mapping,

---

[1] Note that the $O(1)$ time computation described in the appendix was used for the up-remap but a slightly slower algorithm was used for down-moves as they are much more complex to implement, and were more rarely useful.

`FastMultRec` or `MetaEC` found a WGD under the above criterion. This is reassuring: the absence of false positives here shows that when a WGD is detected somewhere, the location may be ambiguous but a large-scale event likely took place.

One and two WGDs. Figure 3 shows the obtained precision and recall for one and two WGDs. The `FastMultRec` results are for duplication cost $\delta = 100$ and loss cost $\lambda = 1$. In the discussion, we justify why the ratio δ/λ should be proportional to the number of gene trees. Precision is generally high in most cases (notice the y-axis starts at 0.9), meaning that the number of false positive WGDs is low. We observe a slightly worse precision when two WGDs are present, but generally speaking getting WGD signals from non-WGDs is unlikely. The phylogenomics approaches tend to slightly improve upon the lca method, while we reckon that `MetaEC` has a better precision in some, but not all, datasets. Recall is also generally high, with an obvious down-spike for the lca-mapping and `MetaEC`. This occurs at the highest duplication rate 1e-6, and low recall indicates that false negatives can occur, probably because the maps are too conservative. We refer to the discussion section for possible explanations of these results.

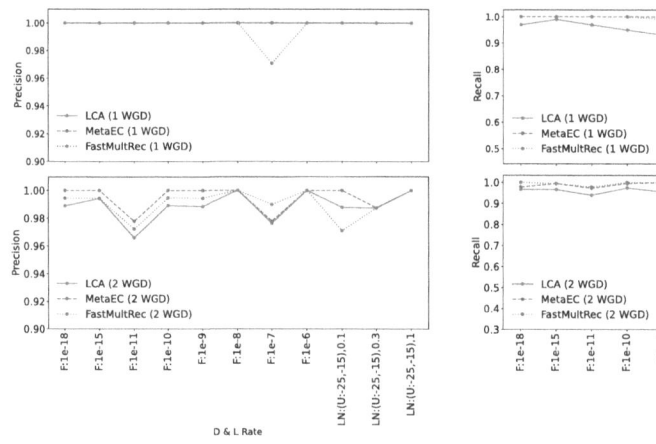
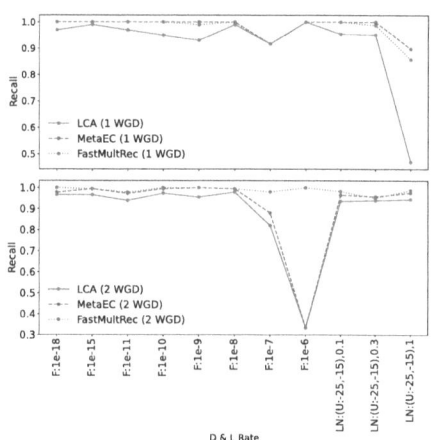

(a) Precision with 1 WGD and 2 WGDs, averaged over 100 runs.

(b) Recall for simulations with 1 WGD and 2 WGDs, averaged over 100 runs.

Fig. 3. Recall and Precision of detected WGDs, with one or two planted WGDs. Each result is an average derived from 100 simulation runs.

As mentioned previously, we also applied NNIs on the simulated trees to test for robustness (see appendix). The results are similar, with a better recall/precision for phylogenomics methods in most (but not all) cases. There are small differences between `MetaEC` and `FastMultRec`, the trend being that `FastMultRec` usually has better recall but a slightly worse precision, which can be explained by the fact that `FastMultRec` considers a larger solution space.

Path Distances. We wanted to validate that our gene-species maps did not worsen those predicted by the lca-mapping. We thus compared our inferred gene-species maps against the lca-mapping, using the simulated maps as our ground truth and the *path distance metric* [37]. Given the true Sim-Phy map s and an inferred map m, the path distance is $pathdist(s,m) = \sum_{v \in V(G)} dist_S(s(v), m(v))$, i.e., the sum of distances between the true and inferred species for each gene. Our values of path distances were similar to the lca-mapping and there is no major difference to report, despite differences in mappings, see appendix.

Perhaps more interesting is Fig. 4. On each dataset, it focuses on the number of gene trees with no errors, i.e., of path distance 0. The appendix shows these numbers for each approach, but Fig. 4 emphasizes the differences: it reports the quantity $noerror(FastMultRec) - noerror(lca)$, where $noerror(x)$ is the number of gene trees at path distance 0 using method x (results are shown for 2 WGDs, they were similar for other WGDs). We see that higher duplication costs have the potential to fix more trees, and that in several settings, a significant number of trees can be fully corrected. On the other hand, there are negative bars at higher duplication rates 1e-7 and 1e-6, indicating that aggressive remaps may perform moves on trees that were correct under the lca mapping. On the other hand, `FastMultRec` achieves better recall on these rates, so there is an inherent tradeoff. Note that `MetaEC` was not included, since its output files focus on WGD locations and not on gene-species maps.

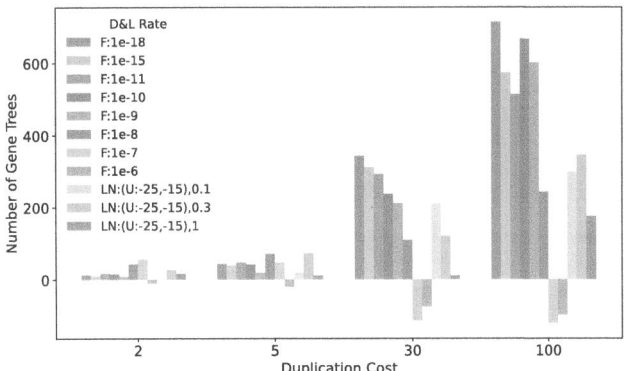

Fig. 4. Difference between the number gene trees with a correct map from `FastMultRec` minus the number from the lca-mapping, for each duplication cost. Note that one dataset has 10,000 trees in total, 100 gene trees for each 100 run.

Results on Real Datasets

We compared our results on three real datasets against the lca-mapping and `MetaEC`. Recall that the latter restricts *Spec* nodes to remain *Spec*, which limits the range of possible remaps for duplications[2]. This did not seem to affect the simulation results much, but significant differences emerge on real datasets. Note that we searched for other implementations, in particular for the Minimum Episodes formulation, but all of [5,8,32,36,42] either provide no link to an implementation, or a dead link. We report the number of gene trees that support a duplication in each of the ancestral species. We adjusted δ to the dataset sizes, see discussion. Note that again, we do not compare these results with those of [12], as this work only used small δ/λ values and, for that reason, gave the same reconciliation as the lca-mapping.

Eukaryotes Dataset. The first dataset consists of 53 gene trees from 16 eukaryote species, introduced by Guigó et al. in [21] and previously analyzed in [5,12,41]. In Fig. 5, each internal node is labeled as A/B, where A is the node identifier, and B represents the number of gene trees supporting duplications at that species. In [21], four segmental duplications were identified at nodes 6, 22, 28, and 30, supported by 6, 17, 9, and 13 gene trees, respectively. Additionally, two consecutive segmental duplications were observed at node 28 (that is, the height of the s-duplication forest is 2, where s is node 28). If we focus on the species in which at least one gene tree supports the duplication, as in [12], Fig. 5b

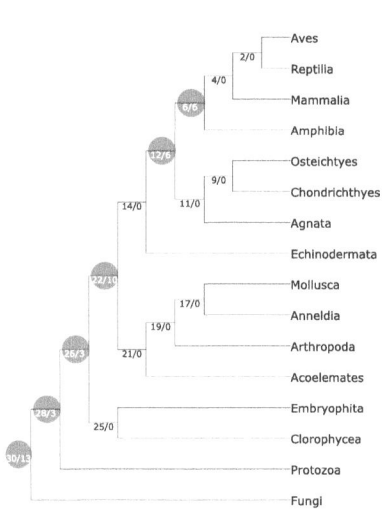

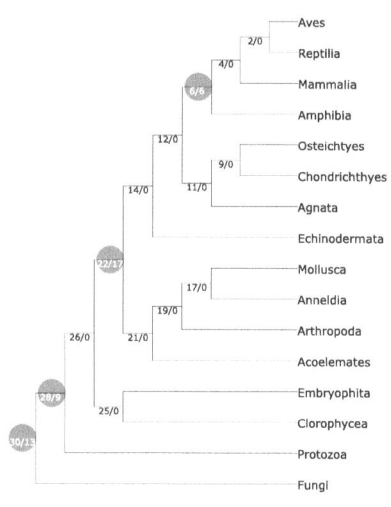

(a) Segmental duplications detected by the lca-mapping shown as gray circles.

(b) Segmental duplications from FastMultRec ($\delta = 50$, $\lambda = 1$). MetaEC yielded the same result as ours.

Fig. 5. Species tree of 16 eukaryotes studied by Guigó et al. for 53 gene trees. Nodes are labeled with two numbers (A/B), with A the node identifier and B the number of gene trees supporting duplications in A.

---

[2] Note that MetaEC can also handle genes from unknown species. We did not use this feature and MetaEC terminates within minutes even on the largest datasets.

shows that our method detects the same segmental duplications at nodes 6, 22, 28 (including two consecutive duplications), and 30, with the same number of supporting gene trees as reported in [21], using a setup where $\delta = 50$ and $\lambda = 1$. On the other hand, Fig. 5a shows that the lca-based method incorrectly identifies nodes 12 and 26 as nodes where one or more segmental duplications may have occurred. Our stochastic implementation gave the same results, and MetaEC also found the same reconciliation as ours, reproducing the results of [21].

Yeast Dataset. We also analyzed the yeast species dataset as presented in [9], restricted to 1,746 gene trees containing all 16 species. To refine unsupported branches, we employed the method detailed in [25] and implemented in ecceTERA [24], using a bootstrap threshold of 0.9 and setting $\delta = \lambda = 1$. Figure 6b demonstrates that our method identifies two significant segmental duplications at nodes 10 and 16, which have been proposed as WGDs in [35] (respectively the "pre-KLE branch" and the "WGD" in Fig. 1 of the paper). Additionally, the two hotspots of duplications at nodes 6 and 29 are suggested by both our algorithm and the lca method. For both nodes, almost all gene trees depicting a duplication on these nodes show apparent duplications (i.e. shared labels between the right and the left child subtrees of the node) so these two hotspots of duplications seem to be robust. Note that node 6 has been indicated in [35] to have a duplication peak (node "n8" in Fig. 1 of the paper). Node 29

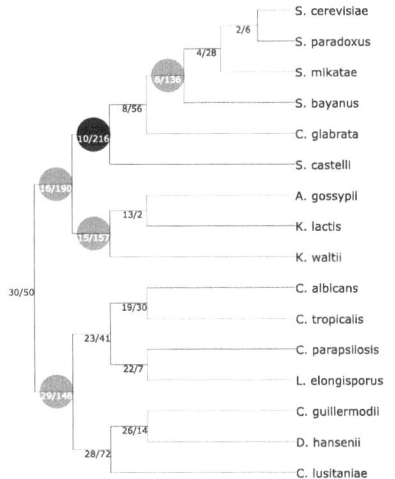
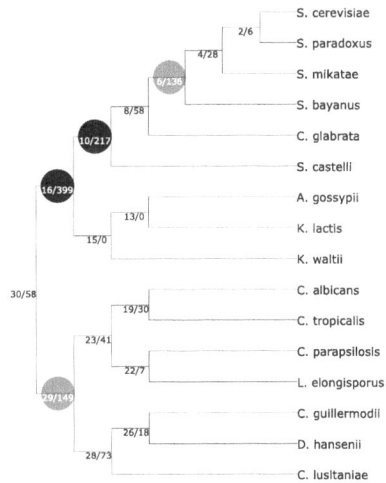

(a) WGD (black cicles) and Segmental duplications (gray circles) detected by lca. MetaEC produced the exact same reconciliation as the lca-mapping.

(b) WGD (black cicles) and Segmental duplications (gray circles) detected by our algorithm where $\delta = 1700$, $\lambda = 1$.

Fig. 6. Species tree of 16 yeast species for 1,746 gene trees. Each node is labeled with two numbers (A/B), where A is the node identifier and B is the count of gene trees supporting duplications in that species.

is not present in the dataset of [35] but several small-scale segmental duplications, often near the telomeres, are known to have happed in these species and can explain the duplication peak we observe for this node. The lca-mapping confirms only one WGD at node 10, see Fig. 6a, and node 15 is identified as a hotspot of duplications by lca but with less than 10% of the gene trees showing apparent duplications; our method relocates these duplications to the parent node, resulting in node 16 being classified as a WGD. Our stochastic version produced almost the same result, but with a slightly stronger support for node 29 (with 222 gene trees). As for MetaEC, it produced the same reconciliation as the lca-mapping and thus provided no additional information. This is likely due to the large number of gene trees, see discussion.

Fish Dataset. The final dataset we analyzed consists of 12,443 gene trees for 14 fish species [23]. In Fig. 7b, two major segmental duplications are identified at nodes 24 and 3, both of which align with previous findings of a WGD at the base of teleost fishes (3R) [3, 26], and a second one at the base of salmonid fish (4R) [2]. Nodes 18 and 12 are identified as segmental duplications by our method but respectively only 3,5% and 0,8% of the gene trees depicting a duplication on these nodes show apparent duplications; the percentage is almost 100% for nodes 24 and 3. We can thus discard nodes 18 and 12 as false positives. Our stochastic version gave the same results. On the contrary, the lca-mapping misses the WGD at node 24 and identifies several hotspots of duplications that are not confirmed

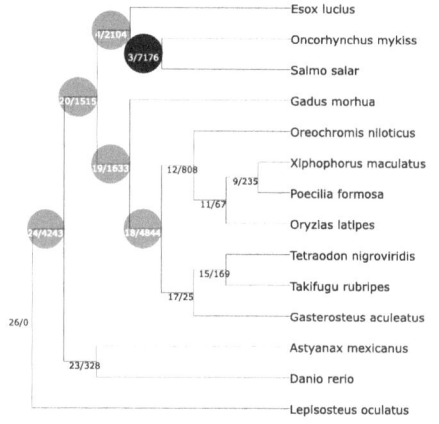
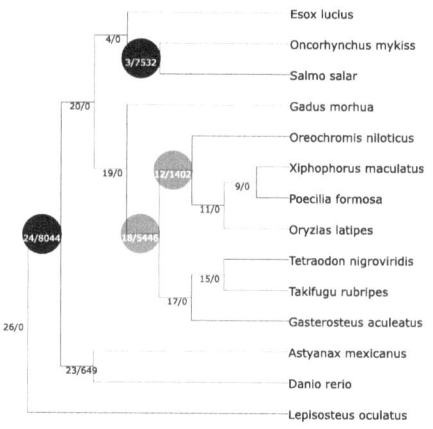

(a) WGD (black cicles) and segmental duplications (gray circles) detected by the lca-mapping. MetaEC produced the exact same reconciliation.

(b) WGD (black cicles) and segmental duplications (gray circles) detected by our algorithm where $\delta = 10000$, $\lambda = 1$.

Fig. 7. Species tree of atlantic salmon for 12,443 gene trees. Each node is labeled with two numbers (A/B), where A is the node identifier and B is the count of gene trees supporting duplications in that species.

by the literature, see Fig. 7a. Again, `MetaEC` produced the same reconciliation as the lca-mapping.

4 Discussion

Our experiments show that when large-scale duplications followed by fractionation occurs, the traditional lca-mapping is prone to mistakes, and alternate scenarios need to be explored. It is important to note that without fractionation, the lca-mapping made much less mistakes in our experiments (results not shown). Therefore, unless fractionation is known to be absent on a dataset, `FastMultRec` should be used as a tool to infer hypotheses on WGD locations that need to be validated through other means. Although our work focused on WGDs, the model could also be used to detect segmental duplications on a smaller scale. As we have seen on real datasets, a significant number of duplications may be identified in some species, but the signal may be too weak to infer a WGD. However, they can be seen as duplication hotpots which require further investigation. It would also be interesting to start with better maps than the lca-mapping to determine the initial speciations, perhaps using extra information such as synteny.

Autopolyploidy vs Allopolyploidy. We must reckon that our model assumes that genome doubling is due autopolyploidy, that is, intraspecies WGD. However, allopolyploidy events can merge the genomes of separate species and produce hybrids with similar doubling effects. In fact, this was hypothesized to have occurred on the yeast dataset that we studied [35]. As the authors argue, when this occurs reconciliation tends to infer duplication hotspots at the lowest common ancestor of the parents of the hybrid. This matches our observations. An interesting question is whether allopolyploidy leaves a signature that can be detected using our approach, perhaps taking inspiration from "segmental hybridation" algorithms that find horizontal gene transfers on whole genomes [27,28].

Importance of Not Restricting the Search Space. Our experiments found that `MetaEC` yielded the same results as the lca-mapping on our two largest real datasets. We did expect the approach to make *some* moves to save duplications, but upon closer inspection of the dataset, this can be explained by the restriction that *Spec* nodes must remain *Spec*. Consider for example the fish dataset, on which we move all the duplications in species 11 to species 12. If, under the lca-mapping, one such duplication v has a parent that is a *Spec* in 12, then v cannot be remapped at all under the EC constraints. This means that a duplication in node 11 *must* be present, making it pointless to move other duplications in 11. We validated that this was indeed the case on the dataset, and that the same phenomenon was present on the other species—as a result, `MetaEC` has no incentive to remap any node. In fact, this is hard to avoid on large datasets, as more gene trees make this more likely to occur. This explains why `MetaEC` performs well on the Guigó et al. dataset with only about 50 gene trees. Note that this constraint is present in both the EC and ME formulations, and it is therefore crucial to allow flexible remappings in phylogenomics reconciliation.

Impact of the Duplication Cost δ. We fixed the loss cost $\lambda = 1$, and δ weighs the importance of duplications versus losses. A small δ infers maps that are close to the lca-mapping, whereas larger δ explore more scenarios. We recommend testing multiple δ's when using `FastMultRec` to evaluate scenarios that cluster more and more duplications together. For WGDs, δ should be proportional to the number of input gene trees. This is because if the lca-mapping has k duplications mapped incorrectly, remapping them will create k losses, which will be considered detrimental if $\delta < k$. This number k of incorrect maps likely increases with the number of gene trees, which is why this quantity should be proportional to δ.

Running Times and Future Improvements. On a desktop computer, FAST-MULTREC took between 30 s and five minutes per SimPhy run, respectively with $\delta = 2$ and 100, and finished in seconds on the Guigó dataset. For yeast and fish, the times are highly dependent on δ. On yeast, a solution is found within two hours when $\delta = 20$, but takes about a day on $\delta = 2000$. On fish, the analysis took almost three days with $\delta = 10000$. Owing to our dynamic programming approach, the time per iteration does not increase much with higher δ, the main bottleneck now being the number of iterations needed to converge.

These technical improvements, coupled with other experiments on real data sets, will be needed to unleash the full potential of `FastMultRec`, which, to date, is the only solution that can infer duplication-loss phylogenomics reconciliation in a model that is synteny-free, that considers gene losses, and that can explore the solution space in an unrestricted manner.

References

1. Albertin, W., Marullo, P.: Polyploidy in fungi: evolution after whole-genome duplication. Proc. Roy. Soc. B Biol. Sci. **279**(1738), 2497–2509 (2012)
2. Allendorf, F.W., Thorgaard, G.H., Turner, B.J.: Evolutionary genetics of fishes. Tetraploidy and the Evolution of Salmonid Fishes, pp. 55–93 (1984)
3. Amores, A., et al.: Zebrafish hox clusters and vertebrate genome evolution. Science **282**(5394), 1711–1714 (1998)
4. Anselmetti, Y., Delabre, M., El-Mabrouk, N.: Reconciliation with segmental duplication, transfer, loss and gain. In: RECOMB International Workshop on Comparative Genomics, pp. 124–145. Springer (2022)
5. Bansal, M.S., Eulenstein, O.: The multiple gene duplication problem revisited. Bioinformatics **24**(13), i132–i138 (2008)
6. Bansal, M.S., Kellis, M., Kordi, M., Kundu, S.: Ranger-dtl 2.0: rigorous reconstruction of gene-family evolution by duplication, transfer and loss. Bioinformatics **34**(18), 3214–3216 (2018)
7. Birchler, J.A., Veitia, R.A.: The gene balance hypothesis: from classical genetics to modern genomics. Plant Cell **19**(2), 395–402 (2007)
8. Burleigh, J.G., Bansal, M.S., Wehe, A., Eulenstein, O.: Locating multiple gene duplications through reconciled trees. In: Vingron, M., Wong, L. (eds.) RECOMB 2008. LNCS, vol. 4955, pp. 273–284. Springer, Heidelberg (2008). https://doi.org/10.1007/978-3-540-78839-3_24
9. Butler, G., et al.: Evolution of pathogenicity and sexual reproduction in eight candida genomes. Nature **459**(7247), 657–662 (2009)

10. De Storme, N., Mason, A.: Plant speciation through chromosome instability and ploidy change: cellular mechanisms, molecular factors and evolutionary relevance. Current Plant Biol. **1**, 10–33 (2014)
11. Delabre, M., et al.: Reconstructing the history of syntenies through super-reconciliation. In: Comparative Genomics: 16th International Conference, RECOMB-CG 2018, Magog-Orford, QC, Canada, October 9-12, 2018, Proceedings 16, pp. 179–195. Springer (2018)
12. Dondi, R., Lafond, M., Scornavacca, C.: Reconciling multiple genes trees via segmental duplications and losses. Algorithms Mol. Biol. **14**, 1–19 (2019)
13. Doyon, J.-P., Scornavacca, C., Yu Gorbunov, K., Szöllősi, G.J., Ranwez, V., Berry, V.: An efficient algorithm for gene/species trees parsimonious reconciliation with losses, duplications and transfers. In: Comparative Genomics: International Workshop, RECOMB-CG 2010, Ottawa, Canada, October 9-11, 2010. Proceedings 8, pp. 93–108. Springer (2010)
14. Duchemin, W., et al.: Decostar: reconstructing the ancestral organization of genes or genomes using reconciled phylogenies. Genome Biol. Evol. **9**(5), 1312–1319 (2017)
15. Durand, D., Halldórsson, B.V., Vernot, B.: A hybrid micro-macroevolutionary approach to gene tree reconstruction. J. Comput. Biol. **13**(2), 320–335 (2006)
16. Gabaldón, T.: Hybridization and the origin of new yeast lineages. FEMS Yeast Res. **20**(5), foaa040 (2020)
17. Goodman, M., Czelusniak, J., William Moore, G., Romero-Herrera, A.E., Matsuda, G.: Fitting the gene lineage into its species lineage, a parsimony strategy illustrated by cladograms constructed from globin sequences. Syst. Biol. **28**(2), 132–163 (1979)
18. Górecki, P., Rutecka, N., Mykowiecka, A., Paszek, J.: Unifying duplication episode clustering and gene-species mapping inference. Algorithms for Molecular Biology **19**(1), 7 (2024)
19. Górecki, P., Tiuryn, J.: Dls-trees: a model of evolutionary scenarios. Theoret. Comput. Sci. **359**(1–3), 378–399 (2006)
20. Gout, J.-F., et al.: Dynamics of gene loss following ancient whole-genome duplication in the cryptic paramecium complex. Molecular biology and evolution **40**(5), msad107 (2023)
21. Guigó, R., Muchnik, I., Smith, T.F.: Reconstruction of ancient molecular phylogeny. Mol. Phylogenet. Evol. **6**(2), 189–213 (1996)
22. Hellmuth, M., Wieseke, N., Lechner, M., Lenhof, H.-P., Middendorf, M., Stadler, P.F.: Phylogenomics with paralogs. Proc. Natl. Acad. Sci. **112**(7), 2058–2063 (2015)
23. Hermansen, R.A., Hvidsten, T.R., Sandve, S.R., Liberles, D.A.: Extracting functional trends from whole genome duplication events using comparative genomics. Biological procedures online **18**, 1–10 (2016)
24. Jacox, E., Chauve, C., Szöllősi, G.J., Ponty, Y., Scornavacca, C.: eccetera: comprehensive gene tree-species tree reconciliation using parsimony. Bioinformatics **32**(13), 2056–2058 (2016)
25. Jacox, E., Weller, M., Tannier, E., Scornavacca, C.: Resolution and reconciliation of non-binary gene trees with transfers, duplications and losses. Bioinformatics **33**(7), 980–987 (2017)
26. Jaillon, O., et al.: Genome duplication in the teleost fish tetraodon nigroviridis reveals the early vertebrate proto-karyotype. Nature **431**(7011), 946–957 (2004)
27. Kloub, L., et al.: Systematic detection of large-scale multigene horizontal transfer in prokaryotes. Molecular Biology Evolution **38**(6), 2639–2659 (2021)

28. Kloub, L., Gosselin, S., Graf, J., Gogarten, J.P., Bansal, M.S.: Investigating additive and replacing horizontal gene transfers using phylogenies and whole genomes. Genome Biol. Evol. **16**(9), evae180 (2024)
29. Kuzmin, E., et al.: Exploring whole-genome duplicate gene retention with complex genetic interaction analysis. Science **368**(6498), eaaz5667 (2020)
30. Lafond, M., Miardan, M.M., Sankoff, D.: Accurate prediction of orthologs in the presence of divergence after duplication. Bioinformatics **34**(13), i366–i375 (2018)
31. Li, Z., et al.: Multiple large-scale gene and genome duplications during the evolution of hexapods. Proc. Natl. Acad. Sci. **115**(18), 4713–4718 (2018)
32. Luo, C.-W., et al.:Linear-time algorithms for the multiple gene duplication problems. IEEE/ACM Trans. Comput. Biol. Bioinform. **8**(1), 260–265 (2009)
33. Lynch, M., Conery, J.S.: The evolutionary fate and consequences of duplicate genes. Science **290**(5494), 1151–1155 (2000)
34. Diego Mallo, Leonardo de Oliveira Martins, and David Posada. Simphy: phylogenomic simulation of gene, locus, and species trees. *Systematic biology*, **65**(2), 334–344 (2016)
35. Marcet-Houben, M., Gabaldón, T.: Beyond the whole-genome duplication: phylogenetic evidence for an ancient interspecies hybridization in the baker's yeast lineage. PLoS Biol. **13**(8), e1002220 (2015)
36. Mettanant, V., Fakcharoenphol, J.: A linear-time algorithm for the multiple gene duplication problem. In: The 12th National Computer Science and Engineering Conference (NCSEC), pp. 198–203 (2008)
37. Moulton, V., Steel, M.: Retractions of finite distance functions onto tree metrics. Discret. Appl. Math. **91**(1–3), 215–233 (1999)
38. Ohno, S.: Evolution by Gene Duplication. Springer-Verlag (1970)
39. Otto, S.P.: The evolutionary consequences of polyploidy. Cell **131**(3), 452–462 (2007)
40. Page, R.D.M.: Maps between trees and cladistic analysis of historical associations among genes, organisms, and areas. Syst. Biol. **43**(1), 58–77 (1994)
41. Roderic, D.M.P., Cotton, J.A.: Vertebrate phylogenomics: reconciled trees and gene duplications. In: Biocomputing 2002, pp. 536–547. World Scientific (2001)
42. Paszek, J., Górecki, P.: Efficient algorithms for genomic duplication models. IEEE/ACM Trans. Comput. Biol. Bioinf. **15**(5), 1515–1524 (2017)
43. Rabier, C.-E., Ta, T., Ané, C.: Detecting and locating whole genome duplications on a phylogeny: a probabilistic approach. Mol. Biol. Evol. **31**(3), 750–762 (2014)
44. Steenwyk, J.L., King, N.: The promise and pitfalls of synteny in phylogenomics. PLoS Biol. **22**(5), e3002632 (2024)
45. Ullah, I., Sjöstrand, J., Andersson, P., Sennblad, B., Lagergren, J.: Integrating sequence evolution into probabilistic orthology analysis. Syst. Biol. **64**(6), 969–982 (2015)
46. Vanneste, K., Van de Peer, Y., Maere, S.: Inference of genome duplications from age distributions revisited. Mol. Biol. Evol. **30**(1), 177–190 (2013)
47. Wolfe, K.H.: Origin of the yeast whole-genome duplication. PLoS Biol. **13**(8), e1002221 (2015)
48. Wood, T.E., Takebayashi, N., Barker, M.S., Mayrose, I., Greenspoon, P.B., Rieseberg, L.H.: The frequency of polyploid speciation in vascular plants. Proc. Natl. Acad. Sci. **106**(33), 13875–13879 (2009)
49. Zhang, L.: On a mirkin-muchnik-smith conjecture for comparing molecular phylogenies. J. Comput. Biol. **4**(2), 177–187 (1997)

A Sankoff-Rousseau-Like Algorithm for Minimizing Lateral Gene Transfers and Losses on Single Origin Characters

Alitzel López Sánchez[1](✉), Guillaume E. Scholz[2], Peter F. Stadler[2,3,4,5,6], and Manuel Lafond[1]

[1] University of Sherbrooke, Sherbrooke, QC, Canada
{alitzel.lopez.sanchez,manuel.lafond}@usherbrooke.ca
[2] University of Leipzig, Leipzig, Saxony, Germany
[3] Max Planck Institute for Mathematics in the Sciences, Leipzig, Saxony, Germany
[4] University of Vienna, Vienna, Austria
[5] Santa Fe Institute, New Mexico, USA
[6] Universidad Nacional de Colombia, Bogotá, Colombia

Abstract. The simple underlying pattern of presence-absence of a character within a species tree provides useful steps to trace complex evolutionary histories. Character-based models such as *Perfect Transfer Networks* and its galled variant aim to leverage this information to predict horizontal gene transfers. Under the assumption that characters have a single origin, are rarely lost and can be transferred horizontally, they remain an efficient inference method for almost tree-like scenarios. Nevertheless, they can sometimes predict overly complicated scenarios and its simplest structural variants are too restrictive for practical uses. With the goal of extending this model to include loss events, we present a Sankoff-Rousseau-like algorithm that aims to recover the simplest possible scenarios that combine gene transfers and losses using solely the single character information already contained in a given species tree. We establish a link between the Small Parsimony Problem and the inference of scenarios with a minimal number of losses and transfers. Allowing losses and transfers to have a user-defined penalization for this end. We also explore the utility of our model for tracing possible highways of gene transfers by presenting a real case study on a dataset of bacterial species and KEGG functions as characters.

Keywords: horizontal gene transfer · LGT networks · character-based methods · homoplasy-free

1 Introduction

Horizontal Gene Transfer (HGT) is the transmission of genetic material between co-existing organisms, and provides an alternative DNA exchange mechanism to the more widely studied parent-offspring relationships. Occurring both between and within all domains of life, HGT constitutes an important contribution to genetic diversity and serves a source of functional innovation through the introduction of novel genes and metabolic pathways. It is one of the most conspicuous features found in bacterial genomes [11], and notably eases the adaptation of organisms to new conditions, which are sometimes extreme and life-threatening [2]. Transfers are also involved in eukaryotes, but large-scale HGT studies are hampered by the complexity of their genomes. They are nevertheless present, with a recent example including the transfer of genes that degrade cellulose from bacteria to beetles [34].

There are currently two basic approaches to infer HGT [26]: *parametric* methods which look for outliers throughout the genomic composition of single individuals, and *phylogenetic* methods which look for unusual evolutionary histories among groups of organisms. The majority of current methods rely on sequence-based information as input. However, these methods become unreliable to predict ancient transfers as sequences are highly divergent [32].

An alternative to sequence-based methods are *character*-based methods. A character is a morphological or molecular trait that a taxon may possess or lack. Character-based approaches may recover phylogenetic signals more accurately when highly divergent sequences are involved. Given a set of taxa and a set of characters, the aim is to explain the character diversity in terms of changes of character states. Several character-based models have been established in the literature, each imposing conditions on how a character may emerge and evolve. For example, a *perfect phylogeny* requires that characters change their state at most once. Here, we will be concerned with a particular type of characters that have only two states: presence and absence. A character transitions between absence to presence is called a *gain*; a transition from presence to absence it is called a *loss*. In this setting, a variant of perfect phylogeny requires a character to be gained once and never lost. Although the perfect phylogeny model has been shown to be too restrictive to explain the evolutionary history of a set of characters, its extensions remain an active area of research with promising applications [6].

Relaxations of character state transition models were subsequently developed, including *Dollo parsimony* [14], where characters can only be gained once but be lost many times. Optimization problems involving the reconstruction of phylogenies under Dollo parsimony were shown to be NP-hard, even for binary characters [12]. More recently, a variant of this model, the *Dollo-k* model, where characters can be lost at most k times, was studied in [7].

An appealing alternative is to consider network-like structures instead of trees to explain characters. A representative model called *Perfect Phylogenetic Networks* (PPNs) was introduced in [24,25], and asks for a tree displayed by the network to be a perfect phylogeny. Unfortunately, this model is difficult to work with, since even deciding whether a known network explains a set of (multi-state) characters is NP-hard. In [22], we introduced a special case of PPNs called *Perfect Transfer Neworks* (PTNs). In this model characters are binary, have a unique origin and cannot be lost once acquired. The biological motivation behind PTNs is to study transferable characters that are difficult to revert such as organelles in endosymbiotic events [1,33] and metabolites [16]. In contrast to PPNs, the tree that dictates the vertical inheritance is fixed in PTNs. This affects the complexity of recognition and optimization problems: Deciding whether a network explains a given set of characters and whether we can augment a tree with transfer arcs to explain the evolution of a set of characters are polynomial-time solvable.

Although easy to compute, PTNs can sometimes predict overly complicated evolutionary scenarios that require a larger number of transfers (note that so-called *galled PTNs* were introduced in [20] to simplify these scenarios, but the structure imposed on the networks was shown to be too restrictive in practice). One of the main reasons that perfect phylogeny models extended to networks is that a single loss within a clade may require adding many transfers within that clade, to ensure that all taxa with the character are connected through the single origin. This leads to the question of whether a balance between transfers and loss events could help us gain a better insight into more plausible biological scenarios. The explicit inclusion of transfer and loss parameters allows us to model diverse biological scenarios. For example, higher loss rates are often associated with the evolution of pathogenic bacteria and symbionts [23]. Conversely, certain genera, such as *Listeria*, exhibit reduced rates of both gene gain and loss [3].

Our Contribution: In this work, our goal is to broaden the definition of PTNs by incorporating loss events as in the Dollo parsimony model. This assumption has been shown to be more suitable for complex characters such as restriction sites and introns [15]. We study the problem of finding the *simplest* possible evolutionary scenario that combines transfer events where the characters are assumed to have originated once (this is known as the "no-homoplasy" condition). We consider the possibility of transfers and loss events having a different cost on the resulting evolutionary scenario. Associating costs to evolutionary events has been previously explored in the context of host-parasite reconciliation networks [10]. We are therefore interested only in scenarios that minimize the total cost provided by transfers and losses.

Through the *Minimum homoplasy-free completion problem* we explore the problem of augmenting a given character-labeled tree with transfer arcs and losses in order to achieve a minimal cost scenario where the transfer arcs are *time-consistent*. We start by illustrating the connection between Dollo Parsimony and the Small Parsimony Problem. We show that through the usage of the well known Sankoff-Rousseau algorithm [27] we can find an inner labeling of the given tree that minimizes the loss and transfer cost.

We conclude with a case study of inferred scenarios on a bacterial dataset consisting of bacterial species and functional characters obtained from the Kyoto Encyclopedia of Genes and Genomes (KEGG) database [18]. We show that the phylogenetic signal of some transfer events reported in the literature is contained within special pairs of nodes in the given species tree known as *transfer highways* [4,5] Note that the full proofs of our results are in Appendix 2.

Related Work: From a stochastic prespective, a model that combines the inference of horizontal transfers with bayesian inference was presented in [19].

From a combinatorial perspective, besides PPNs and variants [31], the model closest to our work, to our knowledge, was presented by Van Iersel, et al. in [30]. In this work the authors provide upper of lower bounds on the number of transfers needed when considering transfer and loss events to explain the evolutionary history of a set of characters. They provide upper and lower bounds for the number of transfers needed to explain a set of characters when we require that every node in the resulting network contains at most k characters. Additionally, they show that finding a character assignment for a given network and a set of characters that satisfies these conditions is NP-complete. Furthermore, the problem of penalizing losses in this context was left as an open problem in their conclusion.

2 Preliminaries

Unless stated otherwise, all graphs in this work are directed and loopless. For a graph N, $V(N)$ and $E(N)$ denote the sets of vertices and arcs of N, respectively, and $L(N)$ denotes the set of leaves, which are the vertices of outdegree 0. The vertices in $V(N) \setminus L(N)$ are called *internal nodes*. For a subset of vertices $X \subseteq V(N)$, $N[X]$ denotes the subgraph of N induced by X, which is the graph with vertex set X and arc set $\{(u,v) : u \in X, v \in X, (u,v) \in E(N)\}$.

A *phylogenetic network*, or simply a *network*, is a directed acyclic graph N with a unique node $\rho(N)$ of indegree zero called the *root*, in which all leaves have indegree 1. A node of indegree one and outdegree 1 is a *subdivision node*, which we allow. We say that a node $v \in V(N)$ *reaches* a node $u \in V(N)$ if there exists a directed path from v to u in N.

A tree T is a network whose underlying undirected graph has no cycles. For $v \in V(T)$, a *child* of v is a node u such that $(v,u) \in E(T)$, and v is the *parent* of u. So edges are oriented from parent to child. More generally, we say that a node $v \in V(T)$ is an *ancestor* of $u \in V(T)$ if v is on the path from $\rho(T)$ to u. In this case, we will call u a *descendant* of v and write $u \preceq_T v$ (or $u \prec_T v$ if we know that $u \neq v$). The ancestor order $\preceq_T$ is a partial order of $V(T)$ and, in particular, $\rho(T)$ is the unique maximal element. We say that two nodes u and v are *comparable* if $u \preceq_T v$ or $v \preceq_T u$ and we say that they are *incomparable* otherwise. We will drop the subscript T when T is clear from the context. For $v \in V(T)$, $T[v]$ denotes the subtree rooted at v.

LGT Networks and Time-Consistency. An *LGT network* (where LGT comes from *Lateral Gene Transfers*) [9] is a network $N = (V, E_S \cup E_T)$, where $\{E_S, E_T\}$ is a specified partition of the arc-set of N, such that the subgraph $\mathcal{T}_N := (V, E_S)$ is a tree with the same set of nodes as N. The tree $\mathcal{T}_N$ is called the *support tree* of N. The arcs in E_S are called *support arcs* and the arcs in E_T are called *transfer arcs*. For a transfer arc $(u,v) \in E_T$, the endpoints u and v are called *transfer nodes* and in particular, u is called the *donor* and v is called the *recipient*. We assume that transfer nodes have exactly one out-neighbor in $\mathcal{T}_N$. The tree obtained from $\mathcal{T}_N$ by suppressing its subdivision nodes[1] is called the

---

[1] Suppressing a subdivision node u with parent p and child v consists of removing u and adding an arc from p to v.

base tree of N. Note, we have not defined a notion of ancestry or partial order between the nodes of an LGT network, as it is not needed. Instead, we always refer to the ancestry relationships between nodes of N in the support tree $\mathcal{T}_N$, using the notation $\preceq_{\mathcal{T}_N}$ (which is well-defined above since $\mathcal{T}_N$ is a tree). We will sometimes refer to the ancestor order of an LGT network N, note that in this case, we will refer to the partial order on its support tree $\preceq_{\mathcal{T}_N}$.

We aim to reconstruct LGT networks that are *biologically-feasible* in terms of time. This implies that the transfers that appear should exist only between ancestral species that co-existed. We define a *time consistent map* over the nodes of an LGT network $N = (V, E_S \cup E_T)$ as a function $\tau : V \to \mathbb{R}$ such that:

- for every $(u, v) \in E_S$, $\tau(u) > \tau(v)$.
- for every $(u, v) \in E_T$, $\tau(u) = \tau(v)$.

A network N is *time-consistent* if there exists a time consistent map of N.

2.1 Homoplasy-Free Scenarios

Let $\mathcal{S}$ be a set of taxa. A *character* γ (on $\mathcal{S}$) is a function $\gamma : \mathcal{S} \to \{0,1\}$, where $\gamma(x) = 1$ indicates that species x possesses the character, while $\gamma(x) = 0$ indicates that it does not. We focus on the evolution of a single character, and discuss in the experiments possible ways to handle multiple characters.

To formalize transfer networks, given an LGT network N, a *labeling* of N is a function $l : V(N) \to \{0,1\}$ that indicates which nodes possess the character or not. Note that $l^{-1}(1)$ denotes the set of nodes of N labeled 1 in l, and $l^{-1}(0)$ those labeled 0. If N and l are clear from the context, the nodes in $l^{-1}(1)$ may be called 1-nodes, and those of $l^{-1}(0)$ may be called 0-nodes.

Definition 1. *Let $\mathcal{S}$ be a set of taxa, let γ be a character on $\mathcal{S}$, and let $N = (V, E_S \cup E_T)$ be an LGT network with leafset $L(N) = \mathcal{S}$. We say that a labeling l of N is an* homoplasy-free fit *of γ if the following conditions hold:*

1. *for each leaf x, $l(x) = \gamma(x)$ (leaves are labeled by their character);*
2. *denote $V_1 = l^{-1}(1)$. Then in $N[V_1]$, the subgraph induced by 1-nodes, there exists a unique node $v \in V_1$ that reaches every node of $N[V_1]$ (single origin).*

Furthermore, we call the pair (N, l) a homoplasy-free scenario *for γ.*

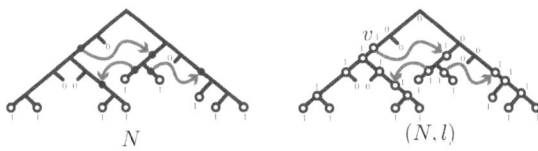

Fig. 1. (Left) A network N and a character γ. (Right) An homoplasy-free scenario for γ, where v is the unique node that reaches all 1-nodes.

The *single origin* condition states that the first node that acquires a character transmits it to every other species that possess the character, through a directed path of 1-nodes. See node v in Fig. 1. This models the "no-homoplasy" condition, i.e., that a character cannot emerge independently in two species.

Transfer and Loss Events. Given an homoplasy-free scenario (N, l) for some character γ, where $N = (V, E_S \cup E_T)$, the *number of transfers* of (N, l) is simply $|E_T|$, the number of transfer arcs. A *loss* is an arc $(u, v) \in E_S$ of the support tree such that $l(u) = 1$ and $l(v) = 0$. This represents failure to transmit the character to a direct vertical descendant.

We emphasize that a loss can only occur on a support tree arc, representing the lack of vertical transmission—a character that is present on the donor of a transfer arc but is not on the recipient is not seen as a loss. Also note that in previous work, losses were entirely forbidden [22], while here we do not restrict the number of losses of a character.

2.2 The Minimum Homoplasy-Free Completion Problem

Let (N, l) be an homoplasy-free scenario for some character γ. Given a transfer cost $cost_T$ and a loss cost $cost_L$, we define the *transfer-loss cost* of (N, l) as:

$$cost_{TL}(N, l) = cost_T \cdot |E_T| + cost_L \cdot |losses(N, l)|$$

where $losses(N, l) = \{(u, v) \in E_S : l(u) = 1, l(v) = 0\}$ is the set of support arcs on which the character was lost. Our main problem of interest is to explain how a set of characters may have evolved along a species tree in an homoplasy-free and time-consistent manner, with transfer and loss events being possible.
The Minimum homoplasy-free completion problem
Input. A character γ on a taxa set $\mathcal{S}$; a tree T with leaf-set $L(T) = \mathcal{S}$; a weight $cost_T$ for transfer events; a weight $cost_L$ for loss events.
Output. An homoplasy-free scenario (N, l) for γ, such that N is time-consistent, has T as base tree, and such that $cost_{TL}(N, l)$ is minimized.

In the following, an homoplasy-free scenario (N, l) for γ that is time-consistent and has T as a base tree will be called a *completion* of T.

3 A Polynomial-Time Algorithm for the Minimum Homoplasy-Free Completion Problem

Our strategy is to infer a labeling λ on the input tree T directly, instead of reconstructing a network N and then labeling it. We call λ an infra-labeling, as it serves as the labeling underlying an intended completion. We argue that the losses inferred on T by the labeling λ can be extended to a completion (N, l) of T with the same number of losses. In a similar manner, the arcs (u, v) of T such that $\lambda(u) = 0$ and $\lambda(v) = 1$ correspond to a *gain*, which must be acquired by a transfer event—except for one such gain that corresponds to the origin of the character. In other words, a labeling λ of T implies the existence of a completion

(N, l) whose number of losses is the number of arcs (u, v) of T that transition from 1 to 0, and the number of transfers is the number of arcs (u, v) of T that transition from 0 to 1, minus one.

3.1 Infra-Labelings and Completions

Formally, for a tree T an *infra-labeling* λ for character γ is a function $\lambda : V(T) \to \{0, 1\}$ such that leaves are labeled by their characters, i.e., $\lambda(x) = \gamma(x)$ for each $x \in L(T)$. The pair (T, λ) is called an *infra-labeled tree* (for γ). Given such a pair (T, λ), an arc $(u, v) \in E(T)$ is a *gain arc* if $\lambda(u) = 0$ and $\lambda(v) = 1$, and we call v a *gain node*, or just a gain for short. For technical reasons, if the root of T satisfies $\lambda(\rho(T)) = 1$, then $\rho(T)$ is also a gain node. We assume henceforth that at least one gain node exists, as otherwise this implies that no leaf has the character. We next define
$losses(T, \lambda) = \{(u, v) \in E(T) : \lambda(u) = 1, \lambda(v) = 0\}$,
$gains(T, \lambda) = \{(u, v) \in E(T) : \lambda(u) = 0, \lambda(v) = 1\}$.

Recall that if (N, l) is a completion of T, then the base tree T of N is obtained by removing transfer arcs and suppressing subdivision nodes. In this way, $V(T) \subseteq V(N)$, and we use $V(N) \cap V(T)$ to explicitly refer to the nodes of N that are also in the base tree. The single origin condition of an homoplasy-free fit is not always satisfied in an infra-labeled tree (T, λ). Nonetheless, we show that regardless of the given infra-labeling, we can always find a completion (N, l) of T such that l and λ assign the same label to nodes of $V(N) \cap V(T)$, that establishes an origin and connects it to the other 1-nodes, in a time-consistent way.

An *island* of λ is an inclusion-maximal subset of nodes $I \subseteq V(T)$ such that: $\lambda(v) = 1$ for all $v \in I$; and between any $u, v \in I$, there exists a path in the underlying undirected graph of T that consists only of nodes w with label $\lambda(w) = 1$. In other words, an island is a connected component of the subgraph of T induced by the nodes labeled 1, where here connected components ignore arc directions. The following follows from this definition.

Lemma 1. *Let (T, λ) be an infra-labeled tree for a character γ, and let I be an island of λ. Then I contains exactly one gain node v. Furthermore, v is the unique node that reaches every node $u \in I$.*

As a consequence, the gain nodes of T are in one-to-one correspondence with the islands. We can now introduce the main idea of our correspondence between infra-labelings and completions: the islands present in the tree already ensure a "vertical" connectivity, thus it suffices to find a suitable ordering of the gains to connect them using transfers. The order of connection matters for time-consistency, but we show that if we add transfers from the "highest" gains to the "lowest" gains in order, and assign times to transfer nodes in decreasing order, then this can be done. Algorithm 1 shows how this can be achieved (Fig. 2).

Algorithm 1 mostly serves as an existential result, i.e., it shows that any infra-labeling can be turned into a completion by adding one particular sequence of transfer arcs. In general there will be many other ways to add such transfers

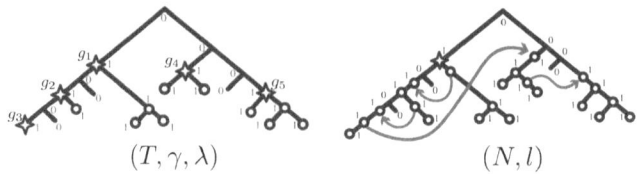

Fig. 2. (Left) An example instance (T, γ, λ), with white nodes and stars representing the nodes whose label is 1, and stars representing gain nodes. (Right) The completion (N, l) returned by Algorithm 1.

to obtain alternate completions, since different orderings of the gains lead to different completions. Thus, the algorithm should be seen as a tool to make the correspondence between the cost of infra-labelings and completions, not as a direct transfer prediction approach. Nonetheless, our experimental analysis shows how the approach can be used to predict transfer highways.

---

1 **function** *IslandCompletion(T, γ, λ)*
2   Let (N, l) be such that $N = T$ and let $l(v) = \lambda(v)$ for all $v \in V(T)$.
3   Let τ^* be a function that assigns a time to each transfer node, which initially assigns no time to any node
4   Let $\{g_1, g_2, \ldots, g_k\}$ be the set of gain nodes in T, ordered so that if $g_i \prec_T g_j$, then $j < i$ (so, ancestors come first)
5   **for** $i \in [k-1]$ **do**
6   **if** g_i *is a leaf* **then**
7 $|$ Subdivide the arc incoming to g_i, which creates a new node w'_i
8   **else**
9 $|$ Let w_i be a child of g_i that is not an ancestor of g_{i+1} in T
10   Subdivide the arc incoming to w_i, which creates a new node w'_i
11   Subdivide the arcs incoming to g_{i+1}, which creates node g'_{i+1}
12   Add the transfer arc (w'_i, g'_{i+1})
13   Set $\tau^*(w'_i) = \tau^*(g'_{i+1}) = |V(T)| - i$
14   Set $l(w'_i) = l(g'_{i+1}) = 1$
15   return(N, l)

Algorithm 1: Finding a completion from an infra-labeling.

Lemma 2. *The network N returned by Algorithm 1 is time-consistent.*

Proof (sketch). Observe that our ordering of the gains ensures that at the moment that a transfer arc (w'_i, g'_i) is created by the algorithm, none of w'_i nor g'_{i+1} has a previously created transfer node as a descendant. Since, at the moment of their creation, these two endpoints are assigned the lowest time so far under τ^*, no time inconsistency is introduced between transfer nodes. One can then fill-in the time of other internal nodes while maintaining time-consistency.
□

Lemma 3. *Let $N = (V, E_S \cup E_T)$ and l be the network and labeling returned by Algorithm 1, on input T, γ, λ. Then T is the base tree of N. Furthermore l is an homoplasy-free fit on γ that satisfies:*

1. $l(v) = \lambda(v)$ for every $v \in V(N) \cap V(T)$.
2. $|losses(N, l)| = |losses(T, \lambda)|$.
3. $|E_T| = |gains(N, \lambda)| - 1$.

Proof (sketch). We know that T is the base tree of N since the algorithm start from T. It also preserves the labels to ensure $l(v) = \lambda(v)$ for $v \in V(T) \cap V(N)$. The labeling l is an homoplasy-free fit on γ since the arcs added by Algorithm 1 connect the island gains together, with g_1 serving as an origin (or its parent in T_N if g_1 is a leaf). The number of losses in l does not change since each transfer node created is made a 1-node. For the last part, the algorithm adds $k-1$ transfer arcs, where k is the number of gain nodes. □

We can finally establish the correspondence between infra-labelings and homoplasy-free scenarios.

Theorem 1. *Let λ be an infra-labeling of T on character γ that minimizes the quantity $cost_T \cdot (|gains(T, \lambda)| - 1) + cost_L \cdot |losses(T, \lambda)|$. Then the pair (N, l) returned by Algorithm 1 on input T, γ, λ is a completion of T that minimizes $cost_{TL}$, among all possible completions of T.*

Proof (sketch). We know by Lemma 2 and Lemma 3 that the pair (N, l) returned by the algorithm is a completion of T. Also by Lemma 3, the value of $cost_{TL}(N, l)$ is the quantity stated in the theorem, and we need to argue that this is the minimum $cost_{TL}$ of any completion of T. To see this, we assume for contradiction that there is an alternate completion (N', l') that achieves a strictly lower cost. We use (N', l') to construct an infra-labeling λ' of T that achieves

$$cost_T(|gains(T, \lambda')| - 1) + cost_L |losses(T, \lambda')|$$
$$< cost_T(|gains(T, \lambda)| - 1) + cost_L |losses(T, \lambda)|.$$

This leads to a contradiction, since we assume that λ minimizes the latter quantity. To see how to convert the hypothetical (N', l') into such a λ', it suffices to define λ' as the restriction of l' to $V(T)$ (i.e., $\lambda'(v) = l'(v)$ for each $v \in V(T)$). With some arguing, we show that if (N', l') has $t - 1$ transfer arcs and q losses, then (T, λ') has at most t gains and at most q losses. Thus the cost of λ' is at most that of $cost_{TL}(N', \lambda')$, which is itself strictly lower than the cost of λ. Since the latter is minimum, this is a contradiction. Therefore, (N', l') cannot exist and the pair (N, l) has minimum cost. □

3.2 Using Sankoff-Rousseau's Algorithm to Find an Infra-Labeling

By Theorem 1, it now suffices to find an infra-labeling λ of T that minimizes the following cost:

$$cost^*_{TL}(T, \lambda) = cost_T \cdot (|gains(T, \lambda)| - 1) + cost_L \cdot |losses(T, \lambda)|.$$

Recalling that $|losses(T, \lambda)|$ counts the number of arcs that transition from 1 to 0, and $|gains(T, \lambda)|$ the arcs that go from 0 to 1, the problem is very similar to the *small parsimony problem*. In the latter, given a tree T, an assignment of character-states to its leaves and a pre-defined cost matrix M where every possible change of state has an assigned weight, the algorithm assigns a label $\lambda(v)$ to each node of T in a way that minimizes the cost $s(T, \lambda) = \sum_{(u,v) \in E(T)} M(\lambda(u), \lambda(v))$. Note that in our case we have only two states $\{0, 1\}$. Since the aim is to penalize the changes from different states, the cost matrix M is defined as $M(1, 0) = cost_L, M(0, 1) = cost_T$, and is 0 otherwise. We show that our problem can be solved by a variant of the Sankoff-Rousseau algorithm [27], which we recall below. We then show how the "minus one" applied on gains in our cost function changes the problem.

For binary labels, given a character γ on $\mathcal{S}$ and a tree T with $L(T) = \mathcal{S}$ we consider the minimum cost $s(T[v], \lambda)$ of a labeling λ of the $T[v]$ subtree, under the condition that $\lambda(v) = a$. The Sankoff-Rousseau algorithm uses a dynamic programming table $C[v, a]$ that stores $s(T[v], \lambda)$ for every $v \in V(T)$ and every label $a \in \{0, 1\}$. If v is a leaf, we have $C[v, \lambda(v)] = 0$ and $C[v, 1 - \lambda(v)] = \infty$. For each internal node v with label $a \in \{0, 1\}$ we have

$$C[v, a] = \sum_{\forall x \in \text{ch}(v)} \min_{a' \in \{0,1\}} \left\{ C[x, a'] + M(a, a') \right\}$$

where $\text{ch}(v)$ represents the set of all children of the node v. Once $C[v, a]$ is computed for all nodes v of T and all labels $a \in \{0, 1\}$, the minimal cost of a labeling of the vertices of T is $\min_{a \in \{0,1\}} C[\rho(T), a]$. A standard backtracking procedure can then reconstruct a labeling of minimum $s(T, \lambda)$ cost.

This cost is *almost* identical to the cost of a completion (N, l) of T, but not quite since one $0 - 1$ arc does not need to be penalized as it represents the origin. This actually matters: even under unit costs $cost_T = cost_L = 1$, there exist optimal scenarii under Sankoff-Rousseau parsimony such that the corresponding number of transfer and loss events is not minimal, and there are scenarii with a minimal number of transfer and loss events that are not minimal under Sankoff-Rousseau, which can therefore not be identified by the Sankoff-Rousseau algorithm. We provide a small example for each of these two problematic situations.

Consider first the tree T depicted in Fig. 3(a) in which two leaves have the character. Then, both labelings depicted in Fig. 3(c) and Fig. 3(d) are optimal in the Sankoff-Rousseau sense (ignoring the subdivision nodes and transfer arc), as both contain exactly two arcs whose ends are given different labels (and these are exactly the minimum-cost scenarii). However, the labeling in Fig. 3(c) corresponds to a scenario with two loss events and no transfer event, while the labeling in Fig. 3 (d) corresponds to a scenario with one transfer event and no loss event. Hence, the second scenario is more parsimonious, in our sense, than the first one. Moreover, if $1 - 0$ arcs have a cost of 2 and $0 - 1$ arcs have a cost of 3, then under Sankoff-Rousseau parsimony, the tree in (d) is preferred since

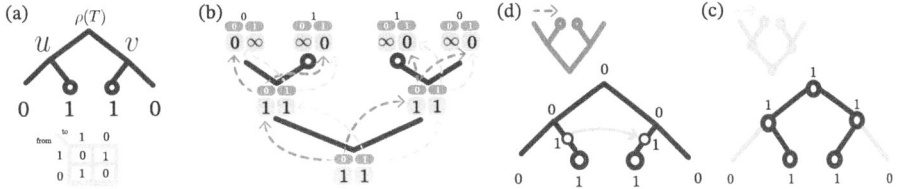

Fig. 3. Examples illustrating that not penalizing one $0-1$ arc changes the space of optimal solutions (see text). (a) Shows a given tree and associated cost matrix. (b) The Sankoff algorithm applied to the tree. (c) and (d) Show two possible optimal solutions in the Sankoff sense and their respective networks.

its cost is 4 whereas the tree on (c) has cost 6. But in terms of homoplasy-free scenarios, the tree on (c) has two losses (cost 4), whereas the network on (d) only needs one transfer (cost 3).

Now, consider the tree T depicted in Fig. 4 (Left). Suppose that $cost_L = cost_T$. One can verify that the labeling in Fig. 4 (Middle) is the only optimal labeling of T in Sankoff-Rousseau's sense. It contains exactly one arc whose ends are given different labels, and it coincides with a scenario with one loss event and no transfer. The labeling depicted in Fig. 4 (Right) has two arcs whose ends are given different labels, so it is not optimal in Sankoff-Rousseau's sense. However, this labeling corresponds to a scenario with one transfer event and no loss event, so the scenario is equally optimal, in our sense, as the first one.

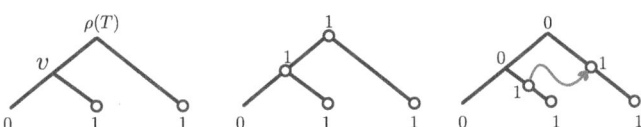

Fig. 4. Another example

3.3 The Genesis Algorithm: a Two-Variable Sankoff-Rousseau

The above shows that we cannot use the Sankoff-Rousseau algorithm as is. We adapt the dynamic programming table by adding an additional dimension that keeps track of where a character could have its origin, if encountered in the current subtree (the originator, thus the *genesis* term). To keep track of the location of the genesis within the subtree, we define a mapping $g : V(T) \to \{0, 1\}$. The value $g(v) = 1$ indicates that the character has possibly originated in the subtree $T[v]$. Conversely, if $g(v) = 0$ it signifies that the character did not originate in $T[v]$. This way, we know that the $0-1$ arc responsible for the origin should not be penalized. Using the same cost matrix as in the previous subsection, we define our dynamic programming table as follows. For $v \in V(T)$, $a \in \{0,1\}, b \in \{0,1\}$:

$$C[v,a,b] = \text{the minimum cost } cost^*(T[v], \lambda) \text{ of a labeling } \lambda \text{ of } T[v],$$
$$\text{given that } l(v) = a \text{ and } g(v) = b.$$

If v is a leaf of T, we set:

$$C[v,1,b] = \begin{cases} 0, & \text{if } l(v) = 1 \\ \infty, & \text{otherwise.} \end{cases} \qquad C[v,0,b] = \begin{cases} 0, & \text{if } l(v) = 0 \text{ and } b = 0. \\ \infty, & \text{otherwise.} \end{cases}$$

The idea is that if v has the character, it could be the origin or not, so both $b \in \{0,1\}$ are allowed, but if v does not it cannot be the origin.

If v is an internal node, then we will have the following cases:

- *Case $C[v,0,0]$:* This corresponds to labeling $l(v) = 0$ and assuming that the origin does not lie in $T[v]$. This implies that all losses and gains in the subtree below will contribute to the cost, hence:

$$C[v,0,0] = \sum_{x \in \text{ch}(v)} \min_{a \in \{0,1\}} \left\{ C[x,a,0] + M(0,a) \right\}$$

- *Case $C[v,1,0]$:* This case corresponds to labeling $l(v) = 1$ and assuming that the origin does not lie within $T[v]$. As in the previous case, all losses and gains in $T[v]$ will contribute to the cost, thus:

$$C[v,1,0] = \sum_{x \in \text{ch}(v)} \min_{a \in \{0,1\}} \left\{ C[x,a,0] + M(1,a) \right\}$$

- *Case $C[v,1,1]$:* This case corresponds to labeling $l(v) = 1$ and assuming that the origin lies in $T[v]$. In this case, v must be the origin: if that origin was a strict descendant of v, it would need to violate time-consistency to transfer to v. Thus, we may assume that no child of v contains the origin, so we have:

$$C[v,1,1] = \sum_{x \in \text{ch}(v)} \min_{a \in \{0,1\}} \left\{ C[x,a,0] + M(1,a) \right\}$$

Note that although the recurrence is identical to the preceding case, the difference lies in the fact that v should be the origin.

- *Case $C[v,0,1]$:* In this case, we assign $l(v) = 0$ and assume that the origin lies within $T[v]$. Note that this implies that *exactly* one child w of v must satisfy that $g(w) = 1$: v itself cannot be an origin since it has label 0, and only one child subtree has the origin. We do need to try every possible w. Note that if the chosen w has label 1, then it will be the origin and no penalty for the state change $M(0,a)$ at the arc (v,w) will be considered. This leads to the following expression:

$$C[v,0,1] = \min_{w \in \text{ch}(v)} \left\{ \min_{a \in \{0,1\}} \left\{ C[w,a,1] \right\} + \sum_{x \in \text{ch}(v) \setminus \{w\}} \min_{a \in \{0,1\}} \left\{ C[x,a,0] + M(0,a) \right\} \right\}$$

We refer the interested reader to Fig. 7 in Appendix 1 for an illustrated example of a binary tree that explores these different cases. Once the optimal score has been computed for the root, the minimal labeling λ can be obtained by *back tracing* in a way that is generic to dynamic programming algorithms in general. Reversing the order of the traversal, i.e. proceeding from the root to the leaves and asking which of the alternative values of label a and child $w \in ch(v)$ achieves a minimum $C[v,p,q]$.

To obtain a completion for a given (T, λ), it suffices to run Algorithm 1 on the resulting infra-labeling. Note that the running time of the Genesis algorithm for a given species tree T on n species is $O(n)$.

4 Experiments on KEGG Characters: a Proof of Concept

We now illustrate our dynamic programming schemes for the inference of transfers that involve gain nodes shared between a set of characters. The whole datasets and implementations used for this part are contained in the following repository: https://github.com/AliLopSan/ptns.

Recall that our strategy is to find the set of gain nodes on a tree and then infer the connections between them a posteriori. Although there may be an exponential number of ways to connect the gains of a character γ (Algorithm 1 just gives one way to do it), there are only $O(|gains(T,\lambda)|^2)$ pairs of gain nodes. Thus it is possible to look for pairs of gains that characters have in common. We use this to predict *transfer highways* [4,5], which are horizontal arcs in a network where a significantly large number of gene (character) transfers have taken place. We now look at a set $\mathcal{C}$ of multiple characters independently, and we say that a pair of gain nodes $\{a, b\}$ in a given tree T is a *transfer highway* if it is present in at least α characters of $\mathcal{C}$.

To build our base species tree T, we took a random subset of 45 species from the bacterial species used in [35], which consists of species that were predicted to be involved in interphylum transfers. We obtained the corresponding species tree from NCBI Taxonomy Browser [28], noting that it is not completely resolved and is therefore non-binary. See Appendix 1. The whole annotated genomes of these species are contained in the KEGG [18] database.

As set of characters, we chose a set of 180 KEGG Ortholgs groups, called KOs, taken from [35], which contain functions that are classified as metabolism-related, information processing and antibiotic resistance. This choice was to ensure that our methods operate on the same input for consistency. We computed various infra-labelings λ using different approaches. The first, which we will refer to as *Basic labeling* described in [22]. This labeling maps every leaf to the character it possesses and an internal vertex v has a 1 label if and only if *all* of its children have a 1 label. The second, which we will refer to as *Sankoff labeling* was derived from the algorithm presented in Sect. 3.2, with transfer/loss cost ratios $cost_T/cost_L \in \{0.25, 0.5, 1\}$. Finally, the third, which we will refer to as *Genesis labeling* was computed using the algorithm from Sect. 3.3, with transfer and loss costs identical to those in the previous approach. For every character γ, we computed a minimum cost infra-labeling λ and obtained the gain nodes. To find the transfer highways, we then look at the pairs of gains that are common to the different sets of gain nodes between the characters. We applied different threshold values: $\alpha \in \{9, 18, 27, 36\}$ corresponding to 5%, 10%, 15%, 20%, and 25% of the total number of characters, respectively. We observed that when loss costs are lower than transfer costs, almost all characters are explained solely by losses. As a result, all characters appear at the root, and no gain arcs are observed throughout the tree. This outcome has been discussed in the literature [13]. In the following we report data on unit costs, $cost_T = cost_L = 1$, as they gave the most balanced results.

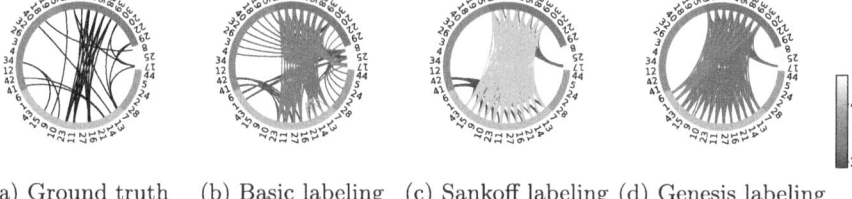

(a) Ground truth   (b) Basic labeling   (c) Sankoff labeling   (d) Genesis labeling

Fig. 5. Transfer highways for different labelings. Every species in T is represented as a segment of the circle, edges represent transfer arcs. For (b), (c), (d), the color bar indicates the number of characters that share the transfer arc and $\alpha = 18$. We used $cost_T = cost_L$. Taxa are numbered and the key can be found in Table 1 in Appendix 1.

HGTs at the Species Level. We contrasted the inferred highways with different transfers found throughout the literature. In Fig. 5 we compare our inference with pairwise interphylum transfers reported in [35], shown in (a), which correspond to a network whose vertices represent the bacterial species and an edge between a pair of species represents a transfer event. This network was built by finding blocks of nearly identical DNA (i.e. more than 500 nucleotides, more than 99% identity) in distantly related genomes (less than 97% of 16rRNA similarity). We conjectured that, interphylum transfers would be more visible to our model than transfers that happen between closely related species. This is because when a transfer happens between closely related species, the *parsimony* criterion implied by our algorithm will most likely explain it through vertical inheritance, rather than transfers. To compare the outcome of our methods at the species level, we take a transfer highway (a, b) and we connect all the leaves in the subtrees $T[a]$ and $T[b]$ between them. In this way we create a graph where nodes represent species and edges represent transfer highways whose weight is proportional to the number of characters that share this highway. We observe that Sankoff and Genesis preserve some of the transfer relationships contained in the ground truth data, especially concerning the interphylum transfers between *Proteobacteria* (purple) and *Actinobacteria* (green) which are shown to be the most popular highways in Sankoff and Genesis and remain faintly in the Basic labeling. Note that contrary to (a), there is a large number of transfers between (17)*Mucispirillum schaedleri* (T08201) and *Proteobacteria* that remained throughout the three models. It has been previously reported in the literature that *Epsilon-* and *Deltaproteobacteria* have shaped the evolution of this genome [21].

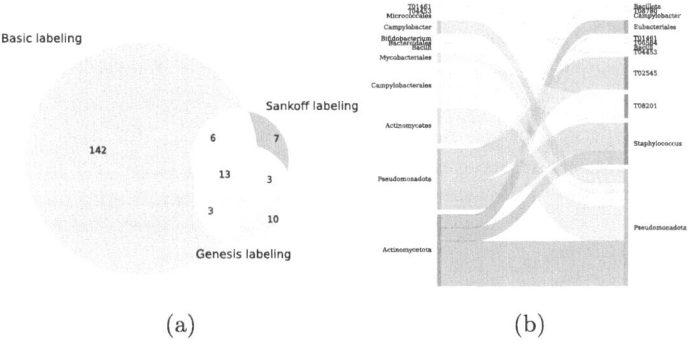

Fig. 6. (a) Quantitative differences between the inferred transfer highways using the three methods with $cost_T = cost_L$ and $\alpha = 5$. (b) A Sankey plot of the highways found with the Genesis labeling. The thickness of the lines is proportional to the number of characters that share the pair. Out of the 45 transfer edges contained in the ground truth graph, the basic labeling found 30, Sankoff labeling 38 and Genesis found 40.

HGTs at Different Taxonomic Ranges. When looking at higher level taxa, as shown in Fig. 6, we see throughout the different types of labeling and parameters used, that the members of the *Pseudomonas* genus remain as a potential gain nodes. This is consistent with the literature, since *Pseudomonas* are known to be not only ecologically versatile pathogens [29], but also implicated in the interphylum transfers of genetic material with members of the *Bacteroidales* order [17]. This shows that our model is biologically sound. It remains to validate other pairs of ancestral species with highly supported transfers, and investigate further the differences between the Sankoff and Genesis outputs.

5 Conclusion

In this work, we have incorporated losses to character-based model on phylogenetic networks. We have shown that a most parsimonious scenario can be found efficiently for a single character, but much remains to be done. Notably, although the transfer network that results from applying Algorithm 1 is time-consistent for the one character case, it does not guarantee that it will remain time consistent when we have more than one character. Because of this, explaining multiple characters with a single scenario whilst minimizing transfers and losses appears challenging—it probably leads to NP-complete problem formulations, but investigating good heuristics or structural restrictions in the resulting networks are interesting future directions. Moreover, in [20], the authors provide examples on which incorporating losses can be used to resolve polytomies in species trees, by looking at resolutions that minimizes our cost criterion. Given that several species phylogenies are only partially resolved, including the NCBI tree used in our experiments, we will look at possible algorithms to resolve them using our model. Finally, we observe that our experiments on real data are preliminary and that the approach can recover transfers that are not far from those found in the literature. It remains to perform experiments at a larger scale and see if our approach can find well-supported and novel transfer highways, especially the ancient ones that are difficult to recover using only sequence comparisons. To make our approach more impactful, one could consider also modeling the adaptability of the transferred genetic material, since there are factors such as the codon usage bias that could lead to certain characters being lost more easily than others [8].

References

1. Anselmetti, Y., El-Mabrouk, N., Lafond, M., Ouangraoua, A.: Gene tree and species tree reconciliation with endosymbiotic gene transfer. Bioinformatics **37**(Suppl._1), i120–i132 (2021)
2. Arnold, B.J., Huang, I.T., Hanage, W.P.: Horizontal gene transfer and adaptive evolution in bacteria. Nat. Rev. Microbiol. **20**(4), 206–218 (2022)
3. den Bakker, H.C., et al.: Comparative genomics of the bacterial genus listeria: genome evolution is characterized by limited gene acquisition and limited gene loss. BMC Genomics **11**, 1–20 (2010)
4. Bansal, M.S., Banay, G., Gogarten, J.P., Shamir, R.: Detecting highways of horizontal gene transfer. J. Comput. Biol. **18**(9), 1087–1114 (2011)
5. Beiko, R.G., Harlow, T.J., Ragan, M.A.: Highways of gene sharing in prokaryotes. Proc. Natl. Acad. Sci. **102**(40), 14332–14337 (2005)
6. Bonizzoni, P., Carrieri, A.P., Vedova, G.D., Dondi, R., Przytycka, T.M.: When and how the perfect phylogeny model explains evolution. Discrete and Topological Models in Molecular Biology, pp. 67–83 (2014)
7. Bouckaert, R., Fischer, M., Wicke, K.: Combinatorial perspectives on dollo-k characters in phylogenetics. Adv. Appl. Math. **131**, 102252 (2021)
8. Callens, M., Scornavacca, C., Bedhomme, S.: Evolutionary responses to codon usage of horizontally transferred genes in pseudomonas aeruginosa: gene retention, amelioration and compensatory evolution. Microbial genomics **7**(6), 000587 (2021)
9. Cardona, G., Pons, J.C., Rosselló, F.: A reconstruction problem for a class of phylogenetic networks with lateral gene transfers. Algorithms for Molecular Biology **10**(1), 28 (2015). https://doi.org/10.1186/s13015-015-0059-z

10. Charleston, M.: Jungles: a new solution to the host/parasite phylogeny reconciliation problem. Mathematical Biosciences **149**(2), 191–223 (1998). https://doi.org/10.1016/S0025-5564(97)10012-8
11. Choi, I.G., Kim, S.H.: Global extent of horizontal gene transfer. Proc. Natl. Acad. Sci. **104**(11), 4489–4494 (2007)
12. Day, W.H., Johnson, D.S., Sankoff, D.: The computational complexity of inferring rooted phylogenies by parsimony. Math. Biosci. **81**(1), 33–42 (1986)
13. Doolittle, W.F., Boucher, Y., Nesbø, C., Douady, C., Andersson, J.O., Roger, A.: How big is the iceberg of which organellar genes in nuclear genomes are but the tip? Philosophical Transactions of the Royal Society of London. Series B: Biological Sci. **358**(1429), 39–58 (2003)
14. Farris, J.S.: Phylogenetic analysis under Dollo's law. Systematic Biol. **26**(1), 77–88 (1977). https://doi.org/10.1093/sysbio/26.1.77
15. Felsenstein, J.: Inferring phylogenies. Sunderland, Mass.: Sinauer Associates (2004)
16. Goyal, A.: Horizontal gene transfer drives the evolution of dependencies in bacteria. iScience **25**(5), 104312 (2022). https://doi.org/10.1016/j.isci.2022.104312
17. Gschwind, R., et al.: Inter-phylum circulation of a beta-lactamase-encoding gene: a rare but observable event. Antimicrobial Agents and Chemotherapy 68(4), e01459–23 (2024). https://doi.org/10.1128/aac.01459-23
18. Kanehisa, M., Sato, Y., Kawashima, M., Furumichi, M., Tanabe, M.: KEGG as a reference resource for gene and protein annotation. Nucleic Acids Res. **44**(D1), D457–D462 (2015). https://doi.org/10.1093/nar/gkv1070
19. Kelly, L.J., Nicholls, G.K.: Lateral transfer in stochastic dollo models. Ann. Appl. Stat. **11**(2), 1146–1168 (2017). https://doi.org/10.1214/17-AOAS1040
20. López Sánchez, A., Lafond, M.: Galled perfect transfer networks. In: Scornavacca, C., Hernández-Rosales, M. (eds.) Comparative Genomics, pp. 24–43. Springer, Cham (2024)
21. Loy, A., et al.: Lifestyle and horizontal gene transfer-mediated evolution of mucispirillum schaedleri, a core member of the murine gut microbiota. Msystems **2**(1), 10–1128 (2017)
22. López Sánchez, A., Lafond, M.: Predicting horizontal gene transfers with perfect transfer networks. In: Boucher, C., Rahmann, S. (eds.) 22nd International Workshop on Algorithms in Bioinformatics. Schloss Dagstuhl - Leibniz-Zentrum für Informatik (2022). https://doi.org/10.4230/LIPICS.WABI.2022.3
23. Moran, N.A.: Microbial minimalism: genome reduction in bacterial pathogens. Cell **108**(5), 583–586 (2002)
24. Nakhleh, L.: Phylogenetic networks. Ph.D. thesis, The University of Texas at Austin (2004)
25. Nakhleh, L., Ringe, D., Warnow, T.: Perfect phylogenetic networks: A new methodology for reconstructing the evolutionary history of natural languages. Language **81**(2), 382–420 (2005), http://www.jstor.org/stable/4489897
26. Ravenhall, M., Škunca, N., Lassalle, F., Dessimoz, C.: Inferring horizontal gene transfer. PLoS Comput. Biol. **11**(5), e1004095 (2015)
27. Sankoff, D., Rousseau, P.: Locating the vertices of a steiner tree in an arbitrary metric space. Math. Program. **9**, 240–246 (1975)
28. Schoch, C.L., et al.: NCBI Taxonomy: a comprehensive update on curation, resources and tools. Database **2020**, baaa062 (2020)
29. Silby, M.W., Winstanley, C., Godfrey, S.A., Levy, S.B., Jackson, R.W.: Pseudomonas genomes: diverse and adaptable. FEMS Microbiology Rev. **35**(4), 652–680 (2011). https://doi.org/10.1111/j.1574-6976.2011.00269.x

30. Van Iersel, L., Semple, C., Steel, M.: Quantifying the extent of lateral gene transfer required to avert a 'genome of eden.' Bull. Math. Biol. **72**, 1783–1798 (2010)
31. Warnow, T., Tabatabaee, Y., Evans, S.N.: Advances in estimating level-1 phylogenetic networks from unrooted SNPs. J. Comput. Biol. **32**(1), 3–27 (2025). https://doi.org/10.1089/cmb.2024.0710, pMID: 39582425
32. Yang, Z., Rannala, B.: Molecular phylogenetics: principles and practice. Nat. Rev. Genet. **13**(5), 303–314 (2012)
33. Zachar, I., Boza, G.: Endosymbiosis before eukaryotes: mitochondrial establishment in protoeukaryotes. Cellular and Molecular Life Sci. **77**(18), 3503–3523 (2020). https://doi.org/10.1007/s00018-020-03462-6
34. Zhang, Y., Tu, C., Bai, J., Li, X., Sun, Z., Xu, L.: Metabolic enhancement contributed by horizontal gene transfer is essential for dietary specialization in leaf beetles. Proceedings of the National Academy of Sciences **122**(1), December 2024. https://doi.org/10.1073/pnas.2415717122
35. Zhou, H., Beltrán, J.F., Brito, I.L.: Functions predict horizontal gene transfer and the emergence of antibiotic resistance. Sci. Adv. **7**(43), eabj5056 (2021)

tMHG-Finder: Tree-Guided Maximal Homologous Group Finder for Bacterial Genomes

Yongze Yin[1], Bryce Kille[1], Huw A. Ogilvie[1], Todd J. Treangen[1,2,3](✉), and Luay Nakhleh[1,4](✉)

[1] Department of Computer Science, Rice University, Houston, TX 77005, USA
{treangen,nakhleh@rice.edu}
[2] Ken Kennedy Institute, Rice University, Houston, TX 77005, USA
[3] Department of Bioengineering, Rice University, Houston, TX 77005, USA
[4] Department of BioSciences, Rice University, Houston, TX 77005, USA

Abstract. A *maximal homologous group*, or MHG, as a group of sequences with a shared evolutionary ancestry, shifts the focus from a gene-centric view to a homology-centric view in comparative genomic studies. Each MHG is formed by identifying and grouping all homologous sequences, which ensures that evolutionary events, such as horizontal gene transfer, gene duplication and loss, or *de novo* sequence evolution, are encapsulated within the same MHG. However, the current MHG computation tool, MHG-Finder, faces challenges in scalability to handle large datasets and lacks the ability to provide detailed insights into intermediate MHGs involving subsets of input genomes. We present tMHG-Finder (https://github.com/yongze-yin/tMHG-Finder), a new method that improves our previous method, MHG-Finder, by utilizing a guide tree to significantly improve scalability and provide more informative biological results. We also introduce a new measure, fractionalization (available at https://github.com/yongze-yin/Fract-Calculator), to assess the accuracy of delineated MHGs compared to ground truth data. Our results show that tMHG-Finder scales linearly with the number of taxa, requiring a small fraction of the computational time of MHG-Finder. Furthermore, according to the fractionalization measure, tMHG-Finder outperforms four state-of-the-art whole-genome aligners on simulated data. Applying tMHG-Finder to a phylum of extreme-environment-resistant bacteria, we validated our results through the encapsulation of 16S rRNA sequences within MHGs. We further investigated how evolutionary rates change with phylogenetic distance and explored the functional roles of genes captured by conserved MHGs, demonstrating the broader utility of tMHG-Finder in uncovering evolutionary insights beyond MHG delineation and phylogenetic relationships.

Supplementary Information The online version contains supplementary material available at https://doi.org/10.1007/978-3-031-94928-9_6.

Keywords: evolution · phylogenetics · phylogenomics · comparative genomics · homologous group · bacterial genomes · graph algorithm

1 Introduction

Bacterial evolution is a complex process affected by frequent genetic events and adaptations to the fast-changing environment [12,23,32]. Homologous groups, each defined as a set of sequences that share a common evolutionary ancestry, are used as loci to understand the evolutionary process. Traditional approaches rely on using functional gene sets as homologous groups [21]. However, this violates the fundamental mathematical assumption that the evolutionary history of each locus should be recombination-free [34]. In contrast, alignment-based methods provide a viable option for studying bacterial evolution without relying on protein annotation. These methods use the inherent genomic similarity of homologous sequences. However, existing whole-genome aligners are challenged at deeper evolutionary timescales where genomes are highly divergent.

Whole-genome alignments play a crucial role in downstream analyses in comparative genomic studies [19]. Particularly, they can serve as the foundation for constructing pan-genomes [8], a field of increasing attention due to its ability to reveal the genetic diversity and key genomic components contributing to bacterial adaptation. Pan-genome studies delineate the core and dispensable genomes, offering insights into bacterial genetic diversity. However, limitations in existing pan-genome construction methods hinder their impact. For example, some methods require the identification and annotation of protein sequences rather than solely relying on genomic sequences as input [25,28,31].

A *maximal homologous group*, or MHG, is defined as a maximal set of maximum-length sequences whose evolutionary history is a single tree [34]. Similar to the property of a locally collinear block (LCB), which denotes a rearrangement-free homologous region, the definition of MHG imposes two additional key optimality constraints. The set is maximal in the sense that no sequences that are not in the set are homologous to those in the set without involving internal rearrangements. Each sequence within the group is also maximal, meaning that flanking base pairs on either side of any sequence in the group do not belong to the homology group without involving internal rearrangements. Our previous alignment-graph-based MHG detector, MHG-Finder [34], is one of the few methods available for identifying MHGs directly. However, it has scalability and information limitation challenges.

To overcome these challenges, we introduce tMHG-Finder, a *de novo* tree-guided MHG finder. tMHG-Finder considers the evolutionary distances between the input genomes to adjust the expected genomic similarity accordingly and computes MHGs for both the complete set and subsets of input genomes. To improve scalability, tMHG-Finder employs a progressive approach by constructing a representative sequence for each MHG at each node in the input tree. As a result, the number of pairwise blastn [2] comparisons drops from $O(n^2)$ to $O(n)$, where n signifies the number of input genomes. Moreover, tMHG-Finder supports parallel processing by utilizing multiple threads to concurrently handle different connected components of the constructed alignment graph.

We assessed the accuracy of tMHG-Finder against four whole-genome aligners: Progressive Cactus, progressiveMauve, SibeliaZ, and Mugsy [3,4,11,22]. Considering each identified alignment as an MHG, we initially employed classical multiple sequence alignment (MSA) set comparison metrics: recall, precision, and F-score. Additionally, we introduced and applied a novel comparison metric, fractionalization. In contrast to existing metrics [6,13] that examine the alignment quality (gaps, mismatches), fractionalization compares the endpoints of the aligned regions to quantify how a reference alignment is fragmented into a query alignment set without considering the alignment quality. When the true alignment set serves as the reference set, fractionalization aids in measuring the quality of identified breakpoints for homologies included in the target alignment. On a set of simulated genomes, aligning MHGs identified by tMHG-Finder, we show that tMHG-Finder and Cactus produce genome alignments with an F-score above 0.90, whereas all other methods produce alignments with an F-score below 0.75. Comparatively, tMHG-Finder yields alignments with higher precision than Cactus, albeit with lower recall. However, alignments generated from tMHG-Finder-identified MHGs exhibit significantly less fragmentation than those identified by Cactus, showcasing superior accuracy.

To explore evolutionary dynamics, we applied tMHG-Finder to an extremophilic bacterial phylum: *Deinococcus-Thermus*. The accuracy of tMHG-Finder was validated by confirming that all 162 copies of near-complete 16S rRNA sequences from 80 studied genomes were encapsulated within a single MHG, indicating that tMHG-Finder effectively captures conserved genomic regions. To assess evolutionary rates, we quantified genomic distances using average pairwise average nucleotide identity (ANI) values. As genomes became more distantly related (lower ANI values), the number of MHGs increases, while the average length of MHGs decreased, consistent with expectations of increased divergence. By tracking MHG results across internal nodes of the guide tree, we observed a reduction in the average number of genes covered by MHGs as genomic distances increased. Focusing on conserved MHGs, defined as MHGs containing exactly one sequence from every studied genome, we conducted a gene ontology (GO) analysis and identified a diverse range of functional categories associated with the genes captured by these conserved MHGs, providing insight into the functional roles of genes under evolutionary constraints. These results highlight the relationship between evolutionary rates and phylogenetic distances, demonstrating the ability of tMHG-Finder to capture genomic evolution and offering insights into the evolutionary dynamics of extremophilic bacteria.

2 Materials and Methods

tMHG-Finder Algorithm

Taking whole-genome nucleotide sequences as the sole input, tMHG-Finder partitions genomes into MHGs (Fig. 1). In contrast to MHG-Finder presented in [34],

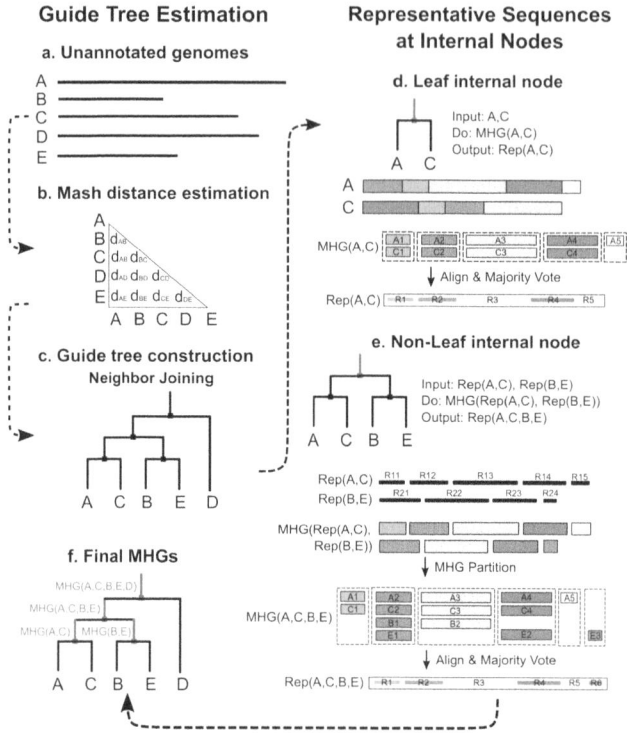

Fig. 1. tMHG-Finder algorithm pipeline. (a): The input comprises a set of two or more genomes. (b): A distance matrix is estimated utilizing Mash. (c): A binary guide tree is constructed using the neighbor-joining method based on estimated pairwise distances by Mash. (d): A leaf internal node is an internal node with both children being leaf nodes. For every leaf internal node in the guide tree, tMHG-Finder utilizes the preceding MHG package MHG-Finder to compute MHGs. Upon obtaining a set of MHGs, tMHG-Finder computes a "representative sequence set" by aligning each MHG and employing majority voting to derive a consensus sequence representing the target MHG. (e): A non-leaf internal node is defined as an internal node with at least one child being an internal node. Instead of performing all-versus-all pairwise alignment for all included taxa under the target internal node, only the two children representative sequence sets are locally aligned and computed for MHGs. (f): The final output comprises an MHG set per internal node, which encompasses all taxa located beneath the respective internal node.

tMHG-Finder utilizes a guide tree that can be either user-provided or estimated. As a result, tMHG-Finder employs a progressive approach by traversing the guide tree from the bottom leaf nodes to the top root node. At each internal node, genomes located at the leaves beneath the target internal node are partitioned into a set of MHGs, and then represented by a set of representative sequences constructed from the identified MHG set. Consequently, an upper-level internal node only needs to compute and partition the homologous matching pairs from the two sets of representative sequences of their direct children. This approach avoids the exhaustive all-versus-all blastn mappings and substantially reduces the blastn comparison complexity from $O(n^2)$ to $O(n)$ where n is the number of input genomes at each internal node. Besides the enhanced computational efficiency compared to MHG-Finder, the final output of tMHG-Finder includes MHGs not only for the entire set of input genomes but also for the subsets along the evolutionary history.

To accurately infer a guide tree, tMHG-Finder leverages the MinHash-based alignment-free distance estimator Mash [24], enabling a fast calculation of pairwise genomic distances between input genomes. Subsequently, neighbor-joining is employed on the distance matrix to sketch a guide tree determining the MHG partition order. Given a guide tree, the MHG partitioning at each internal node involves five steps: (1) blastn for pairwise local alignments, (2) sequence pile-up, (3) initial MHGs and alignment graph construction, (4) alignment graph traversal, and (5) representative sequence calculation. While the first four steps rely on the algorithm in [34], the final novel representative sequence calculation step results in substantial computational savings.

tMHG-Finder categorizes internal nodes into leaf internal nodes and non-leaf internal nodes. A leaf internal node has raw genome assemblies as both children, for which tMHG-Finder relies on MHG-Finder to compute MHGs. Notably, tMHG-Finder computes "representative sequences" for each MHG set to optimize computational efficiency. For single-sequence MHGs, indicating that they are not homologous with any leaf, these sequences are preserved within the parent node's representative sequences for potential higher-level MHG partitioning. For each MHG containing multiple sequences, a representative sequence is derived through MSA using MAFFT [17] and majority voting for the most frequent nucleotide at each site, becoming part of the parent internal node's representative sequence set. In contrast, for a non-leaf internal node for which at least one child is an internal node, it directly utilizes the two representative sequence sets of both children (or one representative sequence set and one raw assembly if a child is a leaf node) to search for pairwise local alignments and partition into MHGs. Each partitioning boundary on a representative sequence (representing a child MHG) is then mirrored back to genome sequences within that MHG. The final representative sequence calculation step eliminates the need for all-versus-all blastn alignments between leaf node genomes, resulting in tMHG-Finder's capability to process a larger input scale.

To determine the homology threshold, tMHG-Finder adopts a dynamic parameter setting that adjusts to the evolutionary distance between genomes. Theoretically, as two species diverge over an extended period of time, their genomes become less similar despite sharing a common ancestor. tMHG-Finder integrates an estimated distance matrix generated by Mash and selectively fine-tunes blastn parameters based on the ground truth of a simulated dataset generated with realistic parameters. Specifically, tMHG-Finder defaults to three-parameter configurations using Mash and blastn. The selection of a configuration is determined by the minimum Mash-estimated pairwise sequence similarity within a given genome set, and each configuration corresponds to a specific task using blastn. Following the guidelines from the blastn manual, if the minimum pairwise similarity within the given genome set exceeds 95%, blastn employs the megablast task due to the high similarity of sequences. For pairwise similarities falling between 65% and 95%, tMHG-Finder relies on blastn to perform dc-megablast to avoid excessively partitioning genomes into small MHGs or being overly strict and identifying too few regions. If the minimum similarity is below 65%, traditional blastn is performed for inter-species comparisons.

Fractionalization: MSA Comparison Statistic

Fractionalization measures the level of diversity or heterogeneity, and we have reformulated it to quantify the level of congruence between two sets of multiple sequence alignments (MSAs). Our proposed fractionalization value, denoted as f, is computed using

$$f(m, C) = 1 - \sum_{i=1}^{n} s_i^2 \tag{1}$$

where $f(m, C)$ represents the fractionalization value for a single MSA m distributed across n MSAs in the comparison set C. Specifically, s_i is the overlapping ratio, calculated as the sum of the lengths of overlapping ranges for sequences in m and C_i, divided by the sum of the sequence lengths in m, providing a nuanced view by considering the extent of overlap. Taking two sets of MSAs as input, because fractionalization is an asymmetric metric, we treat each set alternately as the target and comparison set and compute the fractionalization value for every alignment within each set. The fractionalization value ranges from 0 to 1. As f approaches 0, it indicates high similarity, suggesting there exists an MSA in the comparison set C that closely matches the target alignment m. In contrast, as f approaches 1, it implies a higher degree of fragmentation, indicating that the target alignment m is fragmented into many different MSAs in C. The fractionalization value captures the spectrum of similarity and fragmentation of two sets of MSAs precisely and intuitively (Appendix Fig. S1).

Simulation Study

We generated 30 bacterial genomes using the simulator ALF [10]. We first utilized ALF to build a birth-death model tree containing 30 extent leaves with a birth rate of 0.01 and a death rate of 0.001, and then evolved 30 genomes down the tree. At the root of the model tree, we initialized a genome with 4000 genes, approximately equal to the number of genes in an *E. coli* genome. The gene length was randomly drawn from a Gamma distribution with $k = 3$ and $\theta = 133.8$. Evolutionary events were then introduced along the edges of the tree based on the probability of an event occurring on a single edge. The parameter settings were documented at https://github.com/yongze-yin/tMHG-Finder. Regarding coding versus non-coding length, we maintained a ratio of 85% to 15%, resembling the genomic composition of an *E. coli* genome. The coding regions were simulated using an M2 model, while the non-coding regions were simulated using a TN93 model. Additionally, the simulation provided the ground truth data including true loci, true gene trees, and the true species tree.

To benchmark the scalability and accuracy of tMHG-Finder in identifying MHGs, we compared tMHG-Finder v1.0.0 with four state-of-the-art whole-genome aligners: Cactus v2.3.0 [4], progressiveMauve with a build date of Feb 13, 2015 [11], SibeliaZ v1.2.4 [22], and Mugsy v1.2.3 [3]. Each alignment identified by these methods was regarded as an MHG, enabling a direct accuracy comparison against the ground truth. Progressive Cactus required a guide tree as input. To address this, we provided Cactus with the neighbor-joining topology tree (without branch lengths), which was estimated from the Mash distance matrix as part of the tMHG-Finder output. For SibeliaZ, we set $k = 15$ as recommended by the SibeliaZ manual for bacterial datasets. The remaining parameters of each aligner were set to their default settings. The runtime benchmark was conducted using subsets of 30 simulated genomes, ranging from 2 taxa to the entire dataset. Eight threads were employed if a tool supported using multiple threads; otherwise, the tool was run with a single thread. Notably, when provided with a customized guide tree, tMHG-Finder allows users to obtain MHG sets for specific genome subsets in a single run, eliminating the need for multiple runs for different subsets.

An Extremophilic Phylum of Bacteria

We investigated an extremophilic phylum of bacteria, *Deinococcus-Thermus* or *Deinococcota*, with tMHG-Finder. For each known species within the *Deinococcus-Thermus* phylum, we retrieved the highest-quality assembly from GTDB [26]. To ensure data quality and assembly completeness, We selected assemblies classified as GCF by NCBI and applied an additional filter using CheckM [27], retaining only those with completeness levels exceeding 99%. This curation process yielded a final dataset of 80 assemblies, comprising 52 species from *Deinococcus* and 28 from either *Thermus* or *Meiothermus*.

We applied tMHG-Finder to these 80 genomes and obtained intermediate MHG results for each internal node of the estimated guide tree. Each internal node on the guide tree is annotated with the number and average length of MHGs. To validate the accuracy of these results, we examined gene annotation files (GFF) obtained from NCBI to identify 16S rRNA locations annotated as "16S." Given the highly conserved nature of 16S rRNA, its encapsulation within MHGs serves as a robust validation metric.

To quantify genomic distances, we utilized pyANI [29] which calculates pairwise ANI values using nucleotide BLAST (ANIb). Unlike alignment-free estimators such as fastANI [15], which struggle with ANI values below 80%, pyANI provides accurate results even for divergent genomes. The genomic distance for a given set of genomes was represented as the mean pairwise ANIb value.

We defined a "conserved MHG" as an MHG with exactly one sequence from each studied genome. For genes within conserved MHGs, we developed a pipeline to retrieve UniProt IDs and protein names [9] and query QuickGO [5] for gene ontology (GO) terms, using *Deinococcus radiodurans* (taxonomy ID: 243230) as the reference. Each retrieved GO term was propagated through the gene ontology tree to include all parent terms connected via "is_a" and "part_of" relationships. Functional categorization was achieved by intersecting the resulting GO terms with the Gene Ontology Prokaryote subset [1].

3 Results

3.1 tMHG-Finder Runtime Benchmark

To assess the scalability of tMHG-Finder in comparison to its predecessor and existing whole-genome aligners, we executed all tools using varying thread configurations on subsets of simulated data, gradually scaling up to the entire genome set. This experiment was conducted on an Intel Xeon Gold 5218 processor, utilizing eight threads if the tool supports multiple threads and one thread if not.

Compared to tMHG-Finder, MHG-Finder experienced a substantial increase in runtime when processing more than 6 genomes, taking 54 h for 9 genomes. In contrast, tMHG-Finder processed 30 genomes in 13 h using a single thread and 7 h using eight threads (Fig. 2). Cactus and SibeliaZ took only 30 min to align 30 genomes, exhibiting high speed. Conversely, Mauve took 53 h to process 30 genomes (Fig. 2).

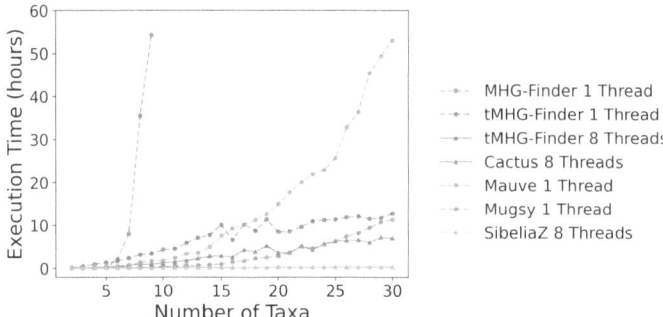

Fig. 2. Execution time comparison. Wall time for computing MHGs using: MHG-Finder with one thread (blue), tMHG-Finder with one thread (red), tMHG-Finder with eight threads (red), Cactus with eight threads (brown), Mauve with one thread (purple), Mugsy with one thread (green), and SibeliaZ with eight threads (orange). Cactus and SibeliaZ overlap at the bottom. The execution time for Cactus does not include guide tree computation and hal-to-maf file conversion time. (Color figure online)

3.2 tMHG-Finder Accuracy Assessment

To assess the accuracy of tMHG-Finder and existing whole-genome aligners at identifying homologous groups, we compared the MHG sets identified by each tool with the ground truth provided by ALF. The simulated dataset comprises 30 taxa and 3989 true MHGs. The number of sequences in each MHG roughly follows a normal distribution with a mean value of 30 (Appendix Fig. S2a). The length of each MHG, defined as the average sequence length, mostly ranges from 500 bp to 1,500 bp, with some extending close to 5,000 bp (Appendix Fig. S2b). The true pairwise Average Nucleotide Identity (ANI), calculated from the ground truth MSA, spans from 65% to 99%, with an average of 72% (Appendix Fig. S2c).

Table 1. Recall, precision, and F-score comparison. Statistics computed using mafComparator comparing the simulated true MHG set with MHG sets identified by Cactus, tMHG-Finder (aligned by MAFFT v7.490), Mauve, Mugsy, and SibeliaZ.

Tool	Recall	Precision	F-score
Cactus	**0.945**	0.987	**0.966**
tMHG-Finder	0.877	**0.994**	0.932
Mauve	0.561	0.991	0.716
Mugsy	0.213	0.984	0.350
SibeliaZ	0.082	0.973	0.151

To quantify the accuracy of each tool, we used mafComparator [13] to calculate the recall, precision, and F-score. mafComparator takes two sets of MSAs as input. For each set, it computes the ratio of the number of aligned nucleotide pairs in both sets to those in the target set. Recall quantifies the percentage of truly aligned pairs successfully identified by a tool, while precision represents the percentage of predicted aligned pairs present in the ground truth.

All tools demonstrated a precision over 97%, indicating their ability to identify aligned pairs correctly. While tMHG-Finder exhibited the highest precision, its recall was slightly lower than Cactus (Table 1). A noticeable observation was that SibeliaZ under-performed in terms of recall. We investigated and found that, unlike other tools with over 95% coverage for all genomes, SibeliaZ's coverage is close to 50% for nine genomes, leading to a suboptimal benchmark for this metric. This is likely due to the high level of divergence and the fact that a 15-mer has <1% chance of remaining unmodified at a distance of 72% ANI [7].

Additionally, we utilized fractionalization values to compare each identified MHG set with the true MHG set. As fractionalization is not symmetric, we computed this value by alternately considering each identified MHG set and the true MHG set as the target set and comparison set. This approach provides insights into whether a tool tends to over-fragment or under-fragment the truth. SibeliaZ (Fig. 3: e and f) and Progressive Cactus (Fig. 3: g and h) exhibit a similar pattern, tending to identify short MHGs. Consequently, most true MHGs have a

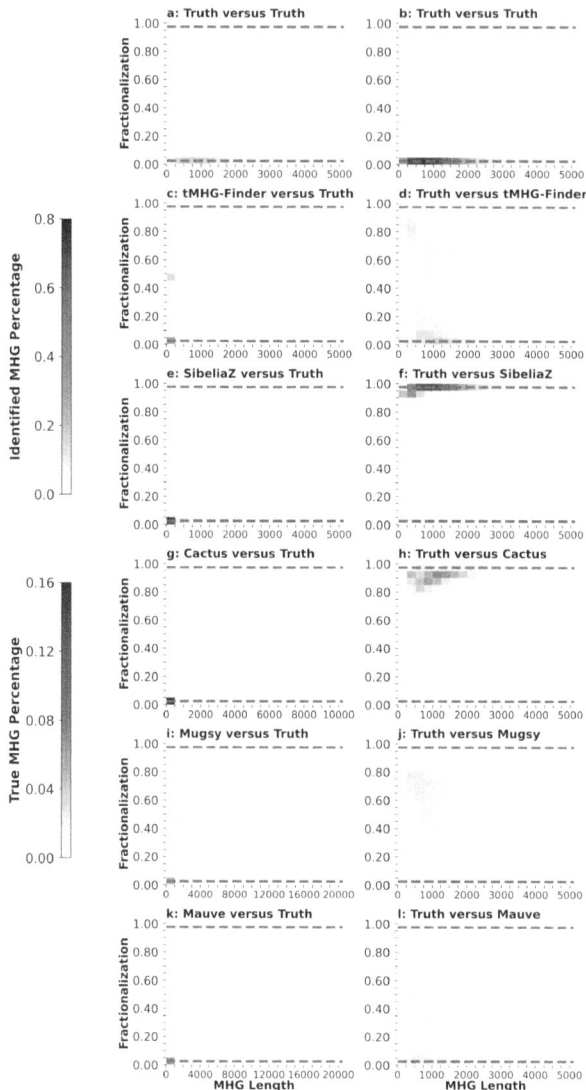

Fig. 3. **Benchmarking identified MHG accuracy with fractionalization.** Each heatmap pair illustrates the distribution of fractionalization values: the left plot demonstrates how identified MHGs are fragmented into true MHGs, while the right plot demonstrates how true MHGs are fragmented into identified MHGs. Proximity to the red line in the left plot implies that the tool under-partitions the genomes, while proximity to the red line in the right plot suggests over-partitioning. Proximity to the blue line in both plots signifies high accuracy. Subplots (a, b): ALF Truth. (c, d): tMHG-Finder. (e, f): SibeliaZ. (g, h): Cactus. (i, j): Mugsy. (k, l): Mauve. (Color figure online)

fractionalization value close to 1, indicating that true MHGs are fragmented into too many identified MHGs. SibeliaZ and Progressive Cactus are overly conservative, even when each identified group is truly homologous, though not necessarily a maximal group.

On the other hand, Mugsy (Fig. 3: i and j) and progressiveMauve (Fig. 3: k and l) identify MHGs ranging up to roughly 20,000 base pairs. As the length of an identified MHG increases, it naturally encompasses a larger number of true MHGs within it. Consequently, this leads to higher fractionalization values for identified MHGs. Notably, both algorithms demonstrate a relaxed approach, displaying a tendency to under-partition MHGs.

tMHG-Finder identifies MHGs ranging up to 4,500 base pairs, aligning closely with the truth. Although tMHG-Finder identified MHGs have some fractionalization values close to 0.5, the performance remains relatively stable as the MHG length increases. Whether considered as the target set or the comparison set, tMHG-Finder identifies MHGs that are closer to the truth, as evidenced by a majority of fractionalization values accumulating around 0 from both perspectives (Fig. 3: c and d).

3.3 Validation in *Deinococcus-Thermus* Genes and Functions

To validate the accuracy of tMHG-Finder, we analyzed 312 sequences annotated as "16S", including both partial and complete 16S rRNA sequences [18,33], ranging from approximately 50 bp to 1,500 bp. Of these, 162 were near-complete (1,400 bp), while the remaining were mostly less than 200 bp. By intersecting the MHGs identified at the root node of the guide tree with the genomic coordinates of 16S sequences, we found that all 162 near-complete 16S sequences were encapsulated within a single MHG, with an average length of 1,256 bp. Shorter sequences were likely artifacts from sequencing or assembly errors, as 16S rRNA sequences are highly conserved, and short fragments often appear as isolated contigs. While the mathematical definition of an MHG assumes all observed homologous relationships result from true biological events, it does not account for such errors. To address this, tMHG-Finder employs heuristics to exclude homologous sequences that are too short, preventing excessive partitioning of MHGs into small units. This approach ensures MHGs reflect biologically meaningful homologous relationships, demonstrating the method's ability to capture true patterns rather than artifacts.

Expanding the analysis beyond 16S rRNA, we examined the relationship between MHG coverage of genes and evolutionary distance. At each internal node of the guide tree, we identified overlaps between MHGs and annotated gene coordinates, defining a gene as "covered" if at least 75% of its length was encompassed by an MHG sequence. To quantify gene coverage, we normalized the number of covered genes by the number of sequences in the MHG, calculating the average gene coverage per MHG sequence.

From leaf nodes containing two genomes to the root node encompassing all 80 genomes, the average number of genes covered per MHG decreased from 4.34 to 0.95. This trend aligns with the observation in Fig. 4 that the average length

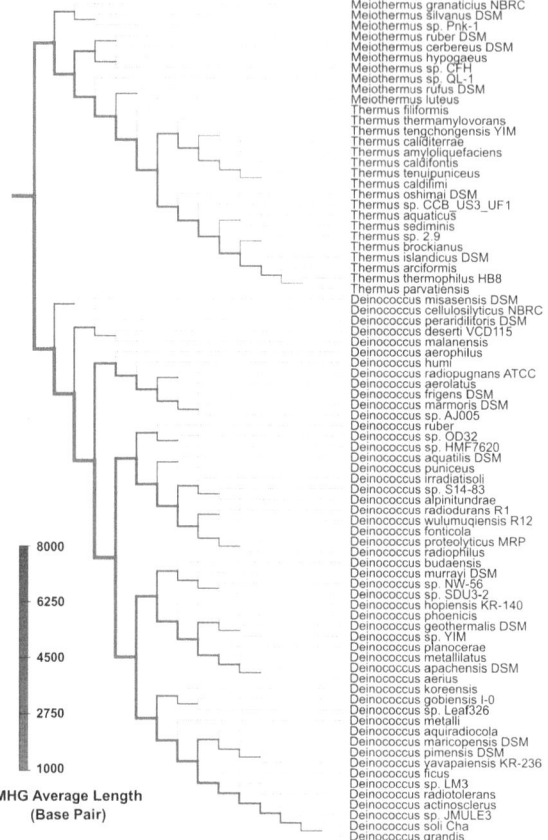

Fig. 4. Annotated guide tree for the *Deinococcus-Thermus* phylum. Each tree branch is annotated with tMHG-Finder results corresponding to the linked internal node. The width of each branch indicates the number of MHGs that contain more than one sequence. A wider branch represents a higher count. The color of each branch, presented as a heatmap, reflects the average length of identified MHGs. The tip branches are represented as dotted lines, symbolizing the initiation of each genome as a single MHG with a length equivalent to the entire genome.

of MHG decreases from approximately 8,000 bp in the leaf nodes to 1,000 bp at the root. At shallow evolutionary distances, MHGs often span multiple genes, exceeding gene boundaries. As the number of genomes increases and evolutionary distances grow, MHGs become shorter, often covering only portions of genes, reflecting reduced homology across their full length.

At the root node, we focused on genes covered by conserved MHGs. A total of 58 conserved MHGs were identified, with average sequence lengths ranging from 1,288 bp to 4,477 bp (Appendix Table S1). Of these, 14 MHGs overlapped with genes lacking annotated names. As expected, most conserved MHGs (38) corresponded to a single gene, with the longest being **rpoC**, which was fully

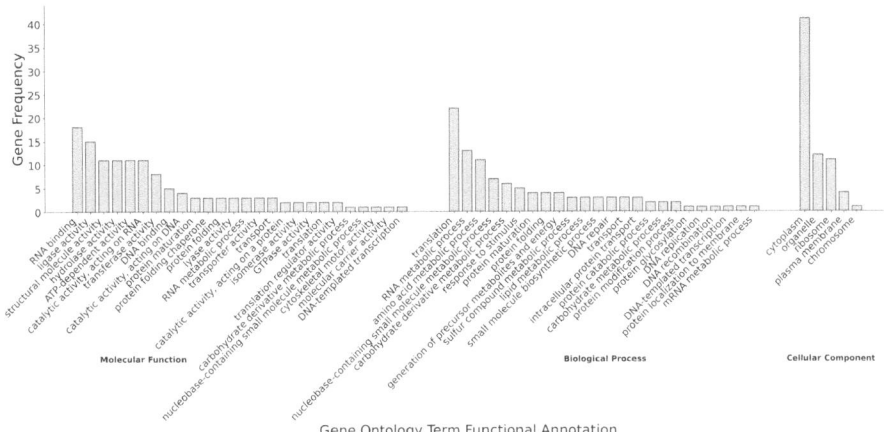

Fig. 5. Gene Ontology Term Distribution Captured by Genes Covered by Conserved MHGs. The frequency distribution is categorized into three main activities: Molecular Function (MF), Biological Process (BP), and Cellular Component (CC). It is based on 54 genes encompassed by 44 conserved MHGs identified at the root node of the phylogenetic guide tree. For each gene, the associated Gene Ontology (GO) terms were propagated to include parent terms and intersected with the GO Slim Prokaryote subset to provide a high-level overview of functional contributions.

encapsulated within its respective MHG. However, five MHGs spanned multiple genes. Among these, three contained multiple highly conserved ribosomal subunits, which are located next to each other and are conserved across all 80 genomes. The other two included **ClpX** and **ClpP**, involved in stress adaptation, and **sucC** and **sucD**, contributing to energy metabolism. An exception was observed with the gene **carB** splitting into two MHGs, each capturing approximately half of the gene. This is not a case of gene fission but rather an instance of over-fragmentation, highlighting a limitation of tMHG-Finder that may reduce its accuracy.

Conserved MHGs extended beyond ribosomal RNAs, prompting an analysis of their broader functional contributions. The genes covered by conserved MHGs were queried against the QuickGO database to retrieve their Gene Ontology (GO) terms, which were propagated through the ontology tree and intersected with the GO Prokaryote subset. While the most frequent terms were generic (e.g., MF: RNA binding, BP: translation, CC: cytoplasm), the GO terms contributed by MHG-covered genes spanned diverse categories (Fig. 5). These included functionalities potentially linked to the extremophilic traits of the studied phylum, such as DNA repair, various metabolic processes, and response to stimulus (Fig. 5). Investigating these functional contributions in greater detail remains a compelling future direction.

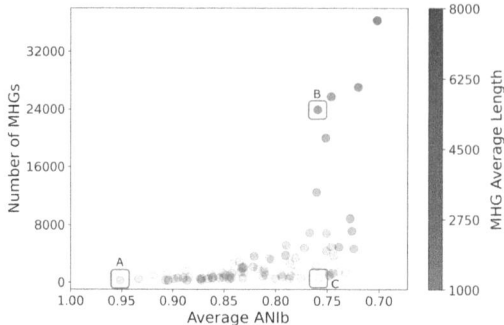

Fig. 6. Correlation between ANI values, genome counts, and MHG statistics. Each dot represents an internal node on the guide tree. The x-axis shows the average ANI value between genome pairs in the left and right child nodes, while the y-axis indicates the number of non-singleton MHGs at the node. Dot color represents the average MHG length as a heatmap, and transparency reflects the number of genomes (ranging from 2 to 80). Points A, B, and C illustrate examples: A and C have similar genome counts but different ANI values, whereas B and C have similar ANI values but different genome counts.

3.4 *Deinococcus-Thermus* Evolutionary Rates

In the analysis of the evolutionary rates of *Deinococcus-Thermus* phylum, the guide tree was constructed by merging two individual trees derived from 52 *Deinococcus* genomes and 28 *Thermus/Meiothermus* genomes, respectively. The guide tree was constructed using the neighbor-joining method based on the distance matrix estimated from Mash for the entire dataset. Since the dataset lacks an outgroup and the neighbor-joining tree is inherently unrooted, the tree was manually rerooted on the branch separating the two clades to provide a meaningful representation of the evolutionary relationships.

tMHG-Finder computed an MHG set at each internal node of the guide tree by including the genomes located at leaf nodes beneath that internal node. Each internal node was annotated with the number of MHGs with more than one sequence and the average length of identified MHGs. From the root to the leaves of the guide tree, there was a decrease in the number of MHGs and an increase in the average MHG length (Fig. 4). At the root node, including the most divergent species pairs within the studied dataset, 36,282 MHGs with an average length of 1,008 bp were identified. In contrast, internal nodes at the bottom level included greater conservation with less than 725 MHGs and a max length of 7,942 bp.

Zooming into specific clades (Fig. 4), the *Deinococcus* clade included 27,066 MHGs with an average length of 1048 bp. In contrast, the *Thermus/Meiothermus* clade, with 8,847 MHGs and an average length of 1010 bp, demonstrated a reduced number of MHGs compared to the *Deinococcus* clade. This observation was consistent with the fact that the *Thermus/Meiothermus* clade included about half the number of species found in the *Deinococcus* clade. Additionally, the average genome length of the *Thermus/Meiothermus* clade (2,599,819 bp) was substantially shorter than that of the *Deinococcus* clade (4,181,147 bp).

Quantifying genomic distance using the average ANIb across all genome pair combinations reveals a clear correlation with MHG statistics. As the genomic distance increases, the number of MHGs increases while their average length decreases. This trend is evident when comparing genome sets of similar size (Fig. 6 A and C). Another key factor influencing MHG statistics is the number of genomes in the compared sets. For example, B and C have similar average genomic distances, but B contains a larger number of genomes (indicated by higher transparency), exhibiting more MHGs with shorter lengths (Fig. 6). In summary, both including a larger number of genomes and increasing genomic distance lead to MHG sets containing a larger number of shorter MHGs.

The tMHG-Finder tree-based analysis demonstrated the complex relationship between evolutionary rates, phylogenetic distance, and homologous groups within the *Deinococcus-Thermus* phylum, providing valuable insights into the evolutionary processes of the extremophilic bacterial phylum.

4 Discussion

The definition of an MHG is rooted in an evolutionary perspective, imposing two crucial "maximal" constraints, ensuring that each MHG incorporates every homologous sequence and all sites for each sequence, sharing a single-tree history. These two constraints distinguish an MHG from a locally collinear block (LCB). It is crucial to note that MHG partitioning does not account for the order of neighboring MHGs. Additionally, MHGs allow for the inclusion of multiple sequences from the same genome, accommodating cases of duplication.

In this study, we present tMHG-Finder as an improvement to MHG-Finder. Leveraging a tree-guided architecture, tMHG-Finder dynamically adjusts the threshold to determine homologous sequences based on phylogenetic distances. tMHG-Finder accelerates blastn mapping with representative sequences and parallelization, enabling efficient processing of large datasets. It supports both user-defined and auto-estimated guide trees, as well as the integration of new genomes by comparing guide tree topologies and retaining MHGs below the insertion point. While guide trees and representative sequences are commonly used in genome alignment [4,11,16,30], they are seamlessly incorporated into tMHG-Finder. Unlike whole-genome alignment methods that prioritize maximizing pairwise alignment scores using the progressive framework, tMHG-Finder aims to optimize the evolutionary breakpoints at each internal node.

In contrast to classical clustering problems with well-defined elements in two compared clusterings, comparing two sets of MSAs in genome partitioning is challenging due to sequences rather than isolated elements in each MSA. Traditional metrics such as recall and precision, akin to the Rand index, treat each nucleotide as an individual element, overlooking that each sequence within an MSA represents a range. Fractionalization, as a novel metric, addresses these limitations by focusing on ranges and quantifying the congruence of breakpoints. It accurately captures the spectrum of similarity and fragmentation compared to existing metrics, providing a comprehensive assessment.

In our simulation study, our results show that tMHG-Finder is not only over 10x faster than MHG-Finder for less than 10 taxa on a single thread but also scales linearly with the number of taxa. Due to the limited scalability and less informative results of MHG-Finder, our evaluation focused on comparing the speed of the two versions of our tools. In contrast, for assessing accuracy, we focused on tMHG-Finder's performance against four existing whole-genome aligners. Both tMHG-Finder and Cactus achieved F-scores above 0.90 (0.932 and 0.966, respectively), while the other methods' F-scores were below 0.75. Cactus outperformed tMHG-Finder in recall (0.945 vs. 0.877), as tMHG-Finder adopts a more conservative approach in defining homologous group boundaries. Aiming to balance the two maximal constraints, the heuristics implemented in tMHG-Finder favor the maximal length constraint, which may result in lower recall but higher precision. However, Cactus exhibited a higher error rate, with 1.3% of its predicted alignments being incorrect, compared to only 0.6% for tMHG-Finder. This is noteworthy as tMHG-Finder's predicted MHGs are significantly less fragmented than those from Cactus. The higher error rate of Cactus can introduce inaccuracies in downstream evolutionary analysis, which is why conservative core-genome alignment methods prioritize precision over recall [14,20].

The validation of tMHG-Finder through the accurate encapsulation of 16S rRNA sequences demonstrates the algorithm's robustness. All near-complete 16S sequences were grouped into a single MHG, underscoring its capability to capture biologically meaningful homologous relationships while minimizing the impact of potential sequencing or assembly artifacts. Evolutionary rate analysis reveals a consistent trend: as species diverge, the number of MHGs decreases while their lengths increase, reflecting reduced conservation of homologous regions over evolutionary time. Additionally, the influence of genomic distance and dataset size on MHG statistics was evident. Larger datasets and greater genomic distances result in a higher number of shorter MHGs, while the average number of genes covered per MHG decreases with increasing evolutionary distance. This transition highlights the shift from MHGs spanning multiple genes at shallow evolutionary distances to covering portions of genes at deeper divergences.

The analysis of conserved MHGs further supports these observations, revealing that most overlap with single genes. These include ribosomal subunits and genes associated with key biological functions, such as stress response and energy metabolism. Notably, conserved MHGs within the *Deinococcus-Thermus* phylum capture genes contributing to extremophilic adaptations, including DNA repair, metabolic processes, and responses to environmental stimuli. These findings illustrate the ability of tMHG-Finder to identify functionally and evolutionarily significant homologous regions across diverse bacterial lineages.

In conclusion, tMHG-Finder provides a robust framework for investigating evolutionary dynamics through MHGs, offering valuable insights into bacterial adaptation to diverse environments. Future research could examine the influence of guide tree selection on MHG computation and assess the applicability of whole-genome aligners for comparative analysis within the *Deinococcus-Thermus* phylum. Additionally, exploring the evolution of key genes and operons through the lens of MHGs may further shed light on extremophilic traits and other adaptive mechanisms shaping bacterial evolution.

Acknowledgements. This research was supported in part by the National Science Foundation, grants DMS/NIGMS-2153704, DBI-2030604, EF-2126387, and IIS-2239114.

Disclosure of Interests. The authors have no competing interests to declare that are relevant to the content of this article.

References

1. Aleksander, S.A., et al.: The gene ontology knowledgebase in 2023. Genetics **224**(1), iyad031 (2023)
2. Altschul, S.F., Gish, W., Miller, W., Myers, E.W., Lipman, D.J.: Basic local alignment search tool. J. Mol. Biol. **215**(3), 403–410 (1990)
3. Angiuoli, S.V., Salzberg, S.L.: Mugsy: fast multiple alignment of closely related whole genomes. Bioinformatics **27**(3), 334–342 (2010). https://doi.org/10.1093/bioinformatics/btq665
4. Armstrong, J., et al.: Progressive cactus is a multiple-genome aligner for the thousand-genome era. Nature **587**(7833), 246–251 (2020)
5. Binns, D., Dimmer, E., Huntley, R., Barrell, D., O'donovan, C., Apweiler, R.: QuickGO: a web-based tool for gene ontology searching. Bioinformatics **25**(22), 3045–3046 (2009)
6. Blackburne, B.P., Whelan, S.: Measuring the distance between multiple sequence alignments. Bioinformatics **28**(4), 495–502 (2012)
7. Blanca, A., Harris, R.S., Koslicki, D., Medvedev, P.: The statistics of k-mers from a sequence undergoing a simple mutation process without spurious matches. J. Comput. Biol. **29**(2), 155–168 (2022)
8. Colquhoun, R.M., et al.: Pandora: nucleotide-resolution bacterial pan-genomics with reference graphs. Genome Biol. **22**, 1–30 (2021)
9. Coudert, E., et al.: Annotation of biologically relevant ligands in UniProtKB using ChEBI. Bioinformatics **39**(1), btac793 (2023)
10. Dalquen, D.A., Anisimova, M., Gonnet, G.H., Dessimoz, C.: ALF—a simulation framework for genome evolution. Mol. Biol. Evol. **29**(4), 1115–1123 (2012)
11. Darling, A.E., Mau, B., Perna, N.T.: progressiveMauve: multiple genome alignment with gene gain, loss and rearrangement. PLoS ONE **5**(6), e11147 (2010)
12. Darling, A.E., Miklós, I., Ragan, M.A.: Dynamics of genome rearrangement in bacterial populations. PLoS Genet. **4**(7), e1000128 (2008)
13. Earl, D., et al.: Alignathon: a competitive assessment of whole-genome alignment methods. Genome Res. **24**(12), 2077–2089 (2014)
14. Fruzangohar, M., Moolhuijzen, P., Bakaj, N., Taylor, J.: CoreDetector: a flexible and efficient program for core-genome alignment of evolutionary diverse genomes. Bioinformatics **39**(11), btad628 (2023)
15. Jain, C., Rodriguez-R, L.M., Phillippy, A.M., Konstantinidis, K.T., Aluru, S.: High throughput ANI analysis of 90k prokaryotic genomes reveals clear species boundaries. Nat. Commun. **9**(1), 5114 (2018)

16. Kaduk, M., Sonnhammer, E.: Improved orthology inference with Hieranoid 2. Bioinformatics **33**(8), 1154–1159 (2017)
17. Katoh, K., Standley, D.M.: Mafft multiple sequence alignment software version 7: improvements in performance and usability. Mol. Biol. Evol. **30**(4), 772–780 (2013)
18. Kembel, S.W., Wu, M., Eisen, J.A., Green, J.L.: Incorporating 16S gene copy number information improves estimates of microbial diversity and abundance. PLoS Comput. Biol. **8**(10), e1002743 (2012)
19. Kille, B., Balaji, A., Sedlazeck, F.J., Nute, M., Treangen, T.J.: Multiple genome alignment in the telomere-to-telomere assembly era. Genome Biol. **23**(1), 182 (2022)
20. Kille, B., Nute, M.G., Huang, V., Kim, E., Phillippy, A.M., Treangen, T.J.: Parsnp 2.0: scalable Core-Genome alignment for massive microbial datasets. Bioinformatics, btae311 (2024). https://doi.org/10.1093/bioinformatics/btae311
21. Lerat, E., Daubin, V., Ochman, H., Moran, N.A.: Evolutionary origins of genomic repertoires in bacteria. PLoS Biology **3**(5) (2005). https://doi.org/10.1371/journal.pbio.0030130
22. Minkin, I., Medvedev, P.: Scalable multiple whole-genome alignment and locally collinear block construction with SibeliaZ. Nat. Commun. **11**(1), 6327 (2020)
23. Nakhleh, L., Ruths, D., Wang, L.-S.: RIATA-HGT: a fast and accurate heuristic for reconstructing horizontal gene transfer. In: Wang, L. (ed.) COCOON 2005. LNCS, vol. 3595, pp. 84–93. Springer, Heidelberg (2005). https://doi.org/10.1007/11533719_11
24. Ondov, B.D., et al.: Mash: fast genome and metagenome distance estimation using MinHash. Genome Biol. **17**(1), 1–14 (2016)
25. Page, A.J., et al.: Roary: rapid large-scale prokaryote pan genome analysis. Bioinformatics **31**(22), 3691–3693 (2015)
26. Parks, D.H., Chuvochina, M., Rinke, C., Mussig, A.J., Chaumeil, P.A., Hugenholtz, P.: GTDB: an ongoing census of bacterial and archaeal diversity through a phylogenetically consistent, rank normalized and complete genome-based taxonomy. Nucleic Acids Res. **50**(D1), D785–D794 (2022)
27. Parks, D.H., Imelfort, M., Skennerton, C.T., Hugenholtz, P., Tyson, G.W.: CheckM: assessing the quality of microbial genomes recovered from isolates, single cells, and metagenomes. Genome Res. **25**(7), 1043–1055 (2015)
28. Perrin, A., Rocha, E.P.: PanACoTA: a modular tool for massive microbial comparative genomics. NAR Genomics Bioinf. **3**(1), lqaa106 (2021)
29. Pritchard, L., Glover, R.H., Humphris, S., Elphinstone, J.G., Toth, I.K.: Genomics and taxonomy in diagnostics for food security: soft-rotting enterobacterial plant pathogens. Anal. Methods **8**(1), 12–24 (2016)
30. Schreiber, F., Sonnhammer, E.L.: Hieranoid: hierarchical orthology inference. J. Mol. Biol. **425**(11), 2072–2081 (2013)
31. Tonkin-Hill, G., et al.: Producing polished prokaryotic pangenomes with the Panaroo pipeline. Genome Biol. **21**, 1–21 (2020)
32. Treangen, T.J., Rocha, E.P.: Horizontal transfer, not duplication, drives the expansion of protein families in prokaryotes. PLoS Genet. **7**(1), e1001284 (2011)
33. Větrovský, T., Baldrian, P.: The variability of the 16s rRNA gene in bacterial genomes and its consequences for bacterial community analyses. PLoS ONE **8**(2), e57923 (2013)
34. Yin, Y., Ogilvie, H.A., Nakhleh, L.: Annotation-free delineation of prokaryotic homology groups. PLoS Comput. Biol. **18**(6), e1010216 (2022)

Phylogenetics

Phylogenetic Network Diversity Parameterized by Reticulation Number and Beyond

Leo van Iersel[1], Mark Jones[1], Jannik Schestag[1], Celine Scornavacca[2], and Mathias Weller[3]

[1] TU Delft, Delft, The Netherlands
{l.j.j.vanIersel,j.t.schestag}@tudelft.nl
[2] ISEM, Université de Montpellier, CNRS, IRD, EPHE, Montpellier, France
celine.scornavacca@umontpellier.fr
[3] LIGM, CNRS, Université Gustave Eiffel, Champs-sur-Marne, France
mathias.weller@u-pem.fr

Abstract. Network Phylogenetic Diversity (Network-PD) is a measure for the diversity of a set of species based on a rooted phylogenetic network (with branch lengths and inheritance probabilities on the reticulation edges) describing the evolution of those species. We consider the MAX-NETWORK-PD problem: given such a network, find k species with maximum Network-PD score. We show that this problem is fixed-parameter tractable (FPT) for binary networks, by describing an optimal algorithm running in $\mathcal{O}(2^r \log(k)(n+r))$ time, with n the total number of species in the network and r its reticulation number. Furthermore, we show that MAX-NETWORK-PD is NP-hard for level-1 networks, proving that, unless P = NP, the FPT approach cannot be extended by using the level as parameter instead of the reticulation number.

1 Introduction

As human activities drive a sixth mass extinction [15], and in the absence of a serious political response to this crisis [11], studying *phylogenetic diversity* (PD) is timely.

Indeed, when experiencing a widespread and rapid decline in Earth's biodiversity, one could wonder where to put our efforts in order to preserve a maximum amount of *biodiversity*, given some temporal and economic constraints [18]. The concept of PD is an attempt at answering this question. The concept has been introduced three decades ago in an impactful paper by Daniel Faith [4]. The

---

L. van Iersel and M. Jones—Partially funded by the Dutch Research Council (NWO) grant OCENW.KLEIN.125 and OCENW.M.21.306.
J. Schestag—Partially funded by the Dutch Research Council (NWO), project "Optimization for and with Machine Learning (OPTIMAL)" OCENW.GROOT.2019.015.
C. Scornavacca—Partially funded by French Agence Nationale de la Recherche through the CoCoAlSeq Project (ANR-19-CE45-0012).

underlying idea is simple: if we want to preserve as much biodiversity as possible within a group X of species and we can rescue at most k species, then we should focus our effort on a size-k subset $S \subseteq X$ of species that showcase, overall, a wide range of features, that is, the distinct traits and qualities covered by the species of S are maximum among all such subsets. This *feature diversity* (FD) of S is often approximated using the PD of S, which is in turn defined as follows: Given a tree T representing the evolution of the species in X, the PD of S (in T) is the sum of the branch lengths of the subtree connecting the root and the species in S. (Note that approximating FD with PD may not always be appropriate, see [20]).

PD has been extensively used in the context of tree-like evolution, and, given a tree T and an integer k, an optimal solution with k species can be found with a greedy algorithm [14,17].

However, when the evolution of the species under interest is also shaped by reticulate events such as hybrid speciation, lateral gene transfer, or recombination, then the picture is no longer as rosy. In the case of reticulate events, a single species may inherit genetic material and, thus, features from multiple direct ancestors, and its evolution should be represented by a phylogenetic network [7] rather than a tree. Several ways of extending the notion of PD for networks have been proposed [2,8,19], one of which is called Network-PD. The optimization problem linked to Network-PD, i.e. computing the maximum Network-PD$_\mathcal{N}$ score over all subsets of species of size at most k for a given phylogenetic network $\mathcal{N}$, is named MAX-NETWORK-PD. Bordewich et al. [2] proved that MAX-NETWORK-PD is NP-hard and cannot be approximated in polynomial time with an approximation ratio better than $1 - \frac{1}{e}$ unless P = NP; furthermore, it remains NP-hard even for the restricted class of phylogenetic networks called "normal" networks.

The contribution of this paper is twofold. First, we present an algorithm for MAX-NETWORK-PD parameterized by the reticulation number of the input network. Herein, we leverage the greedy algorithm for PD on trees [14,17] to efficiently process the subtree below a reticulate event. Surprisingly, we show that this algorithm cannot be generalized to use the "level" as parameter unless P = NP. The level of a network is a measure of its treelikeness, formally defined in the next section, which can be smaller than the reticulation number. More precisely, we prove that MAX-NETWORK-PD is NP-hard even on level-1 networks (which are networks without overlapping cycles), thereby answering an open question of Bordewich et al. [2].

2 Preliminaries

For a positive integer n, denote $[n] := \{1, \ldots, n\}$. Let $(0, 1) := \{x \in \mathbb{R} : 0 < x < 1\}$ and $[0, 1] := (0, 1) \cup \{0, 1\}$. Let $\mathbb{R}_{>0} := \{x \in \mathbb{R} : x > 0\}$ and $\mathbb{R}_{\geq 0} := \mathbb{R}_{>0} \cup \{0\}$. For a set Z and an integer k with $k \leq |Z|$, by $\binom{Z}{k}$ we denote the set of all subsets of Z with exactly k elements. In this paper, we make use of both natural and binary logarithms. We write $\ln x$, and $\log_2 x$, to denote the logarithm of x to the base e and 2, respectively.

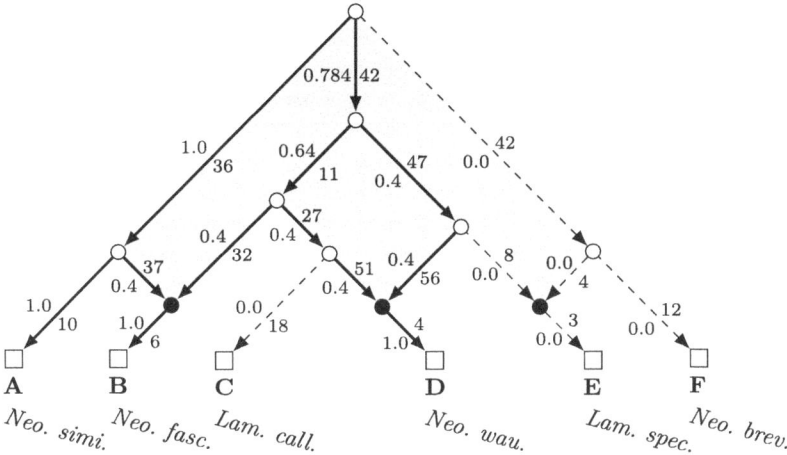

Fig. 1. A hypothesized heritage of several species of fish in a phylogenetic network [9]. We take the inheritance probabilities to be 0.4 for reticulation edges and 1 for other edges. Edge weights are indicated by integers to the right of each edge. *Edge weights and inheritance probabilities are not based on data and for illustrative purposes only.* The three reticulations are depicted as black filled vertices. The biggest subgraph without cut edges is shaded. The level and the reticulation number of the network are 3. It can be shown that the sets $\{B,D\}$ and $\{B,C,D,F\}$ maximize Network-PD$_\mathcal{N}$ among all size-2 and size-4 subsets of taxa, respectively. As an example, we illustrate how to compute the Network-PD score for $Z = \{A,B,D\}$. The decimal numbers left of the edges indicate the $\gamma_Z^p(e)$-values (see Definition 1), leading to a score of Network-PD$_\mathcal{N}^p(Z) = 195.968$. Dashed edges have $\gamma_Z^p(e) = 0$ and hence do not contribute towards the Network-PD score.

Phylogenetics. Consider the example network of Fig. 1. Given a set of taxa X, a *phylogenetic network on X* or *X-network* is a directed acyclic graph $\mathcal{N} = (V, E)$ in which the *leaves*, vertices of indegree 1 and outdegree 0, are bijectively labeled with elements from X, and in which the *root* is the single vertex of indegree 0 and outdegree 2, and in which all other vertices either are *tree vertices* and have indegree 1 and outdegree at least 2 or are *reticulations* and have indegree at least 2 and outdegree 1. Edges incoming at reticulations are *reticulation edges*. When X is clear from context, we refer to an X-network simply as a *network* or *phylogenetic network*. A *phylogenetic tree* $\mathcal{T} = (V, E)$ on X or *X-tree* is an X-network with no reticulations. A network is *binary* if the maximum indegree and outdegree of any vertex is 2.

The *reticulation number* of a network $\mathcal{N}$ is the sum of the indegrees of all reticulations minus the number of reticulations. If $\mathcal{N}$ is binary, then the reticulation number is exactly the number of reticulations. The *level* of $\mathcal{N}$ is the maximum reticulation number among subgraphs with no cut-arcs (arcs whose removal disconnects the network).

For each edge $e = uv$ we say that u is a *parent* of v and v is a *child* of u. For vertices $u, v \in V$, we say u is an *ancestor* of v and v is a *descendant of u* if there is a directed path from u to v in $\mathcal{N}$. If in addition $u \neq v$, we say u is a *strict*

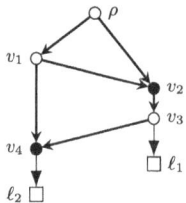

	ρv_1	ρv_2	$v_1 v_2$	$v_2 v_3$	$v_1 v_4$	$v_3 v_4$	$v_3 \ell_1$	$v_4 \ell_2$
$\omega(e)$	50	40	10	5	30	8	4	2
$p(e)$	1	0.4	0.5	1	0.2	0.6	1	1
$\gamma^p_{Z_1}(e)$	0.5	0.4	0.5	1	0	0	1	0
$\gamma^p_{Z_2}(e)$	0.44	0.24	0.3	0.6	0.2	0.6	0	1
$\gamma^p_{Z_3}(e)$	0.6	0.4	0.5	1	0.2	0.6	1	1

Fig. 2. An example for calculating $\gamma^p_Z(e)$. Reticulations are black. The chosen sets are $Z_1 = \{\ell_1\}, Z_2 = \{\ell_2\}, Z_3 = \{\ell_1, \ell_2\}$. Network-PD$^p_{\mathcal{N}}(Z)$ for $Z = Z_1, Z_2, Z_3$ is 55, 50.4, and 72.8, respectively.

ancestor of v and v a *strict* descendant of u. The set of *offspring* of e, denoted off(e), is the set of all $x \in X$ which are descendants of v. Throughout this paper, we use the terms taxon/taxa, species, and leaf/leaves interchangeably.

Diversity. We assume that each edge e in a network $\mathcal{N} = (V, E)$ has an associated positive integer *weight* $\omega(e)$. These weights are used to represent some measure of difference between two species. Given an X-tree $\mathcal{T}$ and a weight function $\omega : E \to \mathbb{N}$, the *phylogenetic diversity* PD$_{\mathcal{T}}(Z)$ of any subset $Z \subseteq X$ is given by PD$_{\mathcal{T}}(Z) := \sum_{e \mid \text{off}(e) \cap Z \neq \emptyset} \omega(e)$. That is, PD$_{\mathcal{T}}(Z)$ is the total weight of all edges in $\mathcal{T}$ that are above some leaf in Z.

The phylogenetic diversity model assumes that features of interest appear along edges of the tree with frequency proportional to the weight of that edge, and that any feature belonging to one species is inherited by all its descendants. Thus, PD$_{\mathcal{T}}(Z)$ corresponds to the expected number of distinct features appearing in all species in Z.

Initially defined only for trees, several extensions of the definition to phylogenetic networks have recently been proposed [2,19]. In this paper, we focus Network-PD$_{\mathcal{N}}$ (defined below), which allows the case that reticulations may not inherit all of the features from every parent. This is modeled via an *inheritance probability* $p(e) \in [0, 1]$ on each reticulation edge $e = uv$. Here, $p(e)$ represents the expected proportion of features present in u that are also present in v; or equivalently, $p(e)$ is the probability that a feature in u is inherited by v. Non-reticulation edges can be considered as having inheritance probability 1.

For a subset of taxa $Z \subseteq X$, the measure Network-PD$_{\mathcal{N}}(Z)$ represents the expected number of distinct features appearing in taxa in Z [2]. For each evolutionary branch uv, this measure is obtained by multiplying the number $\omega(uv)$ of features developed on the branch uv (which is assumed to be proportional to the length of the branch) with the probability $\gamma^p_Z(uv)$ that a random feature appearing in u or developed on uv will survive when preserving Z.

Formally, we define $\gamma^p_Z(uv)$ as follows. Consider an example in Fig. 2.

Definition 1. *Given a network $\mathcal{N} = (V, E)$ with edge weights $\omega : E \to \mathbb{N}$, probabilities $p : E \to [0, 1]$ and a set of taxa $Z \subseteq X$, we define $\gamma^p_Z : E \to [0, 1]$ recursively for each edge $uv \in E$ as follows:*

- If v is a leaf, then $\gamma_Z^p(uv) := p(uv)$ if $v \in Z$, and $\gamma_Z^p(uv) = 0$ otherwise.
  (**Intuition**: *The features of v survive if and only if v is preserved by Z.*)
  In most of the paper, with the notable exception of Sect. 3, $p(uv) = 1$ if v is a leaf.
- If v is a reticulation with outgoing arc vx, then $\gamma_Z^p(uv) = p(uv) \cdot \gamma_Z^p(vx)$.
  (**Intuition**: *v's features are a mixture of features of its parents and the features of u have a certain probability $p(uv)$ of being included in this mix and, thereby, survive in preserved descendants of x.*)
- If v is a tree node with children x_i, then $\gamma_Z^p(uv) = 1 - \prod_i (1 - \gamma_Z^p(vx_i))$. In the special case that v has two children, this is equal to $\gamma_Z^p(vx) + \gamma_Z^p(vy) - \gamma_Z^p(vx) \cdot \gamma_Z^p(vy)$.
  (**Intuition**: *To lose a feature of v, it has to be lost in both children x and y of v, which are assumed to be independent events, since both copies of the feature develop independently.*)

When clear from the context, we will omit the superscript p. Further, we only consider values of p on edges incoming to leaves or reticulations, so we may restrict the domain of p to those edges. We can now define the measure Network-PD$_\mathcal{N}^p(Z)$ for a subset of taxa Z as follows: Network-PD$_\mathcal{N}^p(Z) = \sum_{e \in E} \omega(e) \cdot \gamma_Z^p(e)$.

Since we assume that all weights are non-negative, we observe that both $\gamma_Z^p(e)$ and Network-PD$_\mathcal{N}(Z)$ are monotone on Z, that is, $\gamma_{Z'}^p(e) \leq \gamma_Z^p(e)$ and Network-PD$_\mathcal{N}^p(Z') \leq$ Network-PD$_\mathcal{N}^p(Z)$ for all $Z' \subseteq Z \subseteq X$. We can now formally define the main problem studied in this paper:

MAX-NETWORK-PD
Input: A phylogenetic network $\mathcal{N} = (V, E)$ on X with edge weights $\omega : E \to \mathbb{N}$, inheritance probabilities $p : E \to [0, 1]$, and integers $k, D \in \mathbb{N}$.
Question: Is there a $Z \subseteq X$ with $|Z| \leq k$ and Network-PD$_\mathcal{N}(Z) \geq D$?

Note that, if $p(e) = 1$ for all edges e incoming to leaves (all "preservation projects" succeed with probability 1) and a node v has no reticulation descendants, then $\gamma_Z(uv) = 1$ if off$(e) \cap Z \neq \emptyset$, and otherwise $\gamma_Z(uv) = 0$ (see Lemma 1). In this setting, Network-PD$_\mathcal{N}$ coincides with PD$_\mathcal{N}$ if $\mathcal{N}$ is a tree. This holds even if all leaves are weighted and the total weight of Z must not exceed k.

Throughout the paper, we assume that integers are encoded in binary and that rational numbers p/q (with p and q coprime integers) are encoded using binary encodings of p and q. See Appendix A for details.

3 A Branching Algorithm

In this section, we show that MAX-NETWORK-PD is fixed-parameter tractable with respect to the reticulation number of the input network. To facilitate the explanation of our algorithm, we solve a generalization of MAX-NETWORK-PD, where (a) each leaf ℓ is assigned a cost $c(\ell) \in \{0, 1\}$, (b) the leaf-edges may have inheritance probability $p(v\ell) \leq 1$ (as well as the reticulation edges), with the

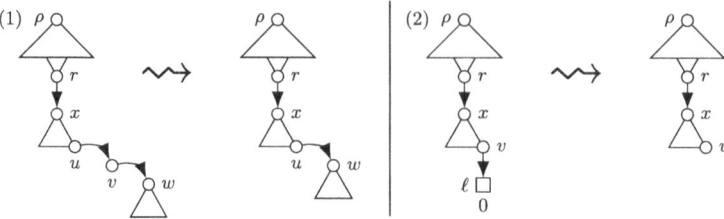

Fig. 3. Examples of Reduction Rules 1 and 2 are depicted on the left and on the right, respectively. White leaves have an inheritance probability of zero. Costs are written below the leaves.

condition that $p(v\ell) = 1$ if $c(\ell) = 1$, and (c) we look for a subset Z of leaves with total cost at most k (instead of cardinality k). We refer to this problem as 0/1-COST MAX-NETWORK-PD and we use $p(\ell)$ instead of $p(v\ell)$ whenever ℓ is a leaf with parent v.

In the following, let $\mathcal{I} := (\mathcal{N}, \omega, p, c, k, D)$ be an instance for 0/1-COST MAX-NETWORK-PD, let r be a lowest reticulation in $\mathcal{N}$ with outgoing edge rx. Our algorithm "guesses" whether or not any cost-1 leaf below r is in a solution Z. If not, then we remove all cost-1 leaves below r and use reduction rules to (a) turn the resulting subtree into a single leaf below r and (b) turn r into two new cost-0 leaves with inheritance probabilities according to $\gamma_Z(rx)$. If some (unknown) cost-1 leaf below r is in a solution, we show that such a leaf can be picked greedily. Then, we decrement k, set the cost of that leaf to zero, and use the knowledge that $\gamma_Z(rx) = 1$ to remove r from the network.

Note that our reduction and branching rules may create nodes with high outdegree, even if the input network is binary. However, the algorithm used to solve the resulting non-binary tree can deal with such polytomies [12].

Reduction. Let r be a lowest reticulation in $\mathcal{N}$ and let E_r be the set of edges below r. The following reduction rules simplify $\mathcal{I}$ by getting rid of cost-0 leaves below r. Note that each rule assumes that $\mathcal{I}$ is reduced with respect to the previous rules. See Fig. 3 for examples of Reduction Rules 1 and 2, and Fig. 4 for examples of Reduction Rules 3 and 4.

Reduction Rule 1. *Let $uv \in E_r$ such that v has a single child w. Then, contract v onto u and set $\omega(uw) := \omega(uv) + \omega(vw)$, $p(uw) = p(vw)$.*

Correctness of Reduction Rule 1. Let $\mathcal{I}' := (\mathcal{N}', \omega', p', c, k, D')$ be the result of applying Reduction Rule 1 to $\mathcal{I}$. Clearly, we have $\gamma_Z^{p'}(e) = \gamma_Z^p(e)$ for any edge e below w and all Z. So by construction $\gamma_Z^{p'}(uw) = \gamma_Z^p(vw)$. Observe that $p(uv) = 1$ since v is not a leaf and r is the lowest reticulation in $\mathcal{N}$; thus, $\gamma_Z^p(uv) = \gamma_Z^p(vw) = \gamma_Z^{p'}(uw)$. This implies that $\gamma_Z^{p'}(e) = \gamma_Z^p(e)$ for all Z and any $e \in E \setminus \{uv, vw\}$. So, Network-PD$_\mathcal{N}(Z)$ − Network-PD$_{\mathcal{N}'}(Z) = \gamma_Z^p(uv) \cdot \omega(uv) + \gamma_Z^p(vw) \cdot \omega(vw) - \gamma_Z^{p'}(uw) \cdot \omega(uw) = \gamma_Z^p(uv) \cdot (\omega(uv) + \omega(vw) - \omega(uw)) = 0$. □

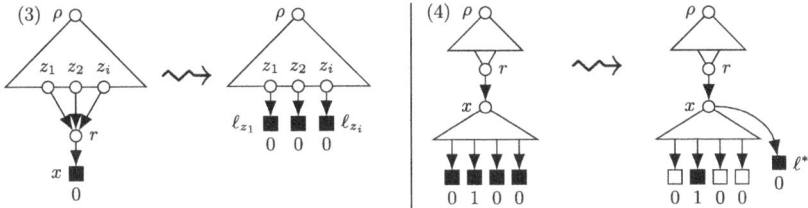

Fig. 4. Examples of Reduction Rules 3 and 4 are depicted on the left and on the right, respectively. Black leaves have a positive inheritance probability. Costs are written below the leaves.

Reduction Rule 2. *Let $v\ell \in E_r$ such that ℓ is a leaf, $v \neq r$, and $p(v\ell) = 0$. Then, remove ℓ.*

Correctness of Reduction Rule 2. Let u be the parent of v, and let $v_1, \ldots, v_t$ denote the children of v with $v_i \neq \ell$. Then $1 - (1 - \gamma_Z^p(v\ell)) \prod_i (1 - \gamma_Z^p(uv_i))$ is the value of $\gamma_Z^p(uv)$ before removing ℓ which equals $1 - \prod_i (1 - \gamma_Z^p(uv_i))$, the value afterward, since $\gamma_Z^p(v\ell) = 0$ for all Z. □

Reduction Rule 3. *Let the unique child x of r be a leaf with cost $c(x) = 0$. Then, for each parent z_i of r, add a new leaf ℓ_{z_i} to z_i with $c(\ell_{z_i}) := 0$ and $p(\ell_{z_i}) := p(z_i r) \cdot p(rx)$ and $\omega(z\ell_{z_i}) := \omega(z_i r)$. Finally, remove r and x and decrease D by $p(rx) \cdot \omega(rx)$.*

Correctness sketch of Reduction Rule 3. As x has cost 0 and γ_Z^p is monotone on Z, every maximal solution for $\mathcal{I}$ contains x. Likewise, every maximal solution for the modified instance $\mathcal{I}'$ contains all ℓ_{z_i}. Then, one can verify that maximal solutions for $\mathcal{I}$ collect exactly the score of rx more than maximal solutions for $\mathcal{I}'$, which is $p(rx) \cdot \omega(rx)$. □

Reduction Rule 4. *Let x be the unique child of r, let Q be the set of cost-0 leaves below r, and let $E_x := E_r \setminus \{rx\}$. Then,*

(1) for each $uv \in E_x$, multiply $\omega(uv)$ by $1 - \gamma_Q^p(uv)$,
(2) for each $\ell \in Q$, set $p(\ell) := 0$,
(3) reduce D by $\sum_{e \in E_x} \gamma_Q^p(e) \cdot \omega(e)$, and
(4) add a new cost-0 leaf ℓ^* as a child of x with $\omega(x\ell^*) = 0$ and $p(\ell^*) = \gamma_Q^p(rx)$.

To prove the correctness of Reduction Rule 4, we use the following lemma.

Lemma 1. *Let uv be an edge in $\mathcal{N}$ such that all descendants of v (including v) are tree nodes and let Z be a leaf set of $\mathcal{N}$. Then, $\gamma_Z^p(uv) = 1 - \prod_{\ell \in \mathrm{off}(uv) \cap Z}(1 - p(\ell))$.*

Proof. We prove the claim by induction on the length of a longest path from v to a leaf. In the induction base, v is a leaf and, thus, $\gamma_Z^p(e) = 1 - \prod_{\ell \in \mathrm{off}(e) \cap Z}(1 - p(\ell))$ since this is $1 - (1 - p(v)) = p(v)$ if $v \in Z$ and 0 otherwise. For the induction step, let v be a tree node with children x_i and assume the claim is true for each edge vx_i. Then,

$$\gamma_Z^p(e) \stackrel{\text{Def. 1}}{=} 1 - \prod_i \left(1 - \gamma_Z^p(vx_i)\right) \stackrel{\text{IH}}{=} 1 - \prod_i \left(1 - \left(1 - \prod_{\ell \in \text{off}(vx_i) \cap Z} (1 - p(\ell))\right)\right)$$

$$= 1 - \prod_i \Big(\prod_{\ell \in \text{off}(vx_i) \cap Z} (1 - p(\ell))\Big) = 1 - \prod_{\ell \in \text{off}(e) \cap Z} (1 - p(\ell)) \qquad \square$$

Correctness of Reduction Rule 4. Let $\mathcal{I}' =: (\mathcal{N}, \omega', p', c, k, D')$ be the result of applying Reduction Rule 4 to $\mathcal{I}$ and let $Q' := Q \cup \{\ell^*\}$. We assume all solutions Z to be maximal, implying that they contain all cost-0 leaves. Note that generality is not lost since Network-PD$_\mathcal{N}^p(Z)$ is monotone on Z. Let Z and Z' be any subsets of leaves of $\mathcal{N}$ and $\mathcal{N}'$, respectively, with $Q \subseteq Z$ and $Z' = Z \cup \{\ell^*\}$. We show that Z is a solution for $\mathcal{I}$ if and only if Z' is a solution for $\mathcal{I}'$.

We consider the contribution of each edge to the diversity score of Z in $\mathcal{N}$ and the diversity score of Z' in $\mathcal{N}'$. If Z (and, thus, also Z') contains a cost-1 leaf ℓ below r, then $p(\ell) = 1$ and, by Lemma 1, we have $\gamma_Z^p(rx) = 1 = \gamma_{Z'}^{p'}(rx)$. Otherwise, $\gamma_Z^p(rx) = \gamma_Q^p(rx) = p'(\ell^*) = \gamma_{\{\ell^*\}}^{p'}(rx) = \gamma_{Z'}^{p'}(rx)$. In both cases, $\gamma_Z^p(e) = \gamma_{Z'}^{p'}(e)$ for all $e \in E \setminus E_r$ since these values only depend on the values of the edges below e. Further, note that $\omega(x\ell^*) \cdot \gamma_{Z'}^{p'}(x\ell^*) = \omega(x\ell^*) \cdot p(\ell^*) = 0$. Thus, it remains to consider the edges in $E_x := E_r \setminus \{rx\}$. For any such edge $e \in E_x$, we observe

$$\gamma_Z^p(e) \stackrel{\text{Lemma 1}}{=} 1 - \prod_{\ell \in \text{off}(e) \cap Z}(1 - p(\ell)) = 1 - \prod_{\ell \in \text{off}(e) \cap Q}(1 - p(\ell)) \cdot \prod_{\ell \in \text{off}(e) \cap Z \setminus Q}(1 - p(\ell))$$

and, since $p(\ell) = 1$ for all $\ell \in Z \setminus Q$ by convention stated in the problem definition, we have

$$\gamma_Z^p(e) = \begin{cases} \underbrace{1 - \prod_{\ell \in \text{off}(e) \cap Q}(1 - p(\ell))}_{\gamma_Q^p(e)} & \text{if off}(e) \cap Z \subseteq Q \\ 1 & \text{otherwise} \end{cases}$$

and the same holds for p' and Z' instead of p and Z since the leaves below e in $\mathcal{N}$ are exactly the leaves below e in $\mathcal{N}'$ (ℓ^* cannot be below e in $\mathcal{N}'$ since $e \in E_x$). Now, since $p'(\ell) = 0$ for all $\ell \in Q$ by construction, we have

$$\gamma_{Z'}^{p'}(e) = \begin{cases} 0 & \text{if off}(e) \cap Z \subseteq Q \\ 1 & \text{otherwise} \end{cases}$$

implying $\gamma_Z^p(e) = \gamma_{Z'}^{p'}(e) \cdot (1 - \gamma_Q^p(e)) + \gamma_Q^p(e)$. Then,

$$\sum_{e \in E} \gamma_Z^p(e) \cdot \omega(e) - \sum_{e \in E \cup \{x\ell^*\}} \gamma_{Z'}^{p'}(e) \cdot \omega'(e) = \sum_{e \in E_x} (\gamma_Z^p(e) \cdot \omega(e) - \gamma_{Z'}^{p'}(e) \cdot \omega'(e))$$

$$\stackrel{\text{Def'n } \omega'}{=} \sum_{e \in E_x} \left(\gamma_Z^p(e) - \gamma_{Z'}^{p'}(e) \cdot (1 - \gamma_Q^p(e))\right) \cdot \omega(e)$$

$$= \sum_{e \in E_x} \gamma_Q^p(e) \cdot \omega(e) = D - D'$$

Thus, $\sum_{e \in E} \gamma_Z^p(e) \cdot \omega(e) \geq D$ if and only if $\sum_{e \in E} \gamma_{Z'}^{p'}(e) \cdot \omega(e) \geq D'$. □

Branching. Observe that, if no reduction rule applies to $\mathcal{N}$, then the subtree below any lowest reticulation r has at least one cost-1 leaf and at most one cost-0 leaf. An important part of the correctness of our branching algorithm is that solutions may be assumed to pick cost-1 leaves "greedily", that is, if a solution chooses any cost-1 leaf below r, then there is also a solution choosing a "heaviest" cost-1 leaf below r instead.

Lemma 2. *Let r be a lowest reticulation in $\mathcal{N}$ and let a be a cost-1 leaf below r in $\mathcal{N}$ maximizing the weight of the r-a-path. Let Z be any set of leaves of $\mathcal{N}$ containing a cost-1 leaf below r. Then, there is a set Z^* of leaves of $\mathcal{N}$ with the same cost as Z with $a \in Z^*$ and Network-PD$_{\mathcal{N}}^p(Z^*) \geq$ Network-PD$_{\mathcal{N}}^p(Z)$.*

Proof. Suppose that $a \notin Z$ as otherwise, the claim is trivial. Let $b \in Z$ be a cost-1 leaf below r such that $u := LCA(a, b)$ is lowest possible (has maximal (unweighted) distance from r), and let $Z^* := (Z \setminus \{b\}) \cup \{a\}$. Let q_a and q_b be the unique paths from u to a and b, respectively, and note that $\omega(q_a) \geq \omega(q_b)$ by choice of a. Furthermore, for each edge uv on q_a, we know that Z contains no leaf below v (by maximality of the r-u-path). Since both a and b are cost-1 leaves, we have $p(a) = p(b) = 1$ by convention stated in the problem definition, implying that $\gamma_Z^p(e_b) = \gamma_{Z^*}^p(e_a) = 1$ for all edges e_a on q_a and e_b on q_b. Thus, Network-PD$_{\mathcal{N}}^p(Z^*)$−Network-PD$_{\mathcal{N}}^p(Z) = \sum_{e \in E} \gamma_{Z^*}^p(e)\omega(e) - \sum_{e \in E} \gamma_Z^p(e)\omega(e) = \omega(q_a) - \omega(q_b) \geq 0$. □

Now, we can present and prove the correctness of our main branching rule, solving 0/1-COST MAX-NETWORK-PD in $O^*(\binom{|R|}{k})$ time, where R is the set of reticulations in the input network and k is the budget.

Branching Rule 1 (See Fig. 5). *Let ρ be the root of $\mathcal{N}$. Let r be a lowest reticulation in $\mathcal{N}$ whose unique child x is not a 0-cost leaf. Let Q be the set of cost-0 leaves below r. Then,*

1. *create the instance $\mathcal{I}_0 := (\mathcal{N}_0, \omega_0, p_0, c_0, k, D)$ by*
   (a) *setting $p_0(t) := 0$ and $c(t) := 0$ for all cost-1 leaves t below r,*
   (b) *replacing rx with ρx, setting $\omega_0(\rho x) := \omega(rx)$ and,*
   (c) *adding a new leaf ℓ to r with $p_0(\ell) := \gamma_Q^p(rx)$ and $c_0(\ell) := \omega_0(r\ell) := 0$, and*

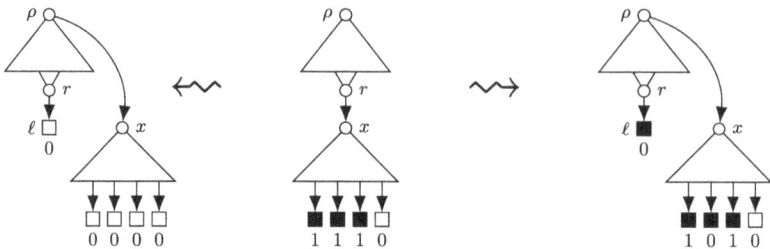

Fig. 5. An example of Branching Rule 1 with $\mathcal{I}_0$ ("do not select a cost-1 leaf below r") on the left and $\mathcal{I}_1$ ("select a cost-1 leaf below r") on the right. Black leaves have an inheritance probability of one. Costs are written below the leaves. Note that the budget for $\mathcal{I}_1$ is $k-1$ and that applying Reduction Rule 4 may change the target diversity.

2. create the instance $\mathcal{I}_1 := (\mathcal{N}_1, \omega_1, p_1, c_1, k-1, D)$ by
  (a) finding a cost-1 leaf a below r maximizing the weight of the r-a-path and setting $c_1(a) := 0$,
  (b) replacing rx with ρx, setting $\omega_1(\rho x) := \omega(rx)$ and
  (c) adding a new leaf ℓ to r with $p_1(\ell) := 1$ and $c_1(\ell) := \omega_1(r\ell) := 0$.

Correctness of Branching Rule 1. Let P denote the set of cost-1 leaves below r in $\mathcal{I}$ and recall that Q contains all cost-0 leaves below r in $\mathcal{I}$, and that $c(Q) = c_0(Q) = 0$. We show that $\mathcal{I}$ has a solution Z if and only if $\mathcal{I}_0$ or $\mathcal{I}_1$ has a solution. Without loss of generality, we may assume solutions to be maximal, that is, they contain all cost-0 leaves. For any leaf-set Z containing all cost-0 leaves in $\mathcal{I}$ and any leaf-set Z_i containing all cost-0 leaves in $\mathcal{I}_i$ for some $i \in \{0, 1\}$, we then have

$$\gamma_{Z_i}^{p_i}(\rho x) = \begin{cases} \gamma_Q^p(rx) & \text{if } i = 0 \\ 1 & \text{if } i = 1 \end{cases} = p_i(\ell) = \gamma_{Z_i}^{p_i}(r\ell) \text{ and } \gamma_Z^p(rx) = \begin{cases} \gamma_Q^p(rx) & \text{if } Z \cap P = \emptyset \\ 1 & \text{if } Z \cap P \neq \emptyset \end{cases} \quad (1)$$

so, under the condition $Z \cap P = \emptyset \iff i = 0$, we have $\gamma_{Z_i}^{p_i}(\rho x) = \gamma_Z^p(rx)$, so

$$\gamma_{Z_i}^{p_i}(\rho x) \cdot \underbrace{\omega_i(\rho x)}_{=\omega(rx)} + \gamma_{Z_i}^{p_i}(r\ell) \cdot \underbrace{\omega_i(r\ell)}_{=0} \stackrel{(1)}{=} \gamma_Z^p(rx) \cdot \omega(rx). \quad (2)$$

Claim 1. Let Z be a leaf-set in $\mathcal{N}$, let $Z' := Z \cup \{\ell\}$, and let $i := \operatorname{sgn}(|Z \cap P|)$. Then, Z is a solution for $\mathcal{I}$ if and only if Z' is a solution for $\mathcal{I}_i$.

Proof. Note that $Z \cap P = \emptyset \iff i = 0$ is satisfied. In the following, we compare the value of Z in $\mathcal{I}$ and the value of Z' in $\mathcal{I}_i$.

First, consider any arc e in $\mathcal{N}$ that is not below r. Since, by (1), we have $\gamma_Z^p(rx) = \gamma_{Z'}^{p_i}(r\ell)$, and since $p(\ell') = p_0(\ell')$ for any leaf $\ell' \neq \ell$, we inductively infer that $\gamma_{Z'}^{p_i}(e) = \gamma_Z^p(e)$ as these values only depend on the edges below e.

Second, by (2), the contribution of the arc rx to the value of the solution Z for $\mathcal{I}$ equals the contribution of ρx and $r\ell$ to the value of the solution Z' for $\mathcal{I}_i$.

It remains to compare the contributions of the arcs e below x in $\mathcal{N}$. In the following, consider such an arc e. If $i = 0$, then Z avoids P and so does Z', so $p(\ell') = p_0(\ell')$ for all $\ell' \in \text{off}(e) \cap Z = \text{off}(e) \cap Z'$. If $i = 1$, then $p(\ell') = p_1(\ell')$ for all leaves in $\text{off}(e)$. Thus, by Lemma 1,

$$\gamma_{Z'}^{p_i}(e) = 1 - \prod_{\ell' \in \text{off}(e) \cap Z'}(1 - p_i(\ell')) = 1 - \prod_{\ell' \in \text{off}(e) \cap Z}(1 - p(\ell')) = \gamma_Z^p(e).$$

Thus, we conclude that Z and Z' score exactly the same in $\mathcal{I}$ and $\mathcal{I}'$, respectively.

Finally, we show that $c(Z) = c_i(Z') - i$. If $i = 0$, then this holds since $c_0(\ell) = 0$. If $i = 1$ then Z intersects P and, by Lemma 2, we can assume that Z contains a. Then, since $c(a) = 1$ and $c_1(\ell) = c_1(a) = 0$, we have $c_1(Z') = c(Z) - 1$. ∎

Now, we can prove the promised equivalence. First, if Z is a solution for $\mathcal{I}$, then $Z' := Z \cup \{\ell\}$ is a solution for $\mathcal{I}_i$ with $i = \text{sgn}(|Z \cap P|)$. Second, if Z_0 is a solution for $\mathcal{I}_0$, then $Z_0' := Z_0 \setminus P$ is also a solution for $\mathcal{I}_0$ since $p_0(\ell') = 0$ for all $\ell' \in P$ and, by Claim 1, $Z := Z_0' \setminus \{\ell\}$ is a solution for $\mathcal{I}$. Third, if Z_1 is a solution for $\mathcal{I}_1$ then we can assume $a \in Z_1$ since $c_1(a) = 0$ so, for $Z := Z_1 \setminus \{\ell\}$, we have $Z \cap P \neq \emptyset$, thereby satisfying the conditions of Claim 1. Thus, Z is a solution for $\mathcal{I}$. □

We can now solve 0/1-COST MAX-NETWORK-PD as follows. If $k = 0$, then the monotonicity of Network-PD$_\mathcal{N}^p(Z)$ in Z implies that "taking" all cost-0 leaves in $\mathcal{N}$ is optimal. Otherwise, we repeatedly find a lowest reticulation r in $\mathcal{N}$, apply all reduction rules, and if r survives, branch into two instances using Branching Rule 1. Note that, in each new instance, r has a leaf child with cost 0. Thus, Reduction Rule 3 will apply and remove r before another branching occurs. If no branching or reduction rules apply, then $\mathcal{N}$ is a tree. In this tree, a slight variation of Reduction Rule 4 can be used to remove all cost-0 leaves, so all remaining leaves have cost 1 and, therefore (by convention), inheritance probability 1. Such an instance can be solved in $\mathcal{O}(n \log k)$ time [12]. Note that the budget k is decreased for one of the two branches and $|R|$ is reduced in each branch, so no more than $\binom{|R|}{k}$ branches need to be explored. Finally, with careful bookkeeping the reduction and branching can be implemented to run in $\mathcal{O}(|E|) = \mathcal{O}(n + r)$ amortized time in total.

Theorem 1. *On binary, n-leaf networks with r reticulations, 0/1-COST MAX-NETWORK-PD and MAX-NETWORK-PD can be solved in $\mathcal{O}(\sum_{i=0}^{\min\{k,r\}} \binom{r}{i} \cdot \log k \cdot (n+r)) \subseteq \mathcal{O}(2^r \cdot \log k \cdot (n+r))$ time, where k is the budget.*[1]

Theorem 1 shows that MAX-NETWORK-PD is fixed-parameter tractable with respect to the number of reticulations. In light of this, one might expect that MAX-NETWORK-PD is also fixed-parameter tractable with respect to the "level" (maximum number of reticulations in any biconnected component ("blob") of the network, since many tractability results for the reticulation number also extend to the level by applying the algorithm separately to each blob, with

---

[1] Note that this running time degenerates to $o(2^r \cdot n)$ if $k \leq r/3$.

minimal adjustment, in such a way that the problem parameterized by level reduces to the problem parameterized by reticulation number. Unfortunately, this approach does not work for MAX-NETWORK-PD– for a given blob, it may be better to pay some diversity score within the blob in order to increase $\gamma_Z^p(e)$ for the incoming edge of that blob. This trade-off means that we need to consider many possible solutions for each blob. Indeed, we will see in the next section that MAX-NETWORK-PD is NP-hard even on level-1 networks.

4 NP-Hardness Results

Complementing the positive result of the previous section, we now show that MAX-NETWORK-PD is NP-hard on level-1 networks, answering an open question in the literature [2, Section 9]. On our way to showing this hardness result, we also show NP-hardness of the following problem, answering an open question of Komusiewicz and Schestag [10]:

UNIT-COST-NAP
Input: A tree $\mathcal{T} = (V, E)$ with leaves L, edge weights $\omega : E \to \mathbb{N}$, success probabilities $p : L \to [0,1]$, and some $k, D \in \mathbb{N}$.
Question: Is there some $Z \subseteq L$ with $|Z| \leq k$ and $\sum_{e \in E} \gamma'_Z(e) \cdot \omega(e) \geq D$, where $\gamma'_Z(e) := (1 - \prod_{x \in \text{off}(e) \cap Z}(1 - p(x)))$?

Note that $\gamma'_Z(e)$ corresponds to the probability that at least one taxa in off(e) survives, under the assumption that every taxon $x \in Z$ survives independently with probability $p(x)$, and every taxon $x \in L \setminus Z$ does not survive. Thus, UNIT-COST-NAP can be viewed as the problem of maximizing the expected phylogenetic diversity on a tree, where each species we choose to save has a certain probability of surviving.

Subset Product. First, we show that the following problem is NP-hard.

SUBSET PRODUCT
Input: A multiset of positive integers $\{v_1, v_2, \ldots, v_m\}$, integers $M, k \in \mathbb{N}$.
Question: Is there any $S \subseteq [m]$ with $|S| = k$ such that $\prod_{i \in S} v_i = M$?

We note that the definition of SUBSET PRODUCT is slightly different here from the formulation of Garey and Johnson [6]. In particular, we assume that the size k of the set S is given and that all integers are positive. This makes the subsequent NP-hardness reductions in this paper slightly simpler.

The NP-hardness of SUBSET PRODUCT is not a new result. It was stated by Garey and Johnson [6] without full proof (the authors indicate that the problem is NP-hard by reduction from EXACT COVER BY 3-SETS (X3C), citing "Yao, private communication") and a full proof appears in [13] and we reproved it for our slightly adapted variant in the appendix.

Lemma 3 ([13]). X3C *reduces to* SUBSET PRODUCT *in polynomial time.*

As X3C is NP-hard [6], so is SUBSET PRODUCT.

Penalty Sum. Komusiewicz and Schestag [10, Theorems 5.3 & 5.4] showed that, if the following problem is NP-hard, then so is UNIT-COST-NAP:

> PENALTY SUM
> **Input**: A set of tuples $\{t_i = (a_i, b_i) \mid i \in [m], a_i \in \mathbb{Q}_+ \cup \{0\}, b_i \in (0,1)\}$, integers k, Q, and a number $D \in \mathbb{Q}_+$.
> **Question**: Is there some $S \subseteq [m]$ with $|S| = k$ such that $\sum_{i \in S} a_i - Q \cdot \prod_{i \in S} b_i \geq D$?

In the following, we reduce SUBSET PRODUCT to PENALTY SUM, to prove the NP-hardness of PENALTY SUM and UNIT-COST-NAP. Afterward, we show that, even on level-1 networks, MAX-NETWORK-PD is NP-hard by a reduction from UNIT-COST-NAP.

4.1 Hardness of Penalty Sum

The full proof of the NP-hardness of PENALTY SUM is given in Appendix B; Here, we give a brief overview of the main ideas. For an instance $(\{v_1, \ldots v_m\}, M, k)$ of SUBSET PRODUCT, we let $Q := M$ and $D := \ln(1/M) - 1$, and let $t_i := (\ln(1/v_i), 1/v_i)$ for each $i \in [n]$. Then, in the instance $(\{t_i \mid i \in [m]\}, k, Q, D)$ of PENALTY SUM, the aim is to find $S \subseteq [m]$ with $|S| = k$ optimizing $\sum_{i \in S} \ln(1/v_i) - M \cdot \prod_{i \in S}(1/v_i) = \ln(\prod_{i \in S} 1/v_i) - M \cdot (\prod_{i \in S} 1/v_i)$. This value maximizes in $\ln(1/M) - 1 = D$, with equality if and only if $\prod_{i \in S} v_i = M$. Thus, $(\{t_i \mid i \in [m]\}, k, Q, D)$ is a yes-instance of PENALTY SUM if and only if $(\{v_1, \ldots v_m\}, M, k)$ is a yes-instance of SUBSET PRODUCT. The full reduction requires additional work in order to ensure that all numbers involved are non-negative rationals.

4.2 Hardness of Network-Diversity

Finally, reducing from UNIT-COST-NAP, we show the following main result.

Theorem 2. MAX-NETWORK-PD *is* NP-*hard even if the input network has level 1, all weights are positive, and the distance between the root and each leaf is 4.*

Proof. Because PENALTY SUM is NP-hard, we know that UNIT-COST-NAP is NP-hard on trees of height 2 [10]. Let $\mathcal{T}$ be an L-tree of height 2 for some L and let an instance $\mathcal{I} = (\mathcal{T}, \omega, q, k, D)$ of UNIT-COST-NAP be given.

We define a leaf-gadget which is illustrated in Fig. 6. Let $\ell \in L$ be a leaf with success-probability $q(\ell)$. Add four vertices $v_\ell^1, v_\ell^2, \ell^*, \ell^-$ and edges $\ell v_\ell^1, \ell v_\ell^2, v_\ell^1 v_\ell^2, v_\ell^1 \ell^-$, and $v_\ell^2 \ell^*$. The only reticulation in this gadget is v_ℓ^2 with incoming edges ℓv_ℓ^2 and $v_\ell^1 v_\ell^2$. We set the inheritance probabilities $p(\ell v_\ell^2) := q(\ell)/(2 - q(\ell))$ and $p(v_\ell^1 v_\ell^2) := q(\ell)/2$ which are both in $[0,1]$ because $q(\ell) \in [0,1]$.

Let $\mathcal{N}$ be the network that results from replacing each leaf of $\mathcal{T}$ with the corresponding leaf-gadget. The leaves of $\mathcal{N}$ are $L' := \{\ell^*, \ell^- \mid \ell \in L\}$. Let d

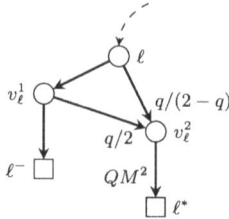

Fig. 6. Illustration of the leaf-gadget. Omitted edge-weights are 1 and $q(\ell)$ is abbreviated to q.

denote the largest denominator in a success-probability $q(\ell)$ of a leaf ℓ of $\mathcal{T}$, so that every $q(\ell)$ is expressible as c'/d' for some pair of integers c, d such that $d' \leq d$. Let M and Q be large integers, such that M is bigger than $\mathrm{PD}_{\mathcal{T}}(L) \geq |L| \geq k$, and $Q \cdot D$ and $Q \cdot d^{-k}$ are both bigger than 3.

Observe that the number of bits necessary to write M and Q is polynomial in the size of $\mathcal{I}$. We set the weight of edges $e \in E(\mathcal{T})$ in $\mathcal{N}$ to $\omega'(e) = kQ \cdot \omega(e)$. For each $\ell \in L$ we set $\omega'(v_\ell^2 \ell^*) := Q \cdot M^2$ and $\omega'(e) := 1$ for $e \in \{\ell v_\ell^1, \ell v_\ell^2, v_\ell^1 v_\ell^2, v_\ell^1 \ell^-\}$.

Finally let $\mathcal{I}' := (\mathcal{N}, \omega', p, k, D' := kQ(M^2 + D))$ be an instance of MAX-NETWORK-PD. Each leaf-gadget is a level-1 network. As the leaf-gadgets are connected by a tree, $\mathcal{N}$ is a level-1 network. Recall that the height of the tree $\mathcal{T}$ is 2, and as such the distance between the root and each leaf in $\mathcal{N}$ is 4.

Before showing that $\mathcal{I}$ and $\mathcal{I}'$ are equivalent, we show that $\gamma_Z^p(e) = q(\ell)$ in the case that $\ell^* \in Z$ but $\ell^- \notin Z$. Indeed because $\ell^- \notin Z$, we conclude that $\gamma_Z^p(\ell v_\ell^2) = p(\ell v_\ell^2) = q(\ell)/(2 - q(\ell))$ and $\gamma_Z^p(\ell v_\ell^1) = \gamma_Z^p(v_\ell^1 v_\ell^2) = p(v_\ell^1 v_\ell^2) = q(\ell)/2$. Subsequently,

$$\gamma_Z^p(e) = 1 - (1 - \gamma_Z^p(\ell v_\ell^1))(1 - \gamma_Z^p(\ell v_\ell^2)) = 1 - \frac{2 - q(\ell)}{2} \cdot \frac{2 - 2q(\ell)}{2 - q(\ell)} = 1 - \frac{2 - 2q(\ell)}{2} = q(\ell). \tag{3}$$

"$\Rightarrow$": Suppose that $\mathcal{I}$ is a yes-instance of UNIT-COST-NAP and that $S \subseteq L$ is a solution of $\mathcal{I}$, that is $|S| \leq k$ and $\mathrm{PD}_{\mathcal{T}}(S) \geq D$. Let $S' := \{\ell^* \mid \ell \in S\}$ be a subset of L'. Clearly $|S'| = |S| \leq k$. Because $\mathcal{T}$ does not contain reticulation edges and $\gamma_Z^p(e) = q(\ell)$ with e being the edge incoming at ℓ, we conclude that

$$\text{Network-PD}_{\mathcal{N}}(S') \geq kQ \cdot \mathrm{PD}_{\mathcal{T}}(S) + k \cdot \omega'(v_\ell^2 \ell^*) \geq kQ \cdot (D + M^2) = D'$$

hence, S' is a solution of $\mathcal{I}'$.

"$\Leftarrow$": Let S' be a solution of $\mathcal{I}'$. Let $S^- = S \cap \{\ell^- \mid \ell \in L\}$ and $S^* = S \cap \{\ell^* \mid \ell \in L\}$. Towards a contradiction, assume $S^- \neq \emptyset$. Then, however, using $3 < Q \cdot D$,

$$\text{Network-PD}_{\mathcal{N}}(S') \leq \sum_{\ell^- \in S^-} (\omega'(v_\ell^1 \ell^-) + \omega'(\ell v_\ell^1)) + |S^*|(QM^2 + 3) + \sum_{e \in E(\mathcal{T})} \omega'(e)$$

$$\leq 2|S^-| + |S^*|(QM^2 + 3) + kQM$$

$$\leq 2 + (k-1)(QM^2 + 3) + kQM$$

$$< kQM^2 - QM^2 + kQM + 3k$$

$$< k(QM^2 + 3) < k(QM^2 + QD) = D'$$

contradicts that S' is a solution. Therefore, we conclude that $S' \subseteq \{\ell^* \mid \ell \in L\}$ and $|S'| = k$. Define $S := \{\ell \mid \ell^* \in S'\}$. Subsequently, with (3) we conclude

$$kQ(M^2 + D) = D' \leq \text{Network-PD}_{\mathcal{N}}(S')$$

$$= k \cdot QM^2 + \sum_{\ell \in S} \underbrace{\left(\frac{q(\ell)}{2} + \frac{q(\ell)}{2} + \frac{q(\ell)}{2 - q(\ell)} \right)}_{\leq 3} + kQ \cdot \text{PD}_{\mathcal{T}}(S).$$

It follows that $\text{PD}_{\mathcal{T}}(S) \geq 1/kQ \cdot (kQ(M + D) - kQM - 3k) = D - 3/Q$.

It remains to show that $\text{PD}_{\mathcal{T}}(S)$ cannot take any values in the range $[D - 3/Q, D)$, i.e. that $\text{PD}_{\mathcal{T}}(S) \geq D - 3/Q$ implies that $\text{PD}_{\mathcal{T}}(S) \geq D$. To this end, let c_ℓ, d_ℓ be the unique positive integers such that $q(\ell) = c_\ell/d_\ell$ for each leaf ℓ in $\mathcal{T}$. Then $q(\ell)$ is a multiple of $1/d_\ell$ by construction, as is $(1 - q(\ell))$. It follows that for any edge e in $\mathcal{T}$, $\gamma'_S(e) = (1 - \prod_{\ell \in \text{off}(e) \cap S}(1 - q(\ell)))$ is a multiple of $1/(\prod_{\ell \in S} d_\ell)$. As all edge weights are integers, we also have that $\text{PD}_{\mathcal{T}}(S)$ is a multiple of $1/(\prod_{\ell \in S} d_\ell)$. It follows that either $\text{PD}_{\mathcal{T}}(S) \geq D$ or $D - \text{PD}_{\mathcal{T}}(S) \geq 1/(\prod_{\ell \in S} d_\ell)$. As $d_\ell \leq d$ for any ℓ, this difference is at least $d^{-k} > 3/Q$. It follows that if $\text{PD}_{\mathcal{T}}(S) \geq D - 3/Q$ then in fact $\text{PD}_{\mathcal{T}}(S) \geq D$.

We conclude $\text{PD}_{\mathcal{T}}(S) \geq D$. Hence with $|S| = |S'| \leq k$ we conclude that S is a solution of $\mathcal{I}$. Thus, $\mathcal{I}$ is a yes-instance of UNIT-COST-NAP. □

5 Discussion

In this paper, we have studied MAX-NETWORK-PD from a theoretical point of view. These results do have some practical implications. In particular, they show that we can only hope to solve MAX-NETWORK-PD efficiently for evolutionary histories that are reasonably tree-like in the sense that the number of reticulate events is small. For this case, we present an algorithm that is theoretically efficient. How well it works in practice is still to be evaluated.

Some open questions on the theoretical front remain. Can MAX-NETWORK-PD be solved in pseudo-polynomial time on level-1 networks? Is MAX-NETWORK-PD polynomial time solvable on level-1 networks if we require the network to be ultrametric, i.e. when all root-leaf paths have the same length? Is MAX-NETWORK-PD FPT when parameterized with the number of selected species k to save plus the level of the network? Is MAX-NETWORK-PD FPT when parameterized with the number of different weights and probabilities? If this is the case, then rounding the weights or probabilities to magnitudes could significantly speed up the running time [5].

From a practical point-of-view however, the most important task is to assess which variants of phylogenetic diversity on networks (see [19]) are biologically most relevant. This could of course depend on the type of species considered and in particular on the type of reticulate evolutionary events. Even if the maximization problem cannot be solved efficiently, having a good measure of phylogenetic diversity can still have great practical use by measuring how diverse a given set of species is.

A A Note About Binary Representation of Rational Numbers

As most of the problems here involve rational numbers as part of the input, it is worth drawing attention to how those numbers are represented, in particular how they affect the input size of an instance. As is standard, we assume that positive integers are represented in binary (so that, for instance, the numbers 3, 4 and 5 are written as 11, 100 and 101 respectively). Thus the number of bits required to represent the integer n is $\mathcal{O}(\log_2(n))$. In the case of rational numbers, we assume throughout that a rational p/q (with p and q coprime integers) can be represented by binary representations of p and q. Thus for example, the number $3/5$ may be written as 11/101. It follows that p/q can be represented using $\mathcal{O}(\log_2(p) + \log_2(q))$ bits.

For rational numbers which are a multiple of a power of 2, (such as $1/8 = 2^{-3}$, or $5/8 = 5 \cdot 2^{-3}$), we can write the number by extending the binary representation 'past the decimal point', so that e.g. $1/8$ would be written as 0.001 and $5/8$ as 0.101. There is also the 'floating point' representation, where the number is expressed as an integer t times 2 to some integer c, and the numbers t and c are expressed in binary. Thus for example $5/8$ would be written as 101 × 2^{-11}. Both of these methods of representing rationals have the drawback that they cannot represent rationals that are not a multiple of a power of 2. The number $1/3$, for instance, cannot be expressed exactly under either method.

This distinction becomes important in Sect. 4.1, where our reduction from SUBSET PRODUCT to PENALTY SUM produces rational numbers that are not multiples of a power of 2. Do our hardness results for PENALTY SUM, UNIT-COST-NAP and MAX-NETWORK-PD still hold when one insists on a different method of representing rationals? This is an interesting question, and we make no attempt to answer it.

B Hardness of Penalty Sum

The reduction from SUBSET PRODUCT to PENALTY SUM can be informally described as follows: For an instance $(\{v_1, \ldots v_m\}, M, k')$ of SUBSET PRODUCT and a big integer A, we let a_i be (a rational close to) $A - \ln v_i$ and let $b_i := 1/v_i$, for each $i \in [m]$. Let $Q := M$, let $k := k'$, and let D be (a rational close to) $kA - \ln Q - 1$.

Note that we cannot set $a_i := A - \ln v_i$ or $D := kA - \ln Q - 1$ exactly, because in general these numbers are irrational and cannot be calculated exactly in finite time (nor stored in finite space). Towards showing the correctness of the reduction, we temporarily forget about the need for rational numbers, and consider how the function $\sum_{i \in S} a_i - Q \cdot \prod_{i \in S} b_i$ behaves when we drop the '(a rational close to)' qualifiers from the descriptions above. In particular we will show that the function reaches its theoretical maximum exactly when S is a solution to the SUBSET PRODUCT instance.

Reduction with Irrational Numbers

Construction 1. *Let $(\{v_1, \ldots v_m\}, M, k)$ be an instance of SUBSET PRODUCT. Let us define the following (not necessarily rational) numbers.*

- *Let $A := \lceil \max_{i \in [m]} (\ln v_i) \rceil + 1$;*
- *Let $a_i^* := A - \ln v_i$ for each $i \in [m]$;*
- *Let $b_i := 1/v_i$ for each $i \in [m]$;*
- *Let $Q := M$;*
- *Let $D^* := kA - \ln Q - 1$.*

Finally, output the instance $(\{(a_i^*, b_i) : i \in [m]\}, k, Q, D^*)$ of PENALTY SUM.

We note that the purpose of A in Construction 1 is simply to ensure that $a_i^* > 0$ for each $i \in [m]$, as required by the formulation of PENALTY SUM. Now, let $f^* : \binom{[m]}{k} \to \mathbb{R}$ be defined by

$$f^*(S) := \sum_{i \in S} a_i^* - Q \cdot \prod_{i \in S} b_i.$$

Lemma 4. *For any $S \in \binom{[m]}{k}$:*

1. *$f^*(S) \leq D^*$, and*
2. *$f^*(S) = D^*$ if and only if $\prod_{i \in S} v_i = Q$.*

Proof. First, observe that given $|S| = k$, the function f^* can be written as

$$f^*(S) = kA - \sum_{i \in S} \ln v_i - Q / \prod_{i \in S} v_i = kA - \ln \left(\prod_{i \in S} v_i \right) - Q / \prod_{i \in S} v_i$$

Letting $x_S := \prod_{i \in S} v_i$, we therefore have $f^*(S) = kA - \ln x_S - Q x_S^{-1}$. Let $g^* : \mathbb{R}_{>0} \to \mathbb{R}$ be defined by $g^*(x) := kA - \ln x - Qx^{-1}$ and note that $f^*(S) = g^*(x_S)$ for any $S \in \binom{[m]}{k}$. Recall that $g^*(x)$ has a critical point at x' when $\frac{dg^*}{dx}(x') = 0$. Since $\frac{dg^*}{dx} = -x^{-1} + Qx^{-2}$, this occurs exactly when $x'^{-1} = Qx'^{-2}$, i.e. when $x' = Q$. Moreover, for $Q > x > 0$, we have $Qx^{-1} > 1$, implying

$$\frac{dg^*}{dx} = -x^{-1} + Qx^{-2} > -x^{-1} + x^{-1} = 0.$$

On the other hand, for $x > Q > 0$, we have $Qx^{-1} < 1$, implying

$$\frac{dg^*}{dx} = -x^{-1} + Qx^{-2} < -x^{-1} + x^{-1} = 0.$$

It follows that $g^*(x)$ is strictly increasing on the range $0 < x < Q$ and strictly decreasing on the range $x > Q$. Thus, $g^*(x)$ has a unique maximum on the range $x > 0$, and this maximum is achieved at $x = Q$. In particular, for all $S \in \binom{[m]}{k}$, we have

$$f^*(S) = g^*(x_S) \leq g^*(Q) = kA - \ln Q - 1 = D^*. \quad (4)$$

With equality if and only if $x_S = \prod_{i \in S} v_i = Q$. □

The above result implies that, abusing terminology slightly, $(\{(a_i^*, b_i) \mid i \in [m]\}, k, Q, D^*)$ is a yes-instance of 'PENALTY SUM' if and only if $(\{v_1, \ldots v_m\}, M, k')$ is a yes-instance of SUBSET PRODUCT.

We are now ready to fully describe the polynomial-time reduction from SUBSET PRODUCT to PENALTY SUM, showing how we can adapt the ideas above to work for rational a_i and D.

Reduction with Rational Numbers. Let $(\{v_1, \ldots v_m\}, M, k')$ be an instance of SUBSET PRODUCT, and let a_i^*, b_i, Q, k, D^* be defined as previously. Then by Lemma 4, $f^*(S) = \sum_{i \in S} a_i^* - Q \cdot \prod_{i \in S} b_i \geq D^*$ if and only if $\prod_{i \in S} v_i = Q = M$ for any $S \in \binom{[m]}{k}$.

Our task now is to show how to replace a_i^* and D^* with rationals a_i and D, in such a way that the same property holds (i.e. that $\sum_{i \in S} a_i - Q \cdot \prod_{i \in S} b_i \geq D$ if and only if $\prod_{i \in S} v_i = M$), and such that the instance $(\{(a_i, b_i) \mid i \in [m]\}, k, Q, D)$ can be constructed in polynomial time. The key idea is to find rational numbers that can be encoded in polynomially many bits but that are close enough to their respective irrationals that the difference between $f^*(S)$ and $f(S)$ (and between D^* and D) is guaranteed to be small. To this end, let us fix a positive integer H to be defined later, and we will require all a_i, b_i, D to be a multiple of 2^{-H}. This ensures that the denominator part of any of these rationals can be encoded using $\mathcal{O}(H)$ bits.

Given any $x \in \mathbb{R}$ and a positive integer H, let $\lfloor x \rfloor_H := r_x/2^H$, where r_x is the largest integer such that $r_x/2^H \leq x$. For example $\lfloor \pi \rfloor_3 = 3.125 = 25/2^3$, because $25/2^3 < \pi < 26/2^3$ (one may think of $\lfloor x \rfloor_H$ as the number derived from the binary representation of x by deleting all digits more than H positions after the binary point. Thus, as the binary expression of π begins 11.00100 10000 11111..., the binary expression of $\lfloor \pi \rfloor_3$ is 11.001). Similarly, let $\lceil x \rceil_H := s_x/2^H$, where s_x is the smallest integer such that $x \leq s_x/2^H$. Finally, let $\delta := 1/2^H$.

Observation 1. *Let $x \in \mathbb{R}$. Then, $x - \delta < \lfloor x \rfloor_H \leq x \leq \lceil x \rceil_H < x + \delta$.*

We can now describe the reduction from SUBSET PRODUCT to PENALTY SUM.

Construction 2. *Let* $(\{v_1, \ldots v_m\}, M, k)$ *be an instance of* SUBSET PRODUCT.

- *Let* $A := \lceil \max_{i \in [m]}(\ln v_i) \rceil + 1$;
- *Let* $a_i := \lceil a_i^* \rceil_H = \lceil A - \ln v_i \rceil_H$ *for each* $i \in [m]$;
- *Let* $b_i := 1/v_i$ *for each* $i \in [m]$;
- *Let* $Q := M$;
- *Let* $D := \lfloor D^* \rfloor_H = \lfloor kA - \ln Q - 1 \rfloor_H$.

Finally, output the instance $\mathcal{I} := (\{(a_i, b_i) \mid i \in [m]\}, k, Q, D)$ *of* PENALTY SUM.

In the following, we show that the two instances are equivalent. To this end, let $f : \binom{[m]}{k} \to \mathbb{R}$ be defined by

$$f(S) := \sum_{i \in S} a_i - Q \cdot \prod_{i \in S} b_i.$$

Note that f is the same as the function f^* defined previously, but with each a_i^* replaced by a_i. Then, $\mathcal{I}$ is a **yes**-instance of PENALTY SUM if and only if there is some $S \in \binom{[m]}{k}$ such that $f(S) \geq D$. The next lemma shows the close relation between f^* and f (and between D^* and D), which will be used in both directions to show the equivalence between **yes**-instances of SUBSET PRODUCT and PENALTY SUM.

Lemma 5. *Let* $S \in \binom{[m]}{k}$. *Then,* $f^*(S) \leq f(S) < f^*(S) + k\delta$ *and* $D^* - \delta < D \leq D^*$.

Proof. Observe that $f(S) - f^*(S) = \sum_{i \in S}(a_i - a_i^*)$ and $|S| = k$. Then, by Observation 1, we have $0 \leq a_i - a_i^* < \delta$ for all $i \in [m]$. Thus, $0 \leq f(S) - f^*(S) < k\delta$, from which the first claim follows. The second claim follows immediately from Observation 1 and the fact that $D = \lfloor D^* \rfloor_H$. □

Corollary 1. *Let* $S \in \binom{[m]}{k}$ *such that* $\prod_{i \in S} v_i = Q$. *Then,* $f(S) \geq D$.

Proof. $D \stackrel{\text{Lem. 5}}{\leq} D^* \stackrel{\text{Lem. 4 (2)}}{=} f^*(S) \stackrel{\text{Lem. 5}}{\leq} f(S)$. □

We now have that $(\{v_1, \ldots v_m\}, M, k')$ being a **yes**-instance of SUBSET PRODUCT implies $\mathcal{I}$ being a **yes**-instance of PENALTY SUM. To show the converse, we show for all $S \in \binom{[m]}{k}$ that $\prod_{i \in S} v_i = Q' \neq Q$ implies $f(S) < D$. Since $f(S) < f^*(S) + k\delta$ and $D^* - \delta < D$, it is sufficient to show that $f^*(S) + k\delta \leq D^* - \delta$, that is $(k+1)\delta \leq D^* - f^*(S)$. To do this, we first establish a lower bound on $D^* - f^*(S')$ in terms of Q, using the following technical lemma, whose proof is deferred to the appendix.

Lemma 6. *Let* $Q, Q' \in \mathbb{N}_+$ *with* $Q \geq 2$ *and* $Q \neq Q'$. *Then,* $\ln Q' - \ln Q + Q/Q' - 1 > Q^{-4}$.

We explicitly note that we use the natural logarithm. For other logarithms, say log_2, this lemma is not true. For example for $Q = 2$ and $Q' = 1$ we have $\log_2(1) - \log_2(2) + 2/1 - 1 = 0 - 1 + 2 - 1 = 0 < 2^{-4}$.

Corollary 2. *Suppose $\prod_{i \in S} v_i = Q' \neq Q$ for some $Q \geq 2$ and $S \in \binom{[m]}{k}$. Then $D^* - f^*(S) > Q^{-4}$.*

Proof. Recall that $D^* = kA - \ln Q - 1$ and that $f^*(S) = kA - \ln(\prod_{i \in S} v_i) - Q/(\prod_{i \in S} v_i) = kA - \ln Q' - Q/Q'$. Then $D^* - f^*(S) = \ln Q' - \ln Q + Q/Q' - 1$. It follows from Lemma 6 that $D^* - f^*(S) > Q^{-4}$. □

Given the above we can now fix a suitable value for H. Given that we wanted $D^* - f^*(S) \geq (k+1)\delta = (k+1)/2^H$ when $\prod_{i \in S} v_i \neq Q$, and assuming without loss of generality that $k < Q$, it is sufficient to set $H = 5\lceil \log_2 Q \rceil$.

Corollary 3. *Let $H = 5\lceil \log_2 Q \rceil$ and $\delta = (1/2^H)$. Then for $(\{(a_i, b_i) \mid i \in [m]\}, k, Q, D)$ constructed as above, it holds that $Q^{-4} \geq (k+1)\delta$.*

Proof. W.l.o.g. we may assume $k < Q$. Then, $(k+1)\delta \leq Q/2^H \leq Q/Q^5 = Q^{-4}$. □

We now have all necessary pieces to reduce SUBSET PRODUCT to PENALTY SUM.

Theorem 3. PENALTY SUM *is* NP-*hard.*

Proof. Given an instance $(\{v_1, \ldots v_m\}, M, k')$ of SUBSET PRODUCT, let $Q := M$, $H := 5\lceil \log_2 Q \rceil$, and $\delta := (1/2^H)$. Construct A, a_i, b_i, k, D as described above, that is: $A := \lceil \max_{i \in [m]}(\ln v_i) \rceil + 1$; $a_i := \lceil a_i^* \rceil_H = \lceil A - \ln v_i \rceil_H$ for each $i \in [m]$; $b_i := 1/v_i$ for each $i \in [m]$; $k := k'$; $D := \lfloor D^* \rfloor_H = \lfloor kA - \ln Q - 1 \rfloor_H$. Let $(\{(a_i, b_i) \mid i \in [m]\}, k, Q, D)$ be the resulting instance of PENALTY SUM.

We first show that $(\{(a_i, b_i) \mid i \in [m]\}, k, Q, D)$ is a **yes**-instance of PENALTY SUM if and only if $(\{v_1, \ldots v_m\}, M, k')$ is a **yes**-instance of SUBSET PRODUCT. Suppose first that $(\{v_1, \ldots v_m\}, M, k')$ is a **yes**-instance of SUBSET PRODUCT. Then there is some $S \in \binom{[m]}{k}$ such that $\prod_{i \in S} v_i = M = Q$. Then by Corollary 1, $f(S) \geq D$ and so $(\{(a_i, b_i) \mid i \in [m]\}, k, Q, D)$ is a **yes**-instance of PENALTY SUM.

Conversely, suppose that $(\{(a_i, b_i) \mid i \in [m]\}, k, Q, D)$ is a **yes**-instance of PENALTY SUM. Then there is some $S \in \binom{[m]}{k}$ such that $f(S) \geq D$. By Lemma 5 and Corollary 3, we have that $f^*(S) > f(S) - k\delta \geq D - k\delta > D^* - (k+1)\delta \geq D^* - Q^{-4}$. Thus $D^* - f^*(S) \leq Q^{-4}$, which by Corollary 2 implies that $\prod_{i \in S} v_i = Q$, and so $(\{v_1, \ldots v_m\}, M, k')$ is a **yes**-instance of SUBSET PRODUCT.

It remains to show that the reduction takes polynomial time. For this, it is sufficient to show that the rationals A, k, D and a_i, b_i for $i \in [m]$ can all be calculated in polynomial time. Observe that $A = \lceil \max_{i \in [m]}(\ln v_i) \rceil$ is the unique integer such that $e^A > \max_{i \in [m]} v_i > e^{A-1}$. Since $\ln v_i < \log_2 v_i$, we have $1 \leq A \leq \lceil \max_{i \in [m]} \log_2 v_i \rceil$ and so we can find A in polynomial time by checking all integers in this range.

For each $i \in [m]$, $a_i = \lceil A - \ln v_i \rceil_H = r_i/2^H$, where r_i is the minimum integer such that $A - \ln v_i \leq r_i/2^H$. Thus, we can compute r_i by checking $e^{A - r_i/2^H} \leq v_i$ with $r_i = 2^H \cdot (A - \lceil \ln v_i \rceil_H)$, setting r_i to its successor if the inequality is not satisfied. Thus we can construct a_i in polynomial time, and a_i can be represented in $\mathcal{O}(\log_2 r + H)$ bits. The construction of D can be handled in a similar way.

For each $i \in [m]$, rational $b_i = 1/v_i$ can be represented in $\mathcal{O}(\log_2 v_i)$ bits (recall that we represent $1/v_i$ with binary representations of the integers 1 and v_i) and takes $\mathcal{O}(\log_2 v_i)$ time to construct. Q and k are taken directly from the instance $(\{v_1, \ldots v_m\}, M, k')$. □

C Omitted Proofs

To prove Lemma 3, we reduce the following problem to SUBSET PRODUCT.

EXACT COVER BY 3-SETS (X3C)
Input: A set X with $|X| = 3n$, a collection $\mathcal{C}$ of subsets of X with $|C| = 3$ for every $C \in \mathcal{C}$.
Question: Is there a collection $\mathcal{C}' \subseteq \mathcal{C}$ such that each element of x appears in exactly one set of $\mathcal{C}'$?

Proof of Lemma 3. Let $(X := \{x_1, \ldots, x_{3n}\}, \mathcal{C} := \{C_1, \ldots, C_m\})$ be an instance of X3C. Let $p_1, \ldots, p_{3n}$ be the first $3n$ prime numbers, so that we may associate each $x_j \in X$ with a unique prime number p_j. For each set $C_i = \{x_a, x_b, x_c\}$, let $v_i := p_a \cdot p_b \cdot p_c$, that is, v_i is the product of the three primes associated with the elements of C_i. Now let $M := \prod_{j=1}^{3n} p_j$, i.e. M is the product of the prime numbers $p_1, \ldots, p_{3n}$. Finally let $k = n$. This completes the construction of an instance $(\{v_1, \ldots v_m\}, M, k)$ of SUBSET PRODUCT.

Now observe that if $\prod_{i \in S} v_i = M$ for some $S \subseteq [m]$, then by uniqueness of prime factorization, every prime number $p_1, \ldots, p_m$ must appear exactly once across the prime factorizations of all numbers in $\{v_i : i \in S\}$. It follows by construction that the collection of subsets $\mathcal{C}' := \{C_i : i \in S\}$ contains each element of X exactly once. Thus, if $(\{v_1, \ldots v_m\}, M, k)$ is a **yes**-instance of SUBSET PRODUCT then $(X, \mathcal{C})$ is a **yes**-instance of X3C. Conversely, if $(X, \mathcal{C})$ is a **yes**-instance of X3C with solution $\mathcal{C}'$, then we can define $S := \{i \in [m] : C_i \in \mathcal{C}'\}$. Since every element of X appears in exactly one $C_i \in \mathcal{C}'$ and $|C_i| = 3$ for all $i \in [m]$, we have that $|\mathcal{C}'| = |X|/3 = n = k$, and $\prod_{i \in S} v_i = p_1 \cdot \ldots \cdot p_{3n} = M$. Thus $(\{v_1, \ldots v_m\}, M, k)$ is a **yes**-instance of SUBSET PRODUCT.

It remains to show that the construction of $(\{v_1, \ldots v_m\}, M, k)$ from $(X, \mathcal{C})$ takes polynomial time. In particular, we need to show that each of the primes $p_1, \ldots p_{3n}$ (and thus the product M) can be constructed in polynomial time. This can be shown using two results from number theory: $p_j < j(\ln j + \ln \ln j)$ for $j \geq 6$, [3,16] and the set of all prime numbers in $[Z]$ can be computed in time $\mathcal{O}(Z/\ln \ln Z)$ [1]. Combining these, we have that the first $3n$ prime numbers can be generated in time $\mathcal{O}(n \ln n / \ln \ln n)$.

Given the prime numbers $p_1, \ldots, p_{3n}$, it is clear that the numbers $\{v_i : i \in [m]\}$ can also be computed in polynomial time. The number M, being the product of $3n$ numbers each less than $3n(\ln 3n + \ln \ln 3n)$, can also be computed in time polynomial in n (though M itself is not polynomial in n). It follows that $(\{v_1, \ldots v_m\}, M, k)$ can be constructed in polynomial time. □

Proof of Lemma 6. We first show that it is enough to consider the cases $Q' = Q + 1$ and $Q' = Q - 1$. Fix an integer $Q \in \mathbb{N}_+$ with $Q \geq 2$. Consider the function $h_Q : \mathbb{R}_{>0} \to \mathbb{R}$ given by

$$h_Q(x) = \ln x - \ln Q + Q/x - 1.$$

So our aim is to show that $h_Q(Q') \geq Q^{-4}$. Similar to the proof of Lemma 4, we can observe that

$$\frac{dh_Q}{dx} = x^{-1} - Qx^{-2} = \frac{1}{x}\left(1 - \frac{Q}{x}\right)$$

is less than 0 when $x < Q$, exactly 0 when $x = Q$, and greater than 0 when $x > Q$. It follows that on the range $x > 0$, the function h_Q has a unique minimum at $x = Q$, and is decreasing on the range $x < Q$ and increasing on the range $x > Q$. Thus in particular $h_Q(Q') \geq h_Q(Q-1)$ if $Q' \leq Q-1$ and $h_Q(Q') \geq h_Q(Q+1)$ if $Q' \geq Q+1$. Since either $Q' \leq Q-1$ or $Q' \geq Q+1$ for any integer $Q' \neq Q$, it remains to show that $h_Q(Q-1) > Q^{-4}$ and $h_Q(Q+1) > Q^{-4}$.

To show $h_Q(Q-1) > Q^{-4}$ for any $Q \in \mathbb{N}_{\geq 2}$: Let $\lambda : \mathbb{R}_{>0} \to \mathbb{R}$ be the function given by

$$\begin{aligned}\lambda(Q) &= h_Q(Q-1) - Q^{-4} \\ &= \ln(Q-1) - \ln Q + Q/(Q-1) - 1 - Q^{-4} \\ &= \ln(Q-1) - \ln Q + 1/(Q-1) - Q^{-4}.\end{aligned}$$

Then

$$\begin{aligned}\frac{d\lambda}{dQ} &= (Q-1)^{-1} - Q^{-1} + (Q-1)^{-2} + 4Q^{-5} \\ &> (Q-1)^{-2} + 4Q^{-5} \\ &> 0.\end{aligned}$$

It follows that λ is a (strictly) increasing function. Since $\lambda(2) = 0 - \ln 2 + 1 - 1/16 \approx 0.244 > 0$, it follows that $\lambda(Q) > 0$ for all $Q \geq 2$, and thus $h_Q(Q-1) > Q^{-4}$.

To show that $h_Q(Q+1) > Q^{-4}$ for all $Q \in \mathbb{N}_{\geq 2}$: First observe that if $Q = 2$, then $h_Q(Q+1) = \ln(3) - \ln(2) + 2/3 - 1 \approx 0.0721 > 0.0625 = 2^{-4}$ and so the claim is true. For $Q \geq 3$, observe that $h_Q(Q+1) = \ln(Q+1) - \ln Q + \frac{Q}{Q+1} - 1 = \ln(\frac{Q+1}{Q}) - \frac{1}{Q+1}$. We use the Mercator series for the natural logarithm:

$$\ln\left(\frac{Q+1}{Q}\right) = \ln\left(1 + \frac{1}{Q}\right) = \sum_{k=1}^{\infty} \frac{(-1)^{k+1}}{kQ^k} = \frac{1}{Q} - \frac{1}{2Q^2} + \frac{1}{3Q^3} - \frac{1}{4Q^4} + \ldots$$

Since $\frac{1}{kQ^k} - \frac{1}{(k+1)Q^{k+1}} > 0$ for all $k > 0$, we can omit all but the first two terms to get

$$\begin{aligned}\ln\left(\frac{Q+1}{Q}\right) &> \frac{1}{Q} - \frac{1}{2Q^2} \\ &= \frac{2Q-1}{2Q^2}.\end{aligned}$$

Then

$$\ln\left(\frac{Q+1}{Q}\right) - \frac{1}{Q+1} > \frac{2Q-1}{2Q^2} - \frac{1}{Q+1}$$
$$= \frac{(2Q-1)(Q+1) - 2Q^2}{2Q^2(Q+1)}$$
$$= \frac{2Q^2 + Q - 1 - 2Q^2}{2Q^2(Q+1)}$$
$$= \frac{Q-1}{2Q^2(Q+1)}$$
$$\geq \frac{1}{Q^2(Q+1)}$$
$$> \frac{1}{Q^4}$$

where the last two inequalities use $Q \geq 3$. □

References

1. Atkin, A.O.L., Bernstein, D.J.: Prime sieves using binary quadratic forms. Math. Comput. **73**, 1023–1030 (2004)
2. Bordewich, M., Semple, C., Wicke, K.: On the complexity of optimising variants of phylogenetic diversity on phylogenetic networks. Theoret. Comput. Sci. **917**, 66–80 (2022)
3. Dusart, P.: The kth prime is greater than $k(\ln k + \ln \ln k - 1)$ for $k \geq 2$. Math. Comput. **68**(225), 411–415 (1999). http://www.jstor.org/stable/2585122
4. Faith, D.P.: Conservation evaluation and phylogenetic diversity. Biol. Cons. **61**(1), 1–10 (1992)
5. Fellows, M.R., Gaspers, S., Rosamond, F.A.: Parameterizing by the number of numbers. Theory Comput. Syst. **50**, 675–693 (2012)
6. Garey, M.R., Johnson, D.S.: Computers and Intractability: A Guide to the Theory of NP-Completeness. W. H. Freeman (1979). ISBN 0-7167-1044-7
7. Huson, D.H., Rupp, R., Scornavacca, C.: Phylogenetic Networks: Concepts, Algorithms and Applications. Cambridge University Press (2010)
8. Jones, M., Schestag, J.: How can we maximize phylogenetic diversity? Parameterized approaches for networks. In: Proceedings of the 18th International Symposium on Parameterized and Exact Computation (IPEC 2023), Schloss-Dagstuhl-Leibniz Zentrum für Informatik (2023)
9. Koblmüller, S., et al.: Reticulate phylogeny of gastropod-shell-breeding cichlids from Lake Tanganyika-the result of repeated introgressive hybridization. BMC Evol. Biol. **7**, 1–13 (2007)
10. Komusiewicz, C., Schestag, J.: A multivariate complexity analysis of the generalized Noah's ark problem. In: Brieden, A., Pickl, S., Siegle, M. (eds.) CTW 2023. AIRO Springer Series, vol. 13, pp. 109–121. Springer, Cham (2023). https://doi.org/10.1007/978-3-031-46826-1_9

11. Merz, J.J., et al.: World scientists' warning: the behavioural crisis driving ecological overshoot. Sci. Prog. **106**(3) (2023). https://doi.org/10.1177/00368504231201372
12. Minh, B.Q., Klaere, S., von Haeseler, A.: Phylogenetic diversity within seconds. Syst. Biol. **55**(5), 769–773 (2006). ISSN 1063-5157. https://doi.org/10.1080/10635150600981604
13. Moret, B.M.: The Theory of Computation. Addison-Wesley (1997)
14. Pardi, F., Goldman, N.: Species choice for comparative genomics: being greedy works. PLoS Genet. **1**(6), e71 (2005). https://doi.org/10.1371/journal.pgen.0010071
15. Ripple, W.J., et al.: World scientists' warning to humanity: a second notice. BioScience **67**(12), 1026–1028 (2017). ISSN 0006-3568. https://doi.org/10.1093/biosci/bix125
16. Rosser, B.: Explicit bounds for some functions of prime numbers. Am. J. Math. **63**(1), 211–232 (1941). http://www.jstor.org/stable/2371291
17. Steel, M.: Phylogenetic diversity and the greedy algorithm. Syst. Biol. **54**(4), 527–529 (2005)
18. Weitzman, M.L.: The Noah's ark problem. Econometrica, 1279–1298 (1998)
19. Wicke, K., Fischer, M.: Phylogenetic diversity and biodiversity indices on phylogenetic networks. Math. Biosci. **298**, 80–90 (2018)
20. Wicke, K., Mooers, A., Steel, M.: Formal links between feature diversity and phylogenetic diversity. Syst. Biol. **70**(3), 480–490 (2021)

On the Robustness to Gene Tree Rooting (or Lack Thereof) of Triplet-Based Species Tree Estimation Methods

Tanjeem Azwad Zaman, Rabib Jahin Ibn Momin, and Md. Shamsuzzoha Bayzid

Department of Computer Science and Engineering, Bangladesh University of Engineering and Technology, Dhaka 1205, Bangladesh
shams_bayzid@cse.buet.ac.bd

Abstract. Species tree estimation is frequently based on phylogenomic approaches that use multiple genes from throughout the genome. This process becomes particularly challenging due to gene tree heterogeneity (discordance), often resulting from Incomplete Lineage Sorting (ILS). Triplet and quartet-based approaches for species tree estimation have gained substantial attention as they are provably statistically consistent in the presence of ILS. However, unlike quartet-based methods, the limitation of rooted triplet-based methods in handling unrooted gene trees has restricted their adoption in the systematics community. Furthermore, since the induced triplet distribution in a gene tree depends on the placement of the root, the accuracy of triplet-based methods depends on the accuracy of gene tree rooting. Despite progress in developing methods for rooting unrooted gene trees, greatly understudied is the choice of rooting technique and downstream effects on species tree inference under realistic model conditions. This study involves rigorous empirical testing with different gene tree rooting approaches to establish a nuanced understanding of the impact of rooting on species tree accuracy.
Link to full version: https://doi.org/10.1101/2024.11.22.624944

Keywords: Species tree · gene tree · gene tree heterogeneity · incomplete lineage sorting (ILS) · triplets · quartets

1 Introduction

The estimation of species trees using multiple loci has become increasingly common. However, combining multi-locus data is difficult, especially in the presence of gene tree discordance. At present, "summary methods", which operate by computing gene trees from different loci and then combining the inferred gene trees into a species tree, are becoming increasingly popular due to drawbacks of the traditional "concatenation methods" [2,6,7,11,12,14,17]. ASTRAL [14,15,21], the most widely used summary method, uses Dynamic Programming to infer the species tree that maximizes the consistency of quartets induced by unrooted gene trees, whereas STELAR [9] seeks to maximize consistency of triplets induced by rooted gene trees. Though both of these methods are statistically consistent, have comparable accuracy, and are scalable to large datasets, STELAR is not as popular due to its limitations in analyzing unrooted gene trees.

Although there is extensive research on approaches for rooting phylogenetic trees, there has been a lack of investigation into their impact on species tree estimation using methods that rely on rooted gene trees. In this study, we report, on an extensive experimental study using a collection of simulated as well as empirical datasets, the performance of STELAR with gene trees rooted by different techniques. Furthermore, we identify different model conditions where STELAR with algorithmically rooted gene trees (i.e. with rooting methods other than outgroup rooting) outperform not only STELAR paired with outgroup rooting, but also the state-of-the-art ASTRAL.

2 Experimental Studies

2.1 Datasets

Simulated Datasets. We used two biologically-based simulated datasets: mammalian (37-taxa) from [18] and avian (48-taxa) from [10]. We also extended our study to two smaller datasets (11-taxa [5] and 15-taxa [1]), as well as larger datasets of 200-taxa and 500 taxa [15]. We varied the number of genes, sequence length per gene and the level of ILS (low[2X], moderate[1X] and high[0.5X]), to obtain a wide range of challenging model conditions for our experiments.

Empirical Datasets. We used an Angiosperm dataset [15,20] of 310 nuclear genes, including 42 species representing all major angiosperm clades. Three gymnosperms (*Picea*, *Pinus*, and *Zamia*) and one lycophyte (*Selaginella*) were the outgroups. We also used the nucleotide (nt) part of an Amniota dataset [4] with 16 species and 248 genes, where the outgroups were *Xenopus* and *Protopterus*.

2.2 Methods Compared

We conducted extensive experimentation using the following rooting methods: Outgroup Rooting (OG), Midpoint Rooting (MP) [8], Minimum Variance rooting (MV) [13], Minimum Ancestor Deviation rooting (MAD) [19], "Search" (RD) and "Exhaustive" (RD-EX) modes of RootDigger [3], and Random rooting (RAND). We refer to STELAR [9] paired with a particular rooting technique using the convention STELAR–⟨quartet-generation-technique⟩ (e.g., STELAR-OG, STELAR-MP, etc.). We benchmark against the latest version of the leading species tree estimation method ASTRAL-III [21], which maximizes quartet consistency.

2.3 Evaluation Metrics

We compared the estimated trees with the model species tree using normalized Robinson-Foulds (RF) Distance [16]. We also compared the triplet and quartet scores of the candidate species trees. We used the term true triplet score when calculating the triplet scores w.r.t true gene trees (no estimation error), and similarly defined true quartet score. To better quantify the errors in root placement, besides the binary-natured "percentage of correct rooting" metric, we computed the non-normalized topological distance between the inferred and true roots [3].

3 Results and Discussion

3.1 Simulated Dataset

In the representative 37-taxa dataset, STELAR with rooting methods like MP, MV, MAD and/or RD-EX performs as well as if not better than with OG in several model conditions, often surpassing even ASTRAL (Fig. 1). Calculating quartet and triplet scores (Table 1), we see that these metrics are closely correlated to and are predictive of the RF scores. Metrics for rooting accuracy reveal that STELAR is robust in predicting the root of the inferred species tree despite being supplied inaccurately rooted gene trees. STELAR becomes more robust to rooting as taxa count increases, evident from the improvement in performance for RAND rooting in higher taxa. RD-EX in 37-taxa (and MAD in 15-taxa), show that certain rootings allow STELAR to mitigate gene tree estimation error.

Table 1. Average triplet score (in hundreds; bold values denote highest in row)

Dataset	Model Cond^n	ASTRAL	STELAR						
			MAD	MP	MV	OG	RAND	RD	RD-EX
37-taxon	0.5X-200-500	9285	**9895**	9895	9891	9893	8741	8431	9818
	1X-200-1000	10511	11584	**11585**	**11585**	11585	9613	9401	11382
	1X-200-500	9814	11014	11017	11016	**11024**	9527	9325	10940
	1X-200-true	13059	**14107**	**14107**	**14107**	**14107**	9988		
	1X-400-500	20013	22027	**22027**	**22027**	22027	19056	19051	21846
	1X-800-500	40046	44160	44160	44160	44160	38107	38097	43763
	2X-200-500	10062	11529	11548	11540	**11549**	9890	9670	11456

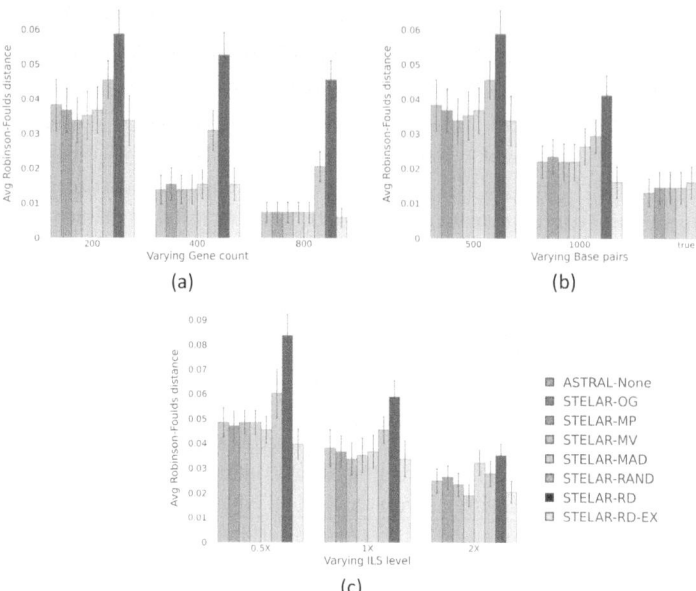

Fig. 1. Comparison of Avg. RF rates with standard error bars over 20 replicates for ASTRAL and STELAR with various rooting methods on 37-taxa simulated dataset. (a) We fixed ILS at moderate (1X) and base pairs at 500bp, and varied gene counts between 200, 400 and 800. (b) We varied sequence length between 500, 1000, and true-gt while keeping moderate ILS (1X) and 200 gene trees. (c) We set gene trees at 200, sequence length at 500bp, and varied ILS [high (0.5X), moderate (1X) and low (2X)]

3.2 Biological Dataset

Analyses on the representative angiosperm dataset aim to address questions in the phylogeny of angiosperms, with a key focus on the placement of *Amborella* relative to other lineages like *Nuphar*. ASTRAL correctly placed *Amborella* as sister to *Nuphar* and the other angiosperms (Fig. 2a). STELAR-MAD correctly placed *Amborella* as sister to the other angiosperms and the clade of all angiosperms as sister to the outgroup (Selaginella, (Zamia, (Pinus, Picea))) (Fig. 2b). Other STELAR variants misplaced *Selaginella*, causing *Amborella* to be placed as a sister to the clade formed by the 3 other outgroups (Fig. 2c). Moreover, all STELAR variants could neither predict *Nuphar* as the sister to the other angiosperms after *Amborella*, nor the clade of (Amborella, Nuphar).

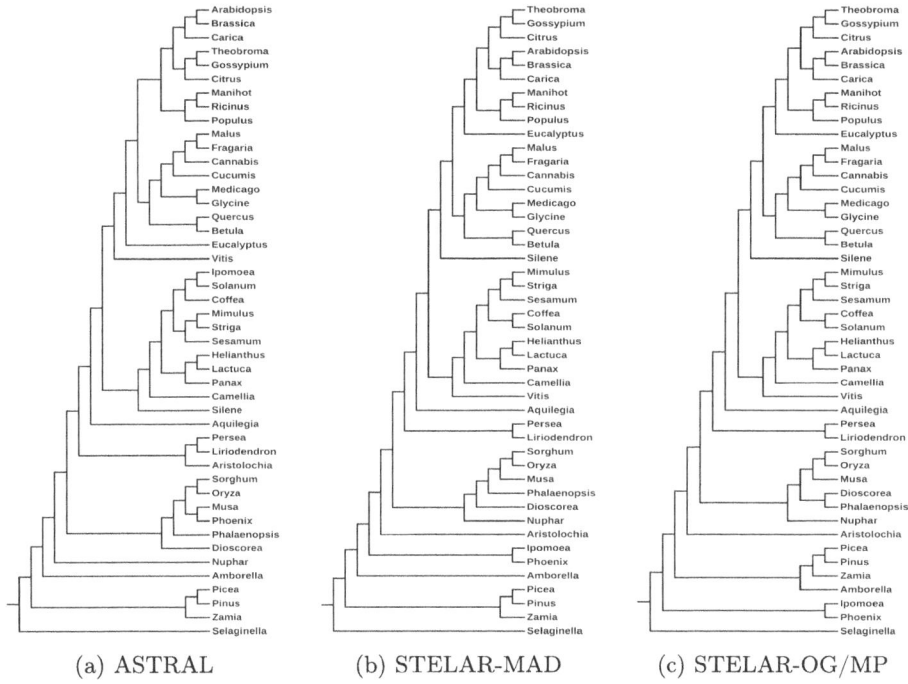

Fig. 2. Estimated Species trees for Angiosperm dataset.

4 Conclusion

In this study, we conducted rigorous empirical testing with different rooting techniques to evaluate their impact on the triplet-based species tree estimation method, STELAR. We evaluated performance in terms of – and found correlations between - key metrics like RF score, triplet and quartet scores. Our experiments showed STELAR to be robust to gene tree rootings across diverse and challenging model conditions spanning both simulated and empirical datasets. We identified several cases where STELAR paired with algorithmic rootings performed as well as, if not better than outgroup rooting, at times even outperforming ASTRAL. This opens up the avenue for further research into the viability of triplet-based methods as competitive alternatives for phylogenetic analysis.

References

1. Bayzid, M.S., Mirarab, S., Boussau, B., Warnow, T.: Weighted statistical binning: enabling statistically consistent genome-scale phylogenetic analyses. PLoS One **10**(6) (2015)
2. Bayzid, M.S., Warnow, T.: Naive binning improves phylogenomic analyses. Bioinformatics **29**(18), 2277–2284 (2013)

3. Bettisworth B., Stamatakis, A.: Root digger: a root placement program for phylogenetic trees. BMC Bioinform. **22**(225) (2021). https://doi.org/10.1186/s12859-021-03956-5
4. Chiari, Y., Cahais, V., Galtier, N., Delsuc, F.: Phylogenomic analyses support the position of turtles as the sister group of birds and crocodiles (archosauria). BMC Biol. **10**(1), 65 (2012)
5. Chung, Y., Ané, C.: Comparing two Bayesian methods for gene tree/species tree reconstruction: a simulation with incomplete lineage sorting and horizontal gene transfer. Syst. Biol. **60**(3), 261–275 (2011)
6. DeGiorgio, M., Degnan, J.H.: Fast and consistent estimation of species trees using supermatrix rooted triples. Mol. Biol. Evol. **27**(3), 552–569 (2009)
7. Edwards, S.V., Liu, L., Pearl, D.K.: High-resolution species trees without concatenation. Proc. Natl. Acad. Sci. **104**(14), 5936–5941 (2007)
8. Farris, J.S.: Estimating phylogenetic trees from distance matrices. Am. Nat. **106**(951), 645–668 (1972)
9. Islam, M., Sarker, K., Das, T., Reaz, R., Bayzid, M.S.: STELAR: a statistically consistent coalescent-based species tree estimation method by maximizing triplet consistency. BMC Genom. **21**(1), 1–13 (2020)
10. Jarvis, E.D., Howard, J.T., et al.: Whole-genome analyses resolve early branches in the tree of life of modern birds. Science **346**(6215), 1320–1331 (2014)
11. Kubatko, L.S., Degnan, J.H.: Inconsistency of phylogenetic estimates from concatenated data under coalescence. Syst. Biol. **56**, 17 (2007)
12. Leaché, A.D., Rannala, B.: The accuracy of species tree estimation under simulation: a comparison of methods. Syst. Biol. **60**(2), 126–137 (2011)
13. Mai, U., Sayyari, E., Mirarab, S.: Minimum variance rooting of phylogenetic trees and implications for species tree reconstruction. PLOS ONE **12**(8), e0182238 (2017). https://doi.org/10.1371/journal.pone.0182238. https://journals.plos.org/plosone/article?id=10.1371/journal.pone.0182238
14. Mirarab, S., Reaz, R., Bayzid, M.S., Zimmermann, T., Swenson, M.S., Warnow, T.: ASTRAL: genome-scale coalescent-based species tree estimation. Bioinformatics **30**(17), i541–i548 (2014)
15. Mirarab, S., Warnow, T.: ASTRAL-II: coalescent-based species tree estimation with many hundreds of taxa and thousands of genes. Bioinformatics **31**(12), i44–i52 (2015)
16. Robinson, D., Foulds, L.: Comparison of phylogenetic trees. Math. Biosci. **53**, 131–147 (1981)
17. Roch, S., Steel, M.: Likelihood-based tree reconstruction on a concatenation of aligned sequence data sets can be statistically inconsistent. Theor. Popul. Biol. **100**, 56–62 (2015)
18. Song, S., Liu, L., Edwards, S.V., Wu, S.: Resolving conflict in eutherian mammal phylogeny using phylogenomics and the multispecies coalescent model. Proc. Natl. Acad. Sci. **109**(37), 14942–14947 (2012)
19. Tria, F.D.K., Landan, G., Dagan, T.: Phylogenetic rooting using minimal ancestor deviation. Nat. Ecol. Evol. **1**(7), 0193 (2017). https://doi.org/10.1038/s41559-017-0193
20. Xi, Z., Liu, L., Rest, J.S., Davis, C.C.: Coalescent versus concatenation methods and the placement of Amborella as sister to water lilies. Syst. Biol. **63**(6), 919–932 (2014)
21. Zhang, C., Rabiee, M., Sayyari, E., Mirarab, S.: ASTRAL-III: polynomial time species tree reconstruction from partially resolved gene trees. BMC Bioinform. **19**(6), 153 (2018)

Exact Counts of Binary Phylogenetic Networks with Two and Three Reticulation Events (Extended Abstract)

Hao Yu and Louxin Zhang[✉]

Department of Mathematics, National University of Singapore, Singapore 119076, Singapore
matzlx@nus.edu.sg

Abstract. Phylogenetic networks are essential models for representing evolutionary history, particularly in studies of genome evolution involving reticulation events. Due to their complex structures, inferring phylogenetic networks for a large number of taxa is highly challenging. To better understand the whole space of phylogenetic networks, we derive closed-form formulas for counting binary phylogenetic networks with two and three reticulation events.

Keywords: Phylogeny · phylogenetic networks · counting

1 Introduction

Phylogenetic networks play a significant role in the study of genome evolution and population biology because of their capability of representing reticulation events (e.g. horizontal gene transfer, introgression or recombination). However, inferring phylogenetic network models is extremely difficult due to the vast space of possible networks. Currently, most algorithmic and software research focuses only on a few restricted classes of networks, such as tree-child networks, reticulation-visible networks, and tree-based networks. The exact counts of unrestricted phylogenetic networks with a few reticulation events remain unknown [1,9]. For example, the count of phylogenetic networks is incorrectly reported in Figure S4 of [1] and in formula (30) of [10]. Both studies underestimate the number of phylogenetic networks with two reticulation events on five taxa and on n taxa (for $n \geq 3$), respectively.

We focus on counting and enumerating binary unrestricted phylogenetic networks with two and three reticulation events. Over the past five years, counting restricted phylogenetic networks has been extensively studied [2–5,11]. Here, we derive closed-form formulas for counting binary phylogenetic networks with two and three reticulation events using the component graph approach [3,6,8], along with exact counts of binary phylogenetic networks with one reticulation event [3,12] and one-component phylogenetic networks [7].

This work was financially supported by the Singapore MOE Academic Research Fund Tier 1 [A-8001951-00-00].

2 Preliminaries

We will work on *binary phylogenetic networks* over n taxa that are defined as rooted, directed acyclic graphs satisfying the following conditions:

- The designated *root* is of indegree 0 and outdegree 1.
- There are n *leaves* which are of indegree 1 and outdegree 0, representing the taxa.
- The non-leaf and non-root nodes are of either indegree 1 and outdegree 2 or indegree 2 and outdegree 1. The former are called *tree nodes*, while latter are called *reticulation nodes*.

Each reticulation node represents a reticulation event. Binary phylogenetic trees are simply those networks without reticulation nodes. In the rest of this paper, binary phylogenetic networks are simply called networks; binary phylogenetic trees are called trees. We will use $\mathcal{R}_{n,k}$ to denote the set of phylogenetic networks with k reticulation nodes on n taxa.

Removing reticulation edges in a network with r reticulation nodes results in a forest consisting of $r + 1$ subtrees whose roots are either the network root or a reticulation node. These obtained subtrees are called *tree components* of the network [6,8]. Each tree component of an unrestricted network may contain only a single reticulation node.

The component graph of a network has the node set consisting of tree components of the network and the arc set consisting of arcs from a component C' to another C'' such that a parent of the root of C'' appears in C'. A network on 5 taxa and its component graph are given in Fig. 1. In general, component graphs are multigraphs in which each non-root vertex is of indegree two. The two parallel arcs between two vertices indicate that the two parents of the root of the child component are both in the parental component.

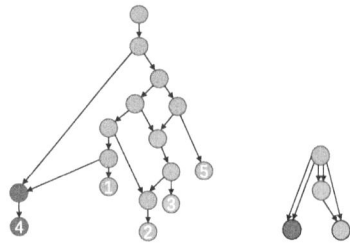

Fig. 1. A binary phylogenetic network and its corresponding component graph. The component graph consists of four nodes, each representing a tree-component of the network, shown in the same color.

Proposition 1. *([3]) (i) There are three possible component graphs for phylogenetic networks with two reticulation nodes, shown in Fig. 2.*

(ii) There are thirteen possible component graphs for phylogenetic networks with three reticulation nodes, shown in Fig. 3.

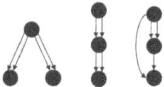

Fig. 2. Three component graphs for networks with 2 reticulation nodes.

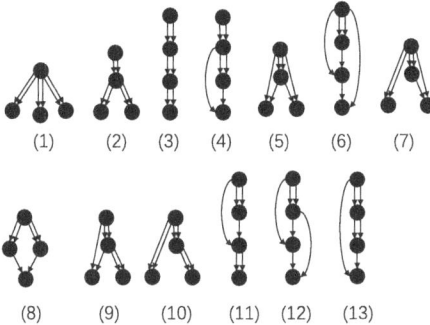

Fig. 3. Thirteen component graphs for networks with 3 reticulation nodes.

3 Count Networks with Two Reticulation Nodes

A network is a *galled network* if the two parents of each reticulation belong to the same tree component. Galled networks correspond to the first or second component graph in Fig. 2, whose count is [3]:

$$\frac{n!}{2^n}\sum_{j=0}^{n-2}\binom{2j}{j}\binom{2n-2j}{n-j}\frac{(j+1)^2(2j+3)}{2n-2j-1} + 2^{n-3}n!(n^2-n) - \frac{(2n)!}{3\cdot 2^n(n-2)!}.$$

By finding the number of networks whose component graph is the third in the figure, we obtain that for $n \geq 1$, the number of networks with 2 reticulation nodes on n taxa is:

$$|\mathcal{R}_{n,2}| = \frac{(2n-2)!}{3\cdot 2^{n-1}(n-1)!}(6n^4 + 19n^3 + 18n^2 - 4n - 6) - 2^{n-1}(n+1)!(2n+3). \quad (1)$$

4 Count of Networks with Three Reticulation Nodes

Similar to the case with two reticulation nodes, by computing the number of networks corresponding to each component graph in Fig. 3, we derive the following closed-form formula for the number of networks on n taxa with three reticulation nodes:

$$|\mathcal{R}_{n,3}| = \frac{(2n-2)!}{3\cdot 2^n(n-1)!}(8n^6 + 88n^5 + 366n^4 + 640n^3 + 325n^2 - 155n - 114)$$
$$- \frac{1}{3}(n+1)!2^{n-4}(48n^3 + 367n^2 + 959n + 840). \quad (2)$$

5 Conclusions

The proof of the formulas (1) and (2) can be found in the full version of this work that appears online on arXiv.

The two closed-form formulas allow us to compute the numbers of networks with 2 and 3 reticulation nodes for large taxon numbers n. For $n = 2, 5, 10, 15$, the number of networks on n taxa with 2 reticulation events is respectively:

$$18, \quad 79,455, \quad 457,518,565,350, \quad 15,146,398,605,413,950,875.$$

Our study provides insights into exploring the full space of phylogenetic networks, which could aid in developing software for network inference. For example, component graphs could be useful for designing random sampling methods in network space. Additionally, they could help compute the distributions of gene coalescence histories in networks.

References

1. Bergström, A., et al.: Origins and genetic legacy of prehistoric dogs. Science **370**(6516), 557–564 (2020)
2. Bouvel, M., Gambette, P., Mansouri, M.: Counting phylogenetic networks of level 1 and 2. J. Math. Biol. **81**(6), 1357–1395 (2020)
3. Cardona, G., Zhang, L.: Counting and enumerating tree-child networks and their subclasses. J. Comput. Syst. Sci. **114**, 84–104 (2020)
4. Fuchs, M., Guan-Ru, Yu., Zhang, L.: On the asymptotic growth of the number of tree-child networks. Eur. J. Comb. **93**, 103278 (2021)
5. Fuchs, M., Guan-Ru, Yu., Zhang, L.: Asymptotic enumeration and distributional properties of galled networks. J. Comb. Theory Ser. A **189**, 105599 (2022)
6. Gunawan, A.D.M., DasGupta, B., Zhang, L.: A decomposition theorem and two algorithms for reticulation-visible networks. Inf. Comput. **252**, 161–175 (2017)
7. Gunawan, A.D.M., Rathin, J., Zhang, L.: Counting and enumerating galled networks. Discret. Appl. Math. **283**, 644–654 (2020)
8. Gunawan, A.D.M., Yan, H., Zhang, L.: Compression of phylogenetic networks and algorithm for the tree containment problem. J. Comput. Biol. **26**(3), 285–294 (2019)
9. Maier, R., Flegontov, P., Flegontova, O., Işıldak, U., Changmai, P., Reich, D.: On the limits of fitting complex models of population history to f-statistics. Elife **12**, e85492 (2023)
10. Mansouri, M.: Counting general phylogenetic networks. arXiv preprint arXiv:2005.14547 (2020)
11. McDiarmid, C., Semple, C., Welsh, D.: Counting phylogenetic networks. Ann. Comb. **19**(1), 205–224 (2015)
12. Zhang, L.: Generating normal networks via leaf insertion and nearest neighbor interchange. BMC Bioinform. **20**(Suppl 20), 642 (2019)

Ancestral Pangenomes and Their Phylogenetic Reconstruction

Xintong Zhou and David Sankoff[✉]

University of Ottawa, Ottawa, Canada
sankoff@uottawa.ca

Abstract. Looking past questions of gene content, we focus on structural variants of the genomes within a pangenome and seek to find a phylogeny where all the ancestral nodes, including the root, are also pangenomes. Representations of pangenomes generally search for compact structures that emphasize common regions or common duplications among the constituent genomes, but necessarily sacrifice some other aspects of gene order. Since the gene order of a monoploid genome is basically just the set of all the gene adjacencies it is composed of, we will consider a pangenome as being made up all the adjacencies of genes appearing in at least one of its constituent genomes. Our key combinatorial tool, *phylogenetic validation*, does not involve optimization, but is simply a filter that removes any adjacencies present in input (extant) pangenomes which are unlikely to have been present in any ancestor, inspired by Dollo's law of irreversible changes. In simulations, this tool turns out to be extraordinarily efficient in retrieving only adjacencies in the original ancestor.

Keywords: pangenome · adjacencies · phylogenetic validation

1 Introduction

Pangenome graphs [1] represent a set of variant genomes of a species as a directed graph with some device for handling differences in gene content and gene orders among these variants. Differences in gene content are usually discussed statistically in terms of core versus non-core genes, while structural variants due to insertion, deletion and duplication (tandem or otherwise) are amenable to several kinds of graphical representation. However, in the models in this paper, for the purposes of focusing on the variability in gene order, we simply assume that gene content is identical across all the genomes, and does not involve paralogy.

The strategy in previous approaches to comparing DAG or DG representations of gene order between two species has been to extract a linear order from each of the two graphs, doing as little violence as possible to the information contained in each of them, in such a way that these two linear orders are optimally similar in terms of rearrangement distance [2,3].

In the context of the phylogenetics of a number of species each represented by a pangenome, it does not seem appropriate to simply reduce each pangenome to a linear order and then proceed with a traditional phylogenetic analysis of these linearized genomes. After all, it is not a new idea that an ancestral population may be more or less heterogeneous with respect to the genomes of individuals or groups. This is explicit in the modern recognition of incomplete lineage sorting [4] but it was understood earlier, such as in the description of species as clouds or quasispecies of more or less closely related individuals [5]. In this paper, then, we explore the notion of phylogenetic analysis of pan-genomes, where the root ancestor and all the intermediate ancestors are also pangenomes. In this initial study, we model the pangenomes and their evolution in the simplest terms.

2 Definitions

Structure. A pangenome consists of a set of related genomes $G = \{g_1, ..., g_\gamma\}$, which are unichromosomal and linear, and which all contain the same n genes. A gene x is denoted by the set of gene ends $\{x^h, x^t\}$, where labels h (heads) and t (tails) are assigned arbitrarily. The distinct identity of each genome g_i resides in its gene order, which is a set Adj_{g_i} or simply Adj_i, of $n+1$ "adjacencies", ordered pairs containing two ends from two different genes (x, y), plus two terminal pairs representing the ends of the genome $(0, x^h)$ (or $(0, x^t)$) and $(0, y^h)$ (or $(0, y^t)$), such that all $2n$ gene ends are in exactly one pair of the $n + 1$ adjacencies in Adj_i. This set contains all information about the structure of a genome.

Evolution. Evolution of the pangenome proceeds by a number of inversions (reversals) affecting each of its constituent genomes independently.

An inversion in genome g_i replaces two adjacencies from Adj_i by two new pairs, with reversed order, as illustrated in Fig. 1.

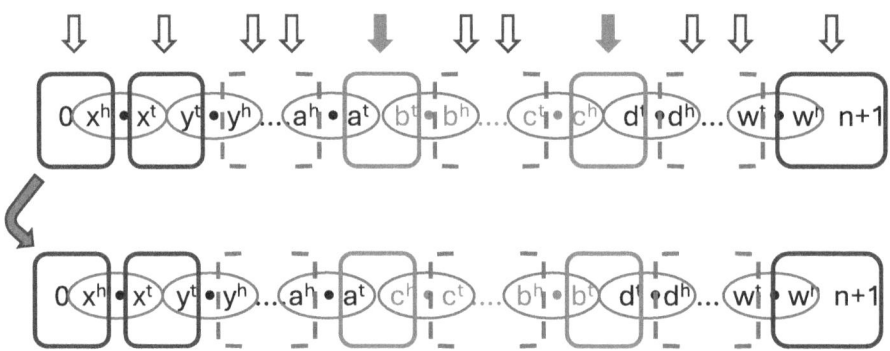

Fig. 1. Inversion involving adjacencies between genes a and b and between c and d. Genes enclosed in ovals, adjacencies in rectangles. Dashed incomplete rectangles contain just one member of an adjacency. Arrows indicate potential break points.

Phylogeny. We will consider evolution on a phylogenetic tree including a root pangenome vertex, three descendant pangenome vertices, each of which may represent a modern (extant) pangenome (terminal vertex - degree 1) or an intermediate ancestor vertex (degree 3). Each intermediate ancestor has two descendants, each of which is either another intermediate ancestor pangenome or a terminal pangenome. Although there is a natural temporal orientation, namely from the root towards the modern genomes, in simulating data, topologically the tree is a binary branching tree, with the root having degree 3, like all other non-terminal vertices.

Measurement. We write $pairs(g_i) = |Adj_i|$ and measure the similarity between two genomes g_i and g_j as $pairs(g_i \cap g_j) = |Adj_i \cap Adj_j|$, and eventually between two pangenomes Y and Z, $pairs(Y \cap Z) = |(\cup Adj_{g_i \in Y}) \cap (\cup Adj_{g_j \in Z})|$, as the number of adjacencies they contain in common. The count of adjacencies takes into account neither the relative order of the two genes in the genome nor the heads/tails identity of the gene ends involved.

3 Generating Divergent Modern Pangenomes from an Ancestral Pangenome

3.1 Generation

The Ancestor. The ancestral pangenome X is simulated by independently generating three genomes X_1, X_2, X_3 from the sequence $1, 2, 3, \cdots, 100$, using r random reversals of lengths sampled from a negative binomial with mean 10 and variance 400, appropriately truncated. We write $X_i \in X$ for $i = 1, 2, 3$, and the set of adjacencies in X is $Adj(X) = \cup_{i=1}^{3} Adj_i$. We explore parameter values $r = 5, 10, 20, 40, 80$ and 120.

The First Generation. Three descendant genomes A_i, B_i, C_i are generated from each genome X_i in pangenome X using r random reversals of lengths sampled from the same negative binomial distribution as before. Then the descendant pangenomes are $A = \{A_1, A_2, A_3\}, B = \{B_1, B_2, B_3\}$ and $C = \{C_1, C_2, C_3\}$ and, for example, $Adj(A) = \cup_{i=1}^{3} Adj_{A_i}$.

Further Branching. If a descendant pangenome D is itself to be considered an (intermediate) ancestor of two other pangenomes F and G, the three genomes in D each produce two further descendants, one which becomes part of F and the other part of G.

4 Inference

4.1 Phylogenetic Validation and the Pangenomic Median

Our key reconstruction technique is based on Dollo's idea that, in certain biological contexts, phylogenetic characters, such as the adjacencies we study here, are gained only once and can never be regained if they are lost [6]. This is realized in an unrooted tree by the property that the set of vertices containing the character are connected. It is a necessary and sufficient condition, valid both for terminal vertices (or degree 1) and internal (ancestral) ones (degree 3 in an unrooted binary tree).

Formally, the connectedness condition for a character can be satisfied by a set of non-terminal nodes of a tree or, trivially, by a single terminal node. For phylogenetic reconstruction, however, we require that an adjacency be present in the genomes associated with at least two terminal vertices, so that by connectedness we can reconstruct that it must have been present in their most recent common ancestor. Otherwise, if it were present only in one terminal set, it could not be inferred as present in any of the non-terminal vertices.

In an unrooted binary branching tree, each non-terminal vertex subtends three subtrees, as in Fig. 2. For an adjacency to be used in constructing a phylogenetic tree and the output sets, clearly each adjacency must be present in at least two of the three subtrees, as illustrated in Fig. 1. More precisely, each adjacency must be present at least in one terminal vertex set in at least two of the three subtrees. We call these adjacencies "phylogenetically validated". The possibility that an adjacency originates twice or more over the phylogenetic time span, so that connectedness is not assured, is not zero, but very small, at least for small or moderate rates of evolution, so that errors in the validation process would be rare.

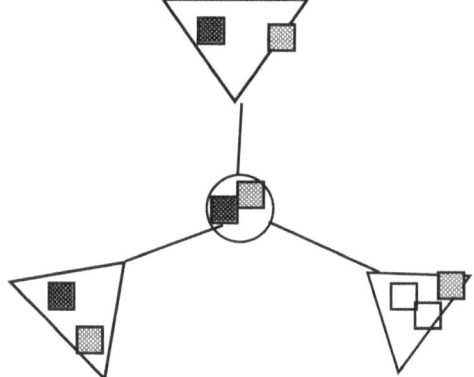

Fig. 2. Necessary condition for adjacencies to appear at an internal vertex associated with an ancestral pangenome of a binary branching phylogenetic tree. Light shaded adjacency (small square) appears in all three trees (triangles) subtended by the internal vertex (circle). Dark shaded adjacency appears in only two of the trees. Unshaded adjacency appears in only one subtree so does not affect internal vertex. The shaded adjacencies are "phylogenetically validated" with respect to the internal vertex. The unshaded one is not validated. Adapted from [7]

The Median. The pairs in common between pangenomes A and B are $Adj_A \cap Adj_B$, between pangenome B and C are $Adj_B \cap Adj_C$ and between C and A are $Adj_C \cap Adj_A$. Then all the pairs in at least two of the three pangemomes are

$$X' = (Adj_A \cap Adj_B) \cup (Adj_B \cap Adj_C) \cup (Adj_C \cap Adj_A). \quad (1)$$

Equation (1) is an expression of the phylogenetic validation criterion, excluding adjacencies that are in only one of the pangenomes A, B or C, as well as adjacencies that are in none of them.

Steinerization. The phylogenies we are modeling and inferring are situated in historical time, with an original "root" vertex representing ancestor X at time zero and all edges directed away from this vertex.

In solving the small phylogeny problem through an iterative "steinerization" process, originally introduced in [8], we first select any three modern pangenomes each located on a different subtree subtended by X. All gene pairs occurring in at least two of these three are considered to form a first estimate

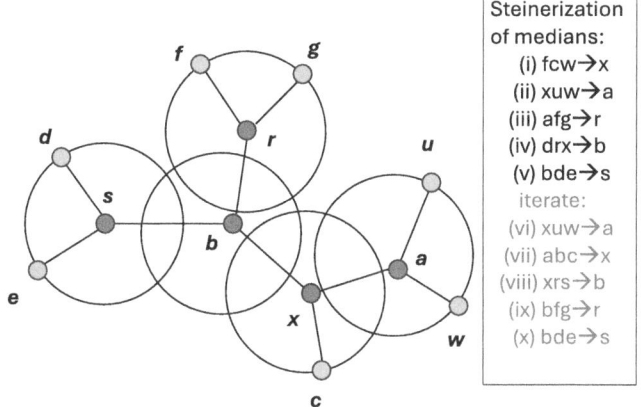

Fig. 3. Calculating the ancestral pangenomes through steinerizing based on the medians.

5 Simulations

The steinerizing process calculates the ancestral (root and intermediate ancestors) pangenomes by iterating the median problem for all non-terminal vertices until convergence which we illustrate in Fig. 3 for seven terminal vertices and four non-terminal vertices.

For our simulations, however, we used the smaller tree in Fig. 4 with only six terminal vertices and three non-terminal vertices. For $r = 5, 10, 20, 40, 80$ and

120, we generated genomes X_1, X_2 and X_3 from the sequence $1, 2, \ldots, 100$ using r inversions for each. We set the ancestor pangenome $X = (X_1, X_2, X_3)$ and generated the intermediate ancestors A, B and C as described in Sect. 3.1 above. From these ancestors we then generated the "extant" pangenomes R, S, V, W, T, Z.

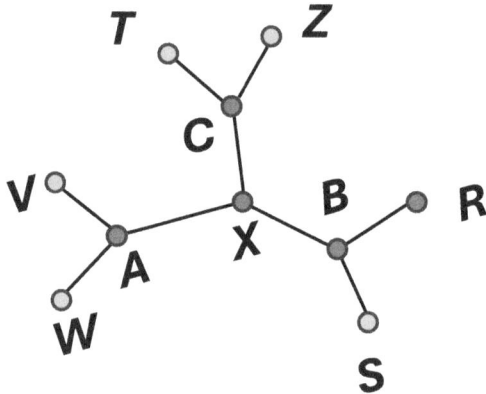

Fig. 4. Phylogeny with ancestor pangenome X used in simulations.

With these simulated data, we could then reconstruct estimated intermediate ancestor pangenomes A', B', C' in terms of the adjacencies in the extant pangenomes that were filtered through the phylogenetic validation criterion.

Finally we used the intermediate ancestors to construct X'. The entire inference procedure was iterated as in Fig. 3 until convergence. Each simulation was repeated 100 times and the mean numbers of adjacencies is reported in Table 1 and Fig. 5. These results are remarkable in that for $r = 5$ and even $r = 10$, almost all the adjacencies in X are recovered in X', and few extraneous adjacencies manage to make it into X'. On the other hand, it is clear that increasing the inversion rate to 40 or higher will defeat the method (Table 2).

Table 1. Results of the inference process. The parameter r measuring the rate of evolution ranges from 5 per time period to 120. The first two columns show that up to $r = 10$, almost all of the adjacencies in X are recovered in X'.

inversions	$|Adj_X \cap Adj_{X'}|$	$|Adj_X \setminus Adj_{X'}|$	$|Adj_{X'} \setminus Adj_X|$
5	125	2	4
10	137	14	8
20	121	69	12
40	44	195	9
80	3	276	3
120	0	290	1

The power of the phylogenetic validity filter is clear from Table 2, when we see the hundreds of adjacencies that are filtered away either in the reconstruction of A', B' and C', or in the final reconstruction of X'.

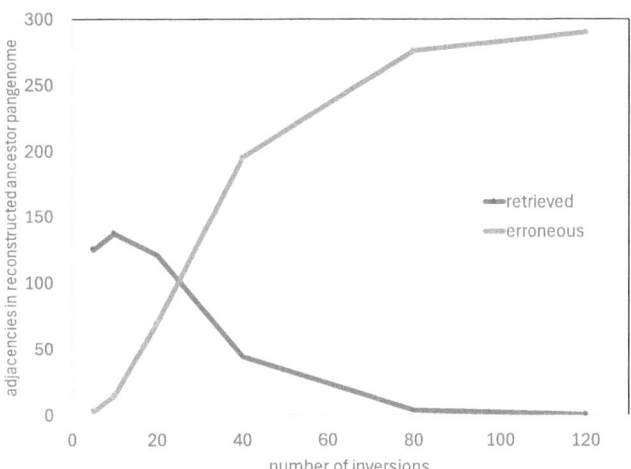

Fig. 5. Effect of evolutionary rate on number of correctly reconstructed adjacencies. With more than 30 inversions per evolutionary period, the number of accurately retrieved adjacencies drops sharply while the number of simulated adjacencies not present in the ancestral pangenome increases.

6 Large Phylogeny

In the pangenomic context, there is a major difference with other phylogenetic problems in the small phylogeny context, namely the use of phylogenetic validation instead of some optimization criterion. In the large phylogeny case, however, there is little difference in the basic intractability of the problem, necessitating exhaustive approaches, heuristics and the like. Here we have six terminal vertices, and only 105 different possible phylogenies. In the present study, however, we simply evaluated one additional tree using the same data. The results of using the second tree to reconstruct the ancestor at the origin of the data were equivocal - slightly fewer original adjacencies recovered and also slightly more extraneous adjacencies in X'. The potential of the method for large phylogenies awaits further investigation (Fig. 6).

Table 2. Adjacencies in extant pangenomes and in intermediate ancestors that are filtered out by the phylogenetic validity criterion.

inversions	union of $(R,S,T,V,W,Z)\backslash X'$	union of $(A,B,C)\backslash X'$
5	229	79
10	424	155
20	780	318
40	1293	600
80	1630	810
120	1691	855

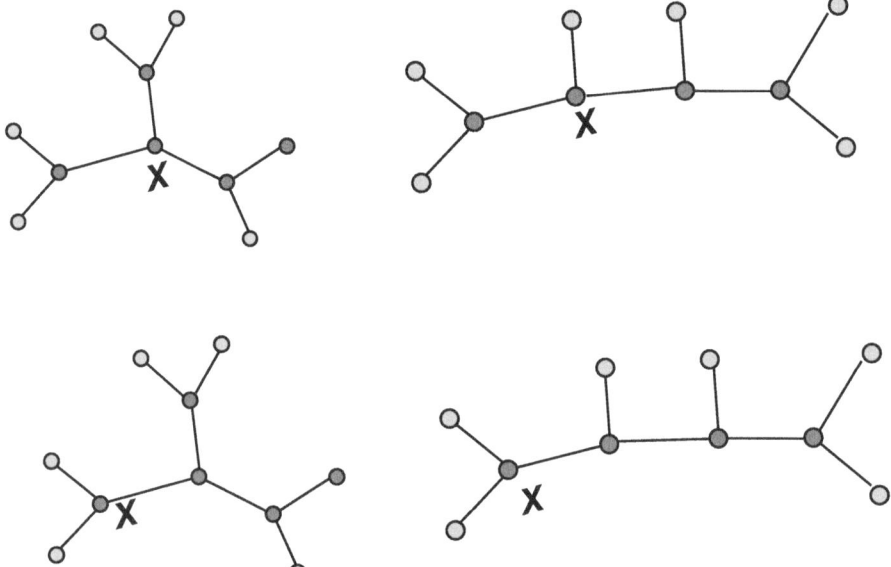

Fig. 6. Various configurations of trees and roots on six leaves.

7 Discussion and Conclusions

The most striking result from this work is the power of the phylogenetic validation criterion based on Dollo's principle to weed out the massive amounts of recently generated adjacency data to preserve the original gene order information in the original pangenome.

Our model is extremely simple. Moreover no suitable data exists to our knowledge for even a more relaxed and parameterized model. Nevertheless, we submit that we have showed proof of principle for a new approach to ancestral pangenome reconstruction, which is itself a new objective.

References

1. Eizenga, J.M., et al.: Pangenome graphs. Annu. Rev. Genomics Hum. Genet. **21**, 139–162 (2020)
2. Zheng, C., Lenert, A., Sankoff, D.: Reversal distance for partially ordered genomes. Bioinformatics **21**, 502–8 (2005)
3. Zheng, C., Sankoff, D.: Genome rearrangements with partially ordered chromosomes.In: International Computing and Combinatorics Conference, pp. 52-62 (2005)
4. Maddison, W.P., Knowles, L., Lacey, T.: Inferring phylogeny despite incomplete lineage sorting. Syst. Biol. **55**, 21–30 (2006)
5. Eigen, M., Schuster, P.: A principle of natural self-organization. Naturwissenschaften **64**, 541–65 (1977)
6. Dollo, L.: Les lois de l'évolution. Bulletin de la Société Belge de Géologie, de Paléontology et d'Hydrolgie **7**, 164–166 (1893)
7. Xu, Q., Sankoff, D.: Gene order phylogeny via ancestral genome reconstruction under Dollo. LNBI, vol. 13883, pp. 100–111 (2023). https://doi.org/10.1007/978-3-031-36911-7_7
8. Sankoff, D., Cedergren, R., Lapalme, G.: Frequency of insertion-deletion, transversion, and transition in the evolution of 5S ribosomal RNA. J. Mol. Evol. **7**, 133–149 (1976)

QT-WEAVER: Correcting Quartet Distribution Improves Phylogenomic Analyses Despite Gene Tree Estimation Error

Navid Bin Hasan(✉), Sohaib, and Md. Shamsuzzoha Bayzid

Computer Science and Engineering, Bangladesh University of Engineering and Technology, Dhaka 1205, Bangladesh
navidhasan0@gmail.com, shams_bayzid@cse.buet.ac.bd

Abstract. Summarizing individual gene trees into species phylogenies using coalescent-based methods has become a standard approach in phylogenomics. However, gene tree estimation error (GTEE) arising from a combination of reasons can potentially impact the accuracy of phylogenomic inference. We, for the first time, introduce the problem of correcting the quartet distribution induced by a set of estimated gene trees, which involves updating the weights of the quartets to better reflect their relative importance within the gene tree distribution. We present QT-WEAVER, which learns the conflicts within the quartet distribution induced by a given set of gene trees and generates an updated quartet distribution by adjusting the weights accordingly. Experimental studies on a collection of simulated and empirical data sets suggest that QT-WEAVER can effectively account for GTEE, which results in a substantial improvement in the species tree accuracy. The full version of this paper is available at https://www.biorxiv.org/content/10.1101/2024.11.11.622962

Keywords: Species tree · Gene tree · Incomplete lineage sorting (ILS) · Quartet distribution · Quartet consistency

1 Introduction

Species tree estimation from multiple genes is commonly done through concatenation, which combines sequence alignments from different loci into a single supermatrix and then computes a tree on the supermatrix. While concatenation can produce accurate species trees when gene trees are concordant, it can be statistically inconsistent [6,29], and produce incorrect trees with high support [14] when gene trees differ from the species tree. As a result, summary methods [4,10,13,16–20,23,24,32] that combine estimated gene trees while accounting for gene tree discordance have drawn great interest. However, summary methods

---

N. B. Hasan and Sohaib—These authors contributed equally to this work.

are prone to gene tree estimation errors (GTEE), which can arise from various reasons including short gene sequences, inaccurate alignments, and the limitations of the models and algorithms used for inferring gene trees from sequence data. As GTEE is a major contributor to inaccuracies of summary methods, there has been great interest in developing tools [2,5,7,9,11,15,25–27,33] to account for GTEE for improved species tree estimations. However, most of these methods require an accurate reference tree, which itself can be hard to obtain due to GTEE. In this study, we address the problem of GTEE, by formulating the *Quartet Distribution Correction* (QDC) problem, where we seek to "correct" the distribution of quartets induced by a given set of estimated gene trees. QDC attempts to account for GTEE without resorting to any reference species tree. We present QT-WEAVER, a novel method that learns the weighted quartet distribution induced by a set of gene trees to identify certain patterns of quartet conflicts and subsequently updates the weights accordingly. We introduced the concept of "quartet conflicts" and proved key analytical and combinatorial results that underpin QT-WEAVER's ability to adjust quartet distributions without relying on any reference species tree or sequence data. Our experimental results, based on a diverse set of simulated and real biological datasets, demonstrate that amalgamating QT-WEAVER's corrected quartets significantly improves species tree inference accuracy.

2 Overview of QT-WEAVER

QT-WEAVER takes a distribution of weighted quartets as input. Then, for each quartet topology, it computes a "conflict score" based on patterns of "quartet conflict", the concept and theoretical backgrounds of which have been introduced in this study. A set of quartets is considered *conflicting* if they cannot coexist in a single tree. Finally, QT-WEAVER scales the weight of each quartet topology based on its conflict score, with higher conflict scores resulting in lower weights. Due to space constraints, a detailed theoretical background is provided only in the full version.

3 Experimental Study

3.1 Datasets

We evaluated the performance of QT-WEAVER using two simulated datasets: a 37-taxon mammalian dataset based on biological data [31] and a 15-taxon dataset, both generated in prior studies [3,22]. We also evaluated QT-WEAVER on a biological avian dataset from Jarvis *et al.* [12] comprising 14,446 genes sampled from 48 birds.

3.2 Species Tree Estimation Methods

We used the well-known quartet amalgamation methods wQFM [20] and wQMC [1], to estimate species trees from weighted quartets. We ran GTF based wQFM and wQMC on uncorrected embedded quartets in the input gene trees with weights reflecting the frequencies of the quartets. wQFM (and wQMC) was also run on the adjusted/corrected weighted quartets generated by QT-WEAVER to demonstrate the impact of quartet weight correction on species tree estimation. We refer to this variant as *wQFM+QT-WEAVER*. We compared wQFM with ASTRAL-III [23,35] (v. 5.7.8), the leading quartet-based species tree method. We also included the weighted variant of ASTRAL (wASTRAL), which utilizes weights derived from the branch support values in gene trees [34].

3.3 Measurements

We evaluated the accuracy of the estimated trees on simulated datasets by comparing them to the model species tree using the normalized Robinson-Foulds (RF) distance [28]. For the biological dataset, we compared the inferred species trees with those reported in the scientific literature. To assess the accuracy of quartet distributions, we compared both corrected and uncorrected distributions to the true quartet distributions derived from the true gene trees using Jansen-Shannon divergence [8].

4 Results and Discussion

4.1 Results on Simulated 37-Taxon Dataset

The results in Fig. 1 demonstrate how wQFM paired with the "corrected" distributions by QT-WEAVER outperforms the rest of the state-of-the-art methods. Even though wQFM is, in general, more accurate than ASTRAL [20,21], the improvements are often not statistically significant. Remarkably, when wQFM leverages the corrected distributions from QT-WEAVER, its improvement over ASTRAL and wASTRAL becomes statistically significant in most model conditions. Moreover, the corrected distributions are less divergent from the true distribution compared to the uncorrected distributions (Fig. 1(d)-(f)), suggesting that QT-WEAVER bridges the gap with the overall nature of the true distribution, thus effectively accounting for GTEE.

4.2 Results on Biological Avian Dataset

We compare trees estimated by ASTRAL and wQFM+QT-WEAVER in Fig. 2. Both trees are highly congruent with the reference binned MP-EST tree presented in [12], with wQFM+QT-WEAVER being more accurate. ASTRAL failed to recover both the Otidimorphae (bustard, turaco, and cuckoo) and the Cursores clade (crane and killdeer), both of which wQFM+QT-WEAVER successfully recovered. Both methods successfully reconstructed the well-established Australaves, Afroaves, Core Waterbird, and Caprimulgimorphae clades, but failed to recover Columbea.

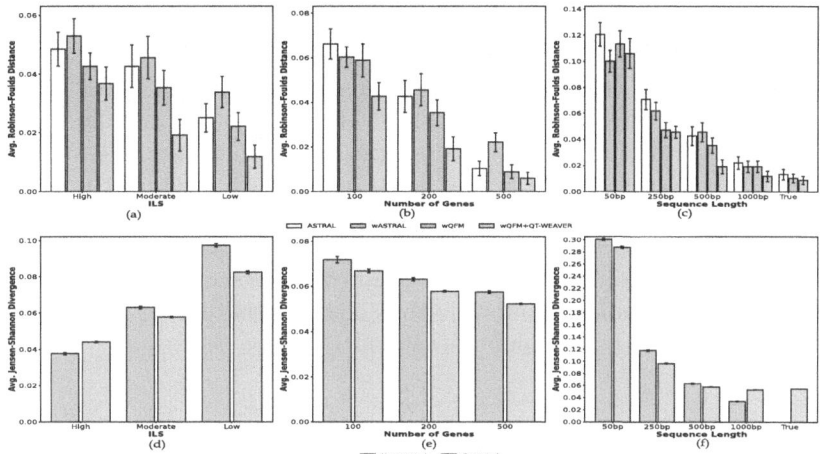

Fig. 1. (a)-(c) Average RF rates with standard errors over 20 replicates for all methods compared. (a) ILS was varied from high to low, keeping the number of genes fixed at 200 and sequence length at 500bp. (b) The number of genes was varied keeping ILS at moderate (1X) and the sequence length at 500bp. (c) Varying sequence lengths with moderate ILS and 200 genes. We also include true gene trees with no GTEE. (d)-(f) Average Jensen-Shannon divergence of the quartet distributions before and after correction with respect to the true quartet distributions.

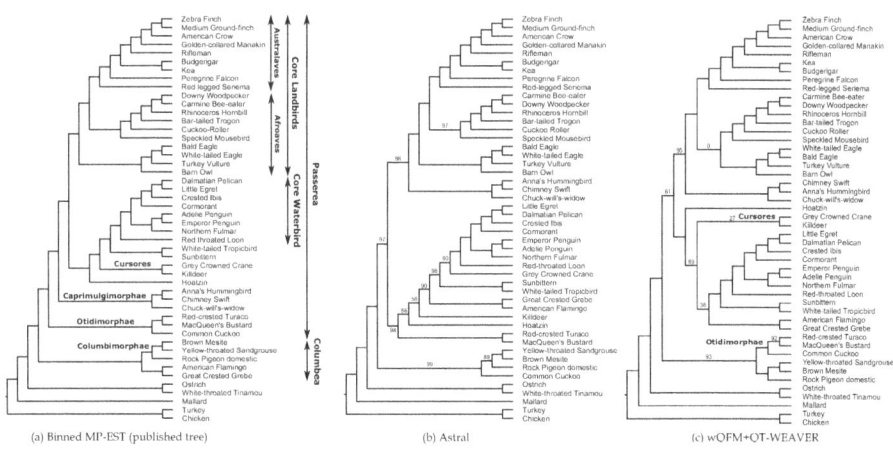

Fig. 2. Analyses of the avian dataset using binned MP-EST, ASTRAL, and wQFM+QT-WEAVER. Branch supports are computed based on quartet-based local posterior probability [30]. All BS values are 100% except where noted.

5 Conclusions

This study, for the first time, introduces the quartet distribution correction problem and shows the impact and clear benefit of using quartet distributions corrected by QT-WEAVER for improved species tree estimations. QT-WEAVER learns the overall quartet distribution based on the pattern of quartet conflicts and seeks to update the weights to better reflect their relative importance. Our experimental study shows that QT-WEAVER may result in substantial improvements over the leading method ASTRAL. Moreover, the concept of quartet conflict and related theoretical results have broad applicability and will be valuable for a range of quartet-based computational methods.

References

1. Avni, E., Cohen, R., Snir, S.: Weighted quartets phylogenetics. Syst. Biol. **64**(2), 233–242 (2015)
2. Bansal, M.S., Wu, Y.C., Alm, E.J., Kellis, M.: Improved gene tree error correction in the presence of horizontal gene transfer. Bioinformatics **31**(8), 1211–1218 (2015)
3. Bayzid, M.S., Mirarab, S., Boussau, B., Warnow, T.: Weighted statistical binning: enabling statistically consistent genome-scale phylogenetic analyses. PLoS ONE **10**(6) (2015)
4. Chifman, J., Kubatko, L.: Quartet from snp data under the coalescent model. Bioinformatics **30**(23), 3317–3324 (2014)
5. Christensen, S., Molloy, E.K., Vachaspati, P., Warnow, T.: Traction: fast non-parametric improvement of estimated gene trees. Leibniz Inter. Proc. Inform. (LIPIcs) **143**(1), 4:1–4:16 (2019)
6. Degnan, J.H., DeGiorgio, M., Bryant, D., Rosenberg, N.A.: Properties of consensus methods for inferring species trees from gene trees. Syst. Biol. **58**, 35–54 (2009)
7. Durand, D., Halldórsson, B.V., Vernot, B.: A hybrid micro–macroevolutionary approach to gene tree reconstruction. J. Comput. Biol. **13**(2), 320–335 (2006)
8. Fuglede, B., Topsoe, F.: Jensen-shannon divergence and hilbert space embedding. In: International Symposium on Information Theory, ISIT 2004. Proceedings, p. 31. IEEE (2004)
9. Górecki, P., Eulenstein, O.: A linear time algorithm for error-corrected reconciliation of unrooted gene trees. Bioinform. Res. Appli. 148–159 (2011)
10. Islam, M., Sarker, K., Das, T., Reaz, R., Bayzid, M.S.: Stelar: a statistically consistent coalescent-based species tree estimation method by maximizing triplet consistency. BMC Genomics **21**(1), 1–13 (2020)
11. Jacox, E., Weller, M., Tannier, E., Scornavacca, C.: Resolution and reconciliation of non-binary gene trees with transfers, duplications and losses. Bioinformatics **33**(7), 980–987 (2017)
12. Jarvis, E.D., et al.: Whole-genome analyses resolve early branches in the tree of life of modern birds. Science **346**(6215), 1320–1331 (2014)
13. Kubatko, L.S., Carstens, B.C., Knowles, L.L.: Stem: species tree estimation using maximum likelihood for gene trees under coalescence. Bioinformatics **25**, 971–973 (2009)
14. Kubatko, L.S., Degnan, J.H.: Inconsistency of phylogenetic estimates from concatenated data under coalescence. Syst. Biol. **56**, 17 (2007)

15. Lafond, M., Chauve, C., El-Mabrouk, N., Ouangraoua, A.: Gene tree construction and correction using supertree and reconciliation. IEEE/ACM Trans. Comput. Biol. Bioinf. **15**(5), 1560–1570 (2017)
16. Larget, B., Kotha, S.K., Dewey, C.N., Ané, C.: BUCKy: gene tree/species tree reconciliation with the Bayesian concordance analysis. Bioinformatics **26**(22), 2910–2911 (2010)
17. Liu, L., Yu, L.: Estimating species trees from unrooted gene trees. Syst. Biol. **60**(5), 661–667 (2011). https://doi.org/10.1093/sysbio/syr027
18. Liu, L., Yu, L., Edwards, S.V.: A maximum pseudo-likelihood approach for estimating species trees under the coalescent model. BMC Evol. Biol. **10**, 302 (2010)
19. Liu, L., Yu, L., Pearl, D.K., Edwards, S.V.: Estimating species phylogenies using coalescence times among sequences. Syst. Biol. **58**(5), 468–477 (2009)
20. Mahbub, M., Wahab, Z., Reaz, R., Rahman, M.S., Bayzid, M.S.: wQFM: highly accurate genome-scale species tree estimation from weighted quartets. Bioinformatics (2021). https://doi.org/10.1093/bioinformatics/btab428
21. Mahbub, S., Sawmya, S., Saha, A., Reaz, R., Rahman, M.S., Bayzid, M.S.: Qt-gild: quartet based gene tree imputation using deep learning improves phylogenomic analyses despite missing data. In: International Conference on Research in Computational Molecular Biology. pp. 159–176. Springer (2022). https://doi.org/10.1007/978-3-031-04749-7_10
22. Mirarab, S., Bayzid, M.S., Boussau, B., Warnow, T.: Statistical binning enables an accurate coalescent-based estimation of the avian tree. Science **346**(6215), 1250463 (2014)
23. Mirarab, S., Reaz, R., Bayzid, M.S., Zimmermann, T., Swenson, M.S., Warnow, T.: ASTRAL: genome-scale coalescent-based species tree estimation. Bioinformatics **30**(17), i541–i548 (2014)
24. Mossel, E., Roch, S.: Incomplete lineage sorting: consistent phylogeny estimation from multiple loci. IEEE/ACM Trans. Comput. Biol. Bioinf. **7**(1), 166–171 (2011)
25. Nguyen, T.H., Doyon, J.-P., Pointet, S., Arigon Chifolleau, A.-M., Ranwez, V., Berry, V.: Accounting for gene tree uncertainties improves gene trees and reconciliation inference. In: Raphael, B., Tang, J. (eds.) WABI 2012. LNCS, vol. 7534, pp. 123–134. Springer, Heidelberg (2012). https://doi.org/10.1007/978-3-642-33122-0_10
26. Noutahi, E, et al.: Efficient gene tree correction guided by genome evolution. PLoS ONE, e0159559 (2016)
27. Rasmussen, M.D., Kellis, M.: A bayesian approach for fast and accurate gene tree reconstruction. Mol. Biol. Evol. **28**(1), 273–290 (2011)
28. Robinson, D., Foulds, L.: Comparison of phylogenetic trees. Math. Biosci. **53**, 131–147 (1981)
29. Roch, S., Steel, M.: Likelihood-based tree reconstruction on a concatenation of aligned sequence data sets can be statistically inconsistent. Theor. Popul. Biol. **100**, 56–62 (2015)
30. Sayyari, E., Mirarab, S.: Fast coalescent-based computation of local branch support from quartet frequencies. Mol. Biol. Evol. **33**(7), 1654–1668 (2016)
31. Song, S., Liu, L., Edwards, S.V., Wu, S.: Resolving conflict in eutherian mammal phylogeny using phylogenomics and the multispecies coalescent model. Proc. Natl. Acad. Sci. **109**(37), 14942–14947 (2012)
32. Vachaspati, P., Warnow, T.: Astrid: accurate species trees from internode distances. BMC Genomics **16**(10), S3 (2015)

33. Wu, Y.C., Rasmussen, M.D., Bansal, M.S., Kellis, M.: Treefix: statistically informed gene tree error correction using species trees. Syst. Biol. **62**(1), 110–120 (2012)
34. Zhang, C., Mirarab, S.: Weighting by gene tree uncertainty improves accuracy of quartet-based species trees. Molecular Biol. Evolut. **39**(12), msac215 (2022)
35. Zhang, C., Rabiee, M., Sayyari, E., Mirarab, S.: Astral-iii: polynomial time species tree reconstruction from partially resolved gene trees. BMC Bioinform. **19**(6), 153 (2018)

Fast Calculation of Cherry Distance on Level-1 Orchard Networks: Optimization, Heuristic and Implementation

Kaari Landry(✉) and Olivier Tremblay-Savard

University of Manitoba, Winnipeg MB, R3T 2N2, Canada
landryk1@cs.umanitoba.ca, olivier.tremblay-savard@umanitoba.ca

Abstract. Phylogenetic networks are increasingly being used to represent more complex evolutionary relationships, such as hybridization and horizontal gene transfer. Calculating distances between networks measures the discrepancies between networks built using different methodologies or reference networks, in the evaluation of construction methods. Here we are interested in the cherry distance, a network distance based on cherry operations. We describe refinements made to a cherry distance algorithm design that takes advantage of a network abstraction that maps reticulated elements between two inputs and performs a preprocessing step. We experimentally show, on a newly implemented and publicly available Rust package, both the improvements this design provides, as well as an exploration of when such an improvement is most effective vis-a-vis the input network topology. Next we present a heuristic strategy to calculate the cherry distance in a non-exact way, and experimentally show how it maintains a very high degree of accuracy while still providing large gains in the runtime efficiency. Finally, we explore particular characteristics of the cherry distance through experiments on real data from the Rose family using another rearrangement operation (rNNI). Specifically, we show how the cherry distance excels at reflecting how many taxa are impacted by changes in the network. We also show a higher degree of sensitivity in cherry distance when compared to a network adaptation of the ubiquitous RF distance on trees, the soft RF distance.

Keywords: Graphs and networks · Trees · Network problems · Applications · Heuristics design · Biology and genetics

1 Introduction

Historically, evolutionary relationships were modelled on phylogenetic trees, however a more general model is required to capture complex evolutionary events, such as hybridization, recombination, and horizontal gene transfer [1]. A more general model comes in phylogenetic networks, which inherit much the same properties of phylogenetic trees with the allowance of a special type of vertex to occur, the *reticulated vertex* (or simply *reticulation*), which represents multi-parental lineage. Of course, with this additional structural entanglement of the networks comes a drastic increase in complexity as network problems tend to be NP-hard, including our problem of interest, *network distance*.

Network distance quantifies the difference between networks and are dominated by two general flavours, rearrangement-based and cluster-based metrics. Some well-established rearrangement operations on trees that also define a distance metric include rooted subtree pruning and regrafting (rSPR) [2], tree bisection and reconnection (TBR) [3], and nearest-neighbour interchange operation (NNI) [4,5]. All three of these rearrangements have fixed-parameter tractable (FPT) algorithms on trees parameterized by the distance themselves [3,6], but none are considered fast enough for practical use. Each of these operations and metrics have been adapted to networks, the subnet prune and regraft (SNPR) on rooted networks [7,8], the TBR on unrooted networks [8], the NNI for general unrooted networks [9], the rooted NNI (rNNI) on rooted networks [10] (no complexity changing moves) and local subnetwork transfer (LST) which defines an rNNI move on relevant classes of networks and two associated complexity-changing moves [11]. The RF metric on trees [12], the symmetric difference on leaf sets that descend vertices between two input trees, by contrast, is computed more efficiently (linear time [12,13]) and is therefore ubiquitous in practical use. Unfortunately, RF has a well-known limitation: it is overly sensitive to small changes, leading to large distances between relatively similar trees. There are many adaptations of RF to networks, however an early adaptation was only shown to be a metric for certain restricted classes of networks [14–16]. Another variation of RF for networks, *soft RF distance*, takes the symmetric difference of leafs sets that descend vertices on trees embedded on the network. The soft RF distance stands apart from RF as being experimentally shown to be more fine-grained that standard RF [17].

Cherry operations were introduced in the context of the minimum hybridization problem [18,19], and have been used with success on other network problems like network construction, network containment and isomorphism [20–22]. In [23], Landry et al. defined four distances based on cherry operations, three of which were proved equal and will be here referred to as *cherry distance*. In [24], Landry et al. suggest a dynamic programming algorithm that solves a problem equivalent to the cherry distance, the MACRS problem. The algorithm presented is fixed-parameter tractable (FPT) in the parameter of the combined number of reticulations in the two input networks.

In this work, we employ various practical strategies to improve its runtime. This includes efficient data preprocessing in an exact algorithm, as well as a heuristic for a fast and accurate calculation. Another main contribution of this paper is the implementation (in Rust) of this program in a comprehensive software package that includes both exact and quick (inexact) calculation of cherry distance, in a user-focused environment that we demonstrate is efficient on both simulated and real data. We further demonstrate how the cherry distance, unlike other well known rearrangement distances, is well-behaved in a few critical aspects; there is a strong correlation between it and how many taxa are affected by changes in the network topology, it is not overly sensitive to a rearrangement move, and gives well distributed distances between random networks. Not many network distances have been implemented to our knowledge, so this software package and accompanying analysis are a meaningful contribution.

Here we focus on level-1 networks. Many difficult problems become tractable on level-1 networks [25] because reticulations are isolated from each other. For example we see results on statistical identifiability of phylogenetic networks, the ability to uniquely determine a network topology based on (for example) sequence data measured at the leaves, that have focused on level-1 networks [26]. Moreover, it is worth noting that many existing network inference methods strictly consider the level-1 class of networks, such as PhyloNetworks [27,28] and NANUQ [29].

The work is organized as follows. After some general definitions (Sect. 2), we introduce the algorithm design including improvements (Sect. 3). Experiments and analysis are presented in Sect. 4 including both simulated and real data sets. We end by articulating possible future work (Sect. 5).

2 Preliminaries

2.1 Network Model Definitions

A network N is an acyclic digraph whose vertices are $V(N)$ and edges are $E(N)$. For a vertex v, let v^- be its in-degree and let $v+$ be it's out-degree. We consider only binary networks, so we say $V(N)$ is composed of

1. a unique root vertex $\rho(N)$ such that $\rho(N)^- = 0$ and $\rho(N)^+ = 2$,
2. the tree vertices such that one, v, has $v^- = 1$ and $v^+ = 2$,
3. the reticulation vertices, $R(N)$, such that one, r, has $r^- = 2$ and $r^+ = 1$,
4. and the leaf vertices such that one, l, has $l^- = 1$ and $l^+ = 0$. Leaves are identified with a label.

Edges are directed from the unique root to the leaves of the network. When there is a directed path from vertex u to vertex v we say v is a descendant of u and that u is an ancestor of v. When the edge uv exists, we say u is a parent of v and that v is a child of u. Additionally, if edge uw exists, then we say v and w are siblings. One subset of $E(N)$ is the *reticulation edges*, $E_R(N)$, edges of the form vr such that $r \in R(N)$.

A biconnected component B is a set of vertices that cannot be disconnected by the removal of an edge (if it exists) $uv|\{u,v\} \in B$. Here we consider biconnected components on the underlying (undirected) graph of the network. A biconnected component B is *non-trivial* when $|B| > 1$. The *level* of a network describes the maximum number of reticulations among all biconnected components of the network. We refer to a network as level-k when its level is $\leq k$. In our case, we assume all networks are level-1.

A cherry is a structure in the network that comes in two flavours, a *simple cherry* and a *reticulated cherry*. First, the simple cherry is a network substructure consisting of two sibling leaves and their parent tree vertex. Second, the reticulated cherry consists of a tree vertex with a leaf child and a reticulation as children, with the reticulation in turn being the parent of a leaf.

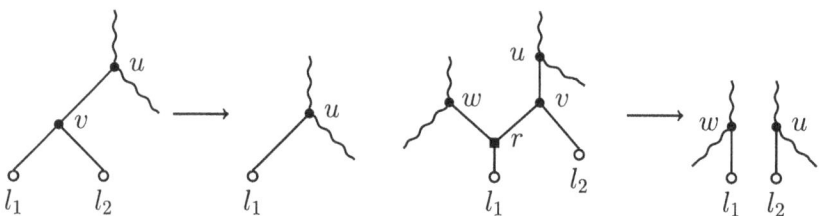

Fig. 1. Showing a simple and reticulated cherry and their reduction. (Left) A simple cherry on v and leaves l_1 and l_2, followed by the result of a simple cherry reduction on l_2. (Right) A reticulated cherry on reticulation edge vr, and leaves l_1 and l_2, followed by the result of the reticulated cherry reduction that removes reticulation r.

There is a *cherry reduction* operation for each cherry flavour that reduces the size of the network. We will not provide a formal definition here, but rather give a sense of the effect on the network. We direct the interested reader to [24] for more detail. It suffices to say that in the simple case one leaf is removed and in the reticulated case a reticulation and reticulation edge is removed. In both cases, another step removes an additional vertex and edge in order to preserve the network definition that precludes a vertex v with $v^- = 1$ and $v^+ = 1$ (see Fig. 1). Consistently, two vertices are removed at every cherry reduction, allowing us to exactly characterize how a sequences of reductions will affect the size of the network. In this paper, we only consider *orchard* networks, i.e. networks that can be completely reduced to a single leaf by cherry reductions.

2.2 Problem Definition

The *maximum agreement cherry-reduced subnetwork* (MACRS) problem asks for the largest common subnetwork of two input networks (say N_1 and N_2) that can be found only by applying cherry reductions. Say the MACRS problem on N_1 and N_2 gives the solution N^*, then we call N^* an MACRS of N_1 and N_2. When obvious from the context, we sometimes drop the specification of the input networks and refer simply to N^* as an MACRS. The minimum number of such cherry reduction operations required to reach the MACRS is the *cherry distance* between N_1 and N_2, characterized as a formula in [23, Thm. 4]. In this way, we treat solving MACRS between two networks as synonymous with finding the cherry distance between them.

3 Methodology

In this section we describe the existing MACRS algorithm, then we elaborate on how the required data enumeration (preprocessing) is improved. We further describe a heuristic strategy that ranks inputs on their promise to deliver a good final result. Finally, with some relevant definitions established, we describe how initial random networks are generated for the experimentation presented in the following section. Pseudocode of the complete algorithm with all elements described here is found in Appendix 5.1.

3.1 Exact Algorithm Design Overview

This section provides a high level summary of the algorithm design from [24]. This work also has a corresponding conference proceeding [30], and open-source version [31]. The original algorithm design summarized here has a time complexity of $O(3^r n^3)$ where r is the combined number of reticulations in each network, and n is the maximum number of vertices in an input network.

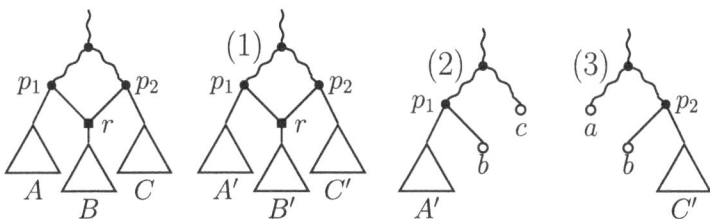

Fig. 2. Showing the three possible options for a reticulation edge in a cherry-reduced network. Let r be a reticulation vertex in this network and r's parents are p_1 and p_2 where p_1 is neither the parent nor the child of p_2 (there would be two choices in that case). (1), (2), (3) depict the three choices for how r may appear on the way to an MACRS (i.e. at some point in the reduction). We show some minimum requirement in the case of (2), (3) or the final MACRS in the case of (1). To simplify the example, subtrees are represented by triangles to emphasize that we assume there are no reticulations in A, B, or C. The prime (') indicates that subtree may or may not be reduced. Small open circles indicate leaf vertices that were required in order to make a cherry on r, assuming $a \in A$, $b \in B$, $c \in C$ in the original network.

The complexity of calculating cherry distance is in the reticulations. So, our strategy is to enumerate through all of the three possible outcomes that each reticulation of the input could take in a calculated MACRS (see Fig. 2). So for an input network N, for every specified set of possible outcomes on $R(N)$ (fixed by choose a disjoint subset of reticulation edges to be reduced), we produce the set of associated minimally cherry-reduced subnetworks of N that accommodates the specification. More formally, we call a subnetwork of N, N', a *reticulation-trimmed subnetwork* of N with respect to $F \subset E_R(N)$ when there exists a sequence of cherry reductions on N that produces N' such that all edges of F are removed, that all edges $E_R(N) \setminus F$ remain, and only the minimum reduction was performed to satisfy this constraint. Refer to Fig. 3 and see [24, Lem. 1] for more details on why these are appropriate structures to produce. We equally refer to such an N' as simply a reticulation-trimmed subnetwork of N, and we can also refer to the set of all reticulation-trimmed subnetworks of N, a set whose members are defined by all appropriate $F \subset E_R(N)$.

The next part of the algorithm involves a new sub-formulation of the MACRS problem, the *simple maximum agreement cherry-reduced subnetwork*, MACRS-SIMPLE, that also asks for the largest common cherry-reduced subnetwork of the input networks, but for one found with only simple cherry reductions. For two input networks N_1 and N_2 we call the solution (if there is one) an MACRS-SIMPLE of N_1 and N_2 or simply an MACRS-SIMPLE.

The algorithm to calculate MACRS consists of two main steps, that is proved to output a correct result in [24, Thm. 2]:

1. enumerate each set of reticulated reduction configurations, resulting in a set of reticulation-trimmed subnetworks for each input network,
2. calculate the MACRS-SIMPLE for each such pair of reduced networks, taking the largest from among them.

The first step can be done in $O(3^r n)$ time, producing a number of pairs reaching $O(3^r)$. Each of those pairs can be handled in the second step in $O(n^3)$

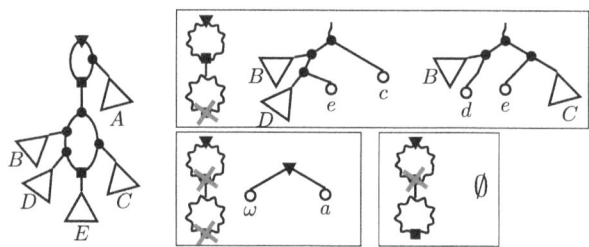

Fig. 3. Showing a set of reticulation-trimmed subnetworks. On the left, we represent a network with two reticulations, one an ancestor of the other. Call the ancestor reticulation r_1 and the descendant reticulation r_2. Subtrees labelled A, B, C, D, and E are just that, subtrees with no reticulations. Let $c \in C$, $d \in D$, $e \in E$ and $\omega \in B \cup C \cup D \cup E$. In each box we show a simplified network representation to the left where the selection of a reticulation edge vr or ur for reduction is denoted with a red 'X' over r. In the right of each box is the resulting combined set of reticulation-trimmed subnetworks, or just the relevant portion in the top case (the upper biconnected component does not change). Notice how on the bottom left example, whether we reduce the "left" or "right" reticulation edge of r_2 is irrelevant: Not every reticulation edge set choice produces a unique reticulation-trimmed subnetwork set. Notice how in the bottom right example, we produce the empty set: the scenario in which only r_1 is slated for reduction cannot produce a reticulation-trimmed subnetwork since r_1 cannot be in any cherry while r_2 is unreduced. (Color figure online)

time. Much of our investigation into runtime improvements of this algorithm concerns reducing the number of pairs produced in the first step, since these serve as the of inputs to the time-consuming dynamic programming in step 2.

See Fig. 3 for a simple description of how not every naive choice of reticulated reductions produce a valid or a unique reticulation-trimmed subnetwork, both due to the bottom-up nature of cherry reductions. This fact, along with an observation regarding a required isomorphism between inputs (described next), is what inspires our optimization: smart enumeration.

3.2 Enumeration of Reticulation-Trimmed Subnetworks

Generating reticulation-trimmed subnetwork pairs can be done, for input networks N_1 and N_2, by taking the product of the two respective sets of their every reticulation-trimmed subnetwork. We call this the naive method of data enumeration (or *naive enumeration*). We compare this method experimentally with the following, improved method in Sect. 4.

For any reticulation r, if we decide that it should remain in the final MACRS i.e. that it will not be reduced to construct a reticulation-trimmed subnetwork, it defines a subnetwork (which we call the *forbidden subnetwork*) above it that can never be a part of a cherry, nor can it be rearranged by any reduction. Similarly, we speak of a forbidden subnetwork defined by a set of reticulations ($R \subseteq R(N)$) that are not reduced. We have previously shown [24, Cor. 1] if there is a solution to MACRS-SIMPLE, then the forbidden subnetwork must form part of an isomorphism (an edge-preserving bijection, since there are no labelled vertices on the forbidden subnetwork) between the two input networks. In this way, we can consider the topological relationship between fixed in-place reticulations as a pre-filter to producing pairs for the MACRS-SIMPLE procedure. We achieve this filter by manipulating a tree structure that we originally defined in [24] for the purpose of counting reticulation-trimmed subnetworks.

For some network N, we define a new graph T, a *dependency tree* of N (or simply a dependency tree), that describes the hierarchy among reticulations of N. Essentially, T is the Hasse diagram on the partially ordered set $R(N) \cup \rho(N)$ defined by the ancestry relation in N. In other words, $V(T)$ is named by $R(N) \cup \{\rho(N)\}$, and the edge uv exists if u covers v. We have previously shown that, by this definition, a dependency tree forms a tree for all level-1 orchard networks [24, Prop. 1]. See Appendix 5.2 for an example. The isomorphism of two dependency trees is edge-preserving as usual, and does **not** preserve labels: labels are network specific, and we seek isomorphisms between dependency trees of different networks. Let T', a subtree of a dependency tree T, be called a *subdependency tree* of T (or simply a subdependency tree when the context is clear) when it contains the root of T. Note that in this way, a subdependency tree of T is a dependency tree of a network reachable by cherry reductions on the network corresponding to T. In particular, we showed that a subdependency tree T' of N leads us to a set of reticulation-trimmed subnetworks on N that have T' as their dependency tree, and critically, that all reticulation-trimmed subnetworks of N can be found in this way [24, Obs. 1].

The final step that we take here is connecting the concept of the subdependency tree generating reticulation-trimmed subnetworks and that of the isomorphism between particular forbidden subnetworks of each input. We assert that there is no MACRS-SIMPLE solution of input networks if there is no isomorphism between the their dependency trees. Observe how the forbidden subnetwork defined by a set of reticulations that remain, R, and a given reticulation-trimmed subnetwork defined strictly by reticulation edges without an endpoint in R (i.e. in which $R(N) \setminus R$ is reduced) both have the same dependency tree. As we enumerate all reticulation-trimmed subnetwork pairs produced by isomorphic (sub)dependency trees, we are only discarding pairs that do not have a possible forbidden subnetwork isomorphism. Put differently, non-isomorphic (sub)dependency trees cannot produce reticulation-trimmed subnetwork pairs that lead to a valid MACRS-SIMPLE solution. It is easy to see that subdependency trees that have a different number of nodes (reticulations) cannot yield a valid MACRS-SIMPLE. In the same way, subdependency trees with the same number of nodes, but with a different topology will produce different forbidden subnetworks that cannot result in a valid MACRS-SIMPLE.

In practice, we implement this optimization as follows. First, for the two input networks N_1 and N_2, we build their corresponding dependency trees T_1 and T_2. Then, we build each possible isomorphic pair between subdependency trees of T_1 and T_2. Next, for two isomorphic subdependency trees T_1' and T_2', we take the product of the two sets of reticulation-trimmed subnetworks corresponding to them. It is these pairs that are the ultimate inputs to the MACRS-SIMPLE algorithm. We call this strategy of generating paired input of reticulation-trimmed subnetworks *smart enumeration*.

Where we expect the smart enumeration strategy to show its strength is when the dependency trees between two input networks are highly divergent in topology, thus reducing the number of possible isomorphisms between them. Importantly, the possibility of seeing such divergence increases with reticulation number, so we expect to get better efficiency gains when they are most appreciated, when network complexity increases. We show this in Sect. 4. However, there are a few angles that show a diminished benefit of the smart enumeration strategy. An interesting but not particularly consequential aspect is that the vertices of a (sub)dependency tree are not completely independent as their treatment would suggest. Siblings in a (sub)dependency tree could be the result of a network topology in which both endpoints of a reticulation edge are the ancestor of distinct reticulations, so they both must be removed for its reduction (Appendix 5.2). We expect this arrangement is uncommon or rare. More significant is the positive association between the number of isomorphisms and the out-degree of dependency tree vertices, which is explored experimentally in Sect. 4.

3.3 Dependency Tree Ranking Heuristic

We acknowledge that the runtime hinges on how many pairs from the algorithm step 1 we check for an MACRS-SIMPLE in algorithm step 2. Here we depart from a guarantee of accuracy in order to make exceptional gains in runtime. As we will see, we do not need to sacrifice much expected accuracy to achieve this. We do so by prioritizing which reticulation-trimmed subnetwork pairs are the most promising in terms of returning an MACRS-SIMPLE that ends up being best from among them all, and discarding tests on those we deem least promising.

Recall how the smart enumeration scheme pre-processes the dependency trees of the input network by finding all isomorphic subdependency tree pairs, and call this set T. In these terms "promise" is measured simply by how large the subdependency tree is, thus the heuristic is quite simply defined. We rank the pairs generated by the smart enumeration scheme by size, largest first/highest priority. Given a threshold $0 \leq t \leq 1$, where $t = c/|T|$, c is defined as follows. When running the MACRS-SIMPLE one-by-one on pairs of reticulation-trimmed subnetworks generated from subdependency trees (where T is sorted), we consider only the first c valid MACRS-SIMPLE solution networks, taking our solution to best from among them as usual. Note that $|T|$ has no correspondence with the resulting number of reticulation-trimmed subnetwork pairs, so even by the same value of t, the value of c will vary by input.

3.4 Generating and Modifying a Network

In acknowledgement of the importance to our experiments of the shape of the dependency tree of a network, we wish to control the dependency tree in our network generator. Therefore, we start with the dependency tree as a base in the following way. First, given a specified number of reticulations r, we generate the set of all isomorphism classes of rooted trees with $r+1$ vertices, representing the possible dependency trees. Sampling equally across the set (except in one experimental case described later), we choose a dependency tree T. Then, we build a base network that contains the appropriate number of non-trivial biconnected components connected in a way that is true to T. We then "grow" a desired number of leaves from this base by randomly choosing edges to be bisected with a pendant leaf.

It is worth noting that when building a base, two sibling vertices in a dependency tree could be the result of many different topological relationships between non-trivial biconnected components in a network. For example, the root of each component could be a sibling, or one component could be descending a non-reticulate vertex in the other. This only gets more complicated with more siblings. For the sake of simplicity, we chose a consistent way to resolve this. We make sibling vertices descendent a common (trivial) parent structure. Using other types of topological relationships between siblings is not expected to alter the speed or accuracy of our approaches in any significant way.

Once the network N has been constructed, we generate another network N' from it by applying d cherry operations on it so that we have a cherry distance of d where d is known. We take half d in cherry reductions, then the remainder in *cherry expansions* (which is the inverse operation to the cherry reduction) with both the simple and reticulated version of the operation. We endeavour to maintain the complexity and apply, if possible, as many reticulated cherry expansions as there were reticulated reductions. Level of the network is maintained by ensuring existing biconnected components are not connected on reticulated cherry expansions. It is not always possible to add a reticulation and preserve the level, so the modified network does not have any guarantee on the total number of reticulations. When we describe experiments on one network in Sect. 4, assume we obtain a pair of inputs by this process.

Also, it is worth noting there is the possible situation in which a reticulated cherry is reduced, then later, before the two edges/leaves are expanded by any other cherries, a reticulated cherry is added back in the same position. These two moves are undetectable, reducing the real distance by two. This may happen in multiple positions, causing the total difference between the requested and real distance to be up to $2r$. While this situation is observed on extremely small networks, we confirmed experimentally that an inaccuracy is exceedingly rare on even small networks, such that it is a non-issue. We avoid this issue entirely for simple cherry expansions by introducing new leaf labels.

4 Results and Discussion

All experiments are conducted on Apple M1 chip with 8GB memory. The software package is implemented in Rust. Networks are input and output in extended Newick format [32]. Our software is fully implemented and available on GitHub.

4.1 Experiments on Simulated Data

Comparing Enumeration Schemes. We hypothesized that the topology of the dependency trees will affect the effectiveness of smart enumeration. This is where we begin our experimentation.

We define the *out-degree* of a dependency tree as the maximum out-degree among all its vertices. We define the *height* of a dependency tree as the number of edges on the longest path from the root to a leaf.

We generate the set of all six-vertex tree isomorphism classes (of which there are 20) representing the possible dependency trees of a network N with $|R(N)| = 5$. In our tests we individually vary both the out-degree and the height, then we show the linear progression between the extrema of these variables, varying both. See Fig. 4 for the results of this experimentation. Additionally, a figure is available in Appendix 5.3 which plots each possible combination of out-degree and height (with 6 vertices), and in which the three experimental tests are highlighted.

Examining Fig. 4, left, we see how indeed there is a much larger percent of pairs that can be ruled out in the smart enumeration method when the dependency tree is more caterpillar-like while the percent of pairs that can be filtered out dramatically lowers as out-degree increases, we attribute this to the fact that there are many more possible isomorphisms between (unordered) sibling sets than on (ordered) ancestral relations. Right, we see the impact in definite terms. For each variable (that is not (1,5)), a not dissimilar absolute number of

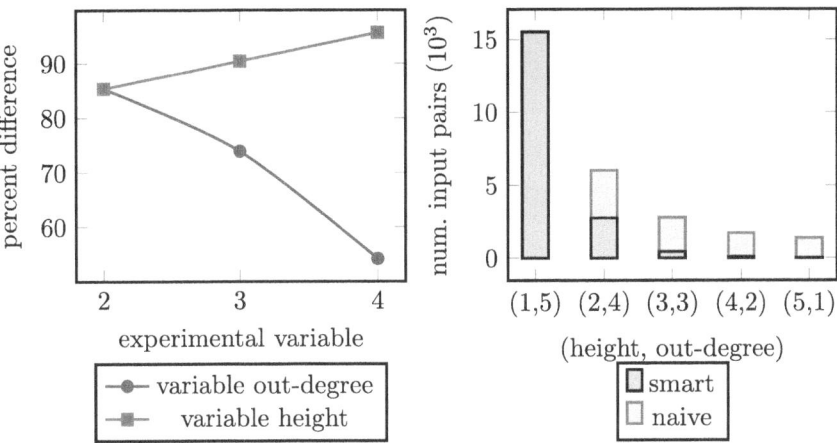

Fig. 4. Comparing the number of dependency tree pairs requiring testing based on dependency tree topology. Both experimental scenarios are averaged over $n = 100$ replicates of 500 leaf networks with distance of 0. (Left) When out-degree is variable, height is set at 2, and vice versa. The y-axis gives the percent (of the naive enumeration) difference of input pairs that is produced by the smart enumeration method i.e. the percent of pairs that is filtered out by smart enumeration. The higher the number, the better. (Right) The pink area of the bar represents how much is filtered out by smart enumeration (bars are overlapping, not stacked).

pairs is filtered out, though what percent of the whole that constitutes varies dramatically. It is obvious to see that the worst case scenario is the *star graph* (height of 1 with the out-degree of 5), which we isolate in the next test.

Next we explore how runtime is affected by other variables of the input networks, number of leaves, of reticulations, and cherry distance. In particular we compare the naive enumeration method with two different scenarios that utilize the smart enumeration scheme, a worst case (a star graph dependency tree) and an average case (any dependency tree). Note that the dependency tree shape may differ between the two input networks after modification by a given cherry distance, especially in the case of larger distances.

See Fig. 5 for results. Overall the smart enumeration scheme may run on par with the naive enumeration scheme when there are less non-isomorphic pairs to rule out, however we still see an improvement in the average case. Indeed, the overbearing source of complexity is the number of reticulations as we see a dramatic rise in runtime as it increases. The size of the input networks also affects runtime, but in a more linear manner. Interestingly, there is an inverse relationship between the runtime and the cherry distance. We attribute this to the functioning of the dynamic programming algorithm, which builds the MACRS-SIMPLE between pairs, a more timely task when the MACRS-SIMPLE is larger as in the case of a lower distance. We return to this fact when analyzing the results of the ranking heuristic.

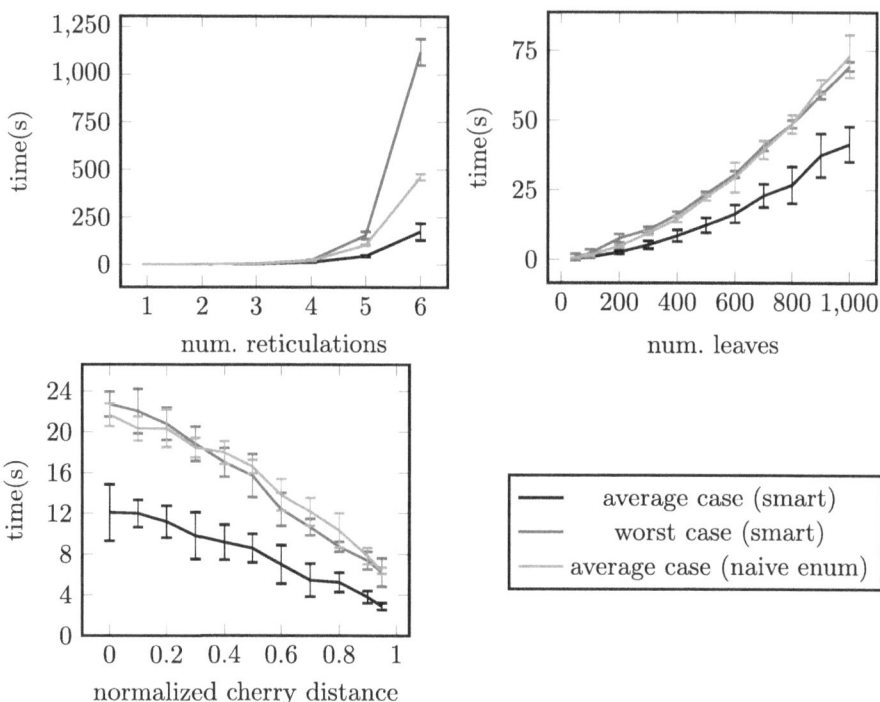

Fig. 5. Showing the runtime under variable network parameters for both smart and naive enumeration methods. In all cases, results are averaged over $n = 100$ replicates and standard error is shown. Number of reticulations step by 1, number of leaves by 100 and normalized cherry distance by 0.1, with the exception of the maximum normalized cherry distance measure being 0.95 as a cherry distance of 1 is trivial to compute. Unless otherwise stated, the default size of the networks is 500 leaves and with 4 reticulations. The distance is 0.05 when experimenting on the number of reticulations, while it is 0 when the number of leaves is varied so as to ensure the number of reticulations is unchanged.

Dependency Tree Ranking Heuristic. We use cherry distance and number of reticulations as the experimental variables. We also show the average case (smart enumeration), as reported in Fig. 5, for comparison. See Fig. 6 for results.

Looking at the $c = 1$ scenario, while it is the fastest scenario (left), the accuracy is not sufficient (bottom right). Interestingly, there is a dip in the accuracy before it rises again, mirrored in the inaccuracy of the worst result (dotted) which rises and then falls towards higher accuracy at high cherry distance. This can be explained by noting that the closer the real distance is to the maximum, the less the heuristic can overshoot the distance. More significant are the results in the $t \geq 0.1$ scenarios, where we see massive increases in accuracy (bottom right) without such dramatic increases in runtime (left). Critically, we see that taking $t = 0.5$ results in nearly perfect accuracy (deviating no more than 10%

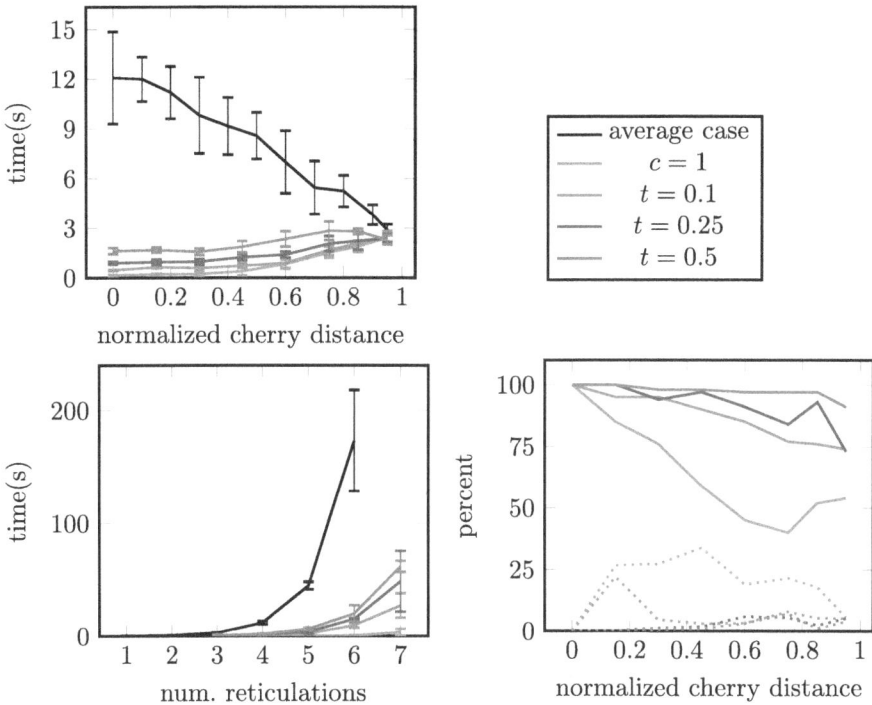

Fig. 6. Showing runtime and accuracy on ranking heuristic method of calculating inexact cherry distance. (Left Upper and Lower) Showing runtime, in both cases results are averaged over $n = 100$ replicates on input networks with 500 leaves and 4 reticulations, standard error is shown. (Bottom right) Solid lines represent the proportion of the correct results in percent (that is, the number of times the distance is correctly returned in 100 replicates). The dotted lines represents the percent difference of the worst solution returned, that is to say it is the percent difference between the correct distance and the largest returned incorrect distance. In an exact program this is always 0. Recall that a threshold t sets the threshold count c, in the case of $c = 1$ we bypass t and simply take the first valid solution to MACRS-SIMPLE.

in both accuracy and in inaccuracy of the worst result) all while maintaining runtime much closer to the $c = 1$ (especially when distance is low) case than to the average case without the ranking heuristic.

Furthermore, we posit that the effectiveness of the ranking heuristic strategy depends on two factors. First, though there are examples to the contrary, we expect the largest MACRS-SIMPLE solution is more likely to have larger (sub)dependency trees. What we see is working in synergy with this factor is that the dynamic programming algorithm in step two runs slower when cherry distance is low, as previously illuminated, because it builds the entire MACRS-SIMPLE. Taken together, we put forward that the small subdependency tree pairs (of which there are many more possible than large ones) end up having a complete MACRS-SIMPLE being built, which ends up dominating the run time of the exact method.

4.2 Experiments Based on Real Data

Now we show properties of the cherry distance through experiments on real data. Specifically, we obtain six networks on a 27 taxa sample of the apple tribe, Maleae, of the Rosaceae (Rose) family [33]. This sample includes genera containing apple, crabapple, pear, quince, hawthorn, and others. The six species networks have each a number of reticulations between 0 and 5, and are inferred from Species Networks applying Quartets (SNaQ) [27,28] network analysis.

First, we verify the accuracy of the ranking heuristics on this real data. In nearly all cases, across a variety of number of reticulations and of cherry distances tested, the resulting accuracy is 100% with $t = 0.5$. In many cases that deviate from this perfect accuracy, it still remains above 95%. There is one case in which accuracy drops to 87%, however in this case the distance calculated was only one "step" away from the real distance. See Appendix 5.4 for tabular results.

Next, we single out a rearrangement operation that is an earlier adaptation of the rNNI operation which, together with two complexity changing moves, constitutes the rLST operation [11]. This version is selected as, in its definition, level is maintained at 1. In short, the rNNI operation selects two edges who exchange a child subnetwork, with specifications as to which edges may be appropriately picked so that level is maintained. An rNNI move is illustrated in Appendix 5.5.

In Fig. 7, left, the experiment performs a single rNNI move, doing so 1000 times for each Maleae network for 6000 total replicates. We measure both the resulting cherry distance, as well as the number of leaves below any vertex affected by the move. Noticing the gap in results on the real data, we also perform the test for simulated data of the same size, 27 leaves and with a random number of reticulations between 0 and 5, we generate 1000 networks in this way, performing the same experiment (Appendix 5.6). We confirm the gap results from the specific network topology.

In analyzing the results, first note how, in each test, a single rNNI move results in a fairly uniform distribution of resulting cherry distances measured, which indicates that the cherry distance is valuable in differentiating the changed networks. Then, note how there is a strong correlation between the cherry distance and the proportion of leaves being affected. This is a strength of the cherry distance, it is accentuated by how many taxa have a lineage change by an incongruence. In this way, superficial changes to the network make a weak contribution to cherry distance while changes closer to the root make a stronger contribution.

Let us contrast this with the *soft RF* distance on networks which more or less (aside from rare special cases) corresponds directly to the number of rNNI moves, regardless of where that rearrangement takes place, confirmed through testing whose results are not shown here. We contend that this fact favours cherry distance as a more accurate reflection of the extent that a change propagates through lineages in the network. We choose the soft RF as comparison since it is (to our knowledge) one of the very few real implementations of a network

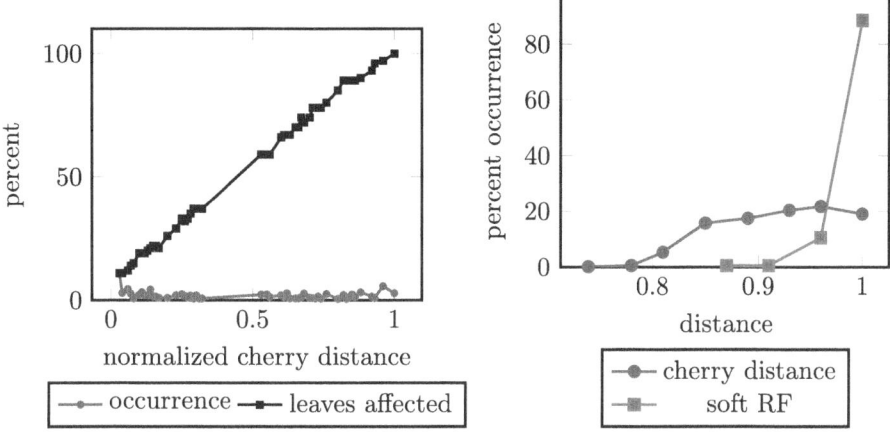

Fig. 7. (Left) After one rNNI move, we compare the resulting cherry distance and the number of leaves affected by the move, measured across 6000 replicates. Concerning the y-axis, the legend displays what percent means in each case. (Right) This plot shows the percent occurrence that each distance is measured on a pair that consists of a base network from the real, Malus, data set and on a random network of the same leaf label set and number of reticulations. The original networks include a random removal of two leaves to keep a consistent leaf set size of 25. Tests include an equal mix of network pairs on a number of reticulations $r = \{0, 1, 2, 3\}$. Cherry distance is normalized by the diameter while soft RF is normalized by the maximum result, as a formal diameter on level-1 orchards does not yet exist to our knowledge.

distance, in the publicly available PhyloNetwork software [17]. Despite being able to handle large numbers of reticulations, PhyloNetwork does not handle large leaf sets well, its runtime being exponential in the number of leaves. Due to this, the experiment using PhyloNetwork (right in Figure 7) has a reduced number of replicates. A network is randomly generated on the same leaf set and number of reticulations as a real network from the Malus data set, then the distance (soft RF and cherry) is taken. We see how soft RF does not have the sensitivity of cherry distance. Due to these differences in efficiency and characteristics, our definition and implementation of the cherry distance offer an interesting avenue for advancing the study and practice of network comparison.

5 Conclusion and Future Work

In summary, we have developed and implemented an optimized version of an exact algorithm for the cherry distance on level-1 orchard networks, and a fast and accurate heuristic. Our experimentation demonstrated some favourable characteristics of the cherry distance, namely a good distribution of values on random networks, a correlation with the number of leaves affected by a network modification operation, and a reasonable running time for the exact method when the number of reticulations is not too high, with the heuristic being much faster.

In terms of increased efficiency, calculations on independent biconnected components can be parallelized, an improvement to the software that would be especially helpful in the worst case. It is also worth asking how often we encounter something like the worst case in real data. Is there a common "naturally occurring" dependency tree shape or shapes?

Next, we look to investigate the broader applications of the dependency tree analysis and ranking heuristic that has proven so effective in our algorithm implementation. The smart enumeration application would be relevant to any problem on a directed graph in which there is a bottom-up dependency on processing. Furthermore, any problem in which a time consuming calculation is repeated a number of times on (perhaps) similar input may benefit from a ranking strategy.

Finally, we show the strong effect of reticulate complexity on runtime, emphasizing the need to develop an algorithm running in a smaller complexity parameter such as network level. In the same vein, we understand that generalizing our algorithm to level-k networks is critical in expanding the application of our algorithm to a broader use. This acknowledges a level of dependency on reticulations that are on overlapping cycles, but to achieve such a runtime biconnected components must be handled independently.

Acknowledgements. This work was funded by the NSERC Discovery (grant number RGPIN-2016-06051) program.

Disclosure of Interests. Authors have no competing interests to declare.

Appendix

5.1 Algorithm pseudocode

The following pseudocode is repeated directly from [24]. This is the naive form of enumeration.

Algorithm 1. MACRS Finder

  Input Two networks N_1 and N_2
  Output A MACRS of N_1 and N_2
1: $\tilde{N} \leftarrow$ empty network
2: **for each** reticulation-trimmed subnetwork N'_1 of N_1 **do**
3:    **for each** reticulation-trimmed subnetwork N'_2 of N_2 **do**
4:       Let N' be a MACRS-SIMPLE of N'_1 and N'_2
5:       **if** N' exists and $|V(N')| > |V(\tilde{N})|$ **then** $\tilde{N} \leftarrow N'$
6:    **end for**
7: **end for**
8: return $\tilde{N}$

The next pseudocode incorporates all of the improvements discussed in this work. This includes the smart enumeration in step 1 using subdependency trees, as well as the heuristic ranking that filters out less important tests in step 2.

Algorithm 2. MACRS Finder with Smart Enumeration and Ranking Heuristic

Input Two networks N_1 and N_2, and a ranking threshold $0 < t \leq 1$
Output A MACRS of N_1 and N_2

1: let T_1 be the dependency tree of N_1
2: let T_2 be the dependency tree of N_2
3: $P = \{(T_1', T_2') | T_1' \simeq T_2', T_1'$ a subdependency tree of T_1, T_2' a subdependency tree of $T_2\}$
4: Order **P** by size
5: $\mathbf{N} \leftarrow \emptyset$
6: **for each** $(T_1', T_2') \in P$ **do**
7: $\quad \mathcal{N}_1 \leftarrow$ all reticulation-trimmed subnetworks corresponding to T_1'
8: $\quad \mathcal{N}_2 \leftarrow$ all reticulation-trimmed subnetworks corresponding to T_2'
9: $\quad \mathbf{N}.append($all pairs between $\mathcal{N}_1$ and $\mathcal{N}_2)$
10: **end for**
11: $c \leftarrow |T| * t$
12: $\tilde{N} \leftarrow$ empty network
13: **while** $c > 0$ **do**
14: $\quad (N_1', N_2') \leftarrow \mathbf{N}.pop()$ $\quad \triangleright$ Pop the largest list element
15: $\quad$ let N' be a MACRS-SIMPLE of N_1' and N_2'
16: $\quad$ **if** N' exists **then**
17: $\quad\quad$ **if** $|V(N')| > |V(\tilde{N})|$ **then**
18: $\quad\quad\quad \tilde{N} \leftarrow N'$
19: $\quad\quad$ **end if**
20: $\quad\quad$ c--
21: $\quad$ **end if**
22: **end while**
23: return $\tilde{N}$

5.2 Dependency tree

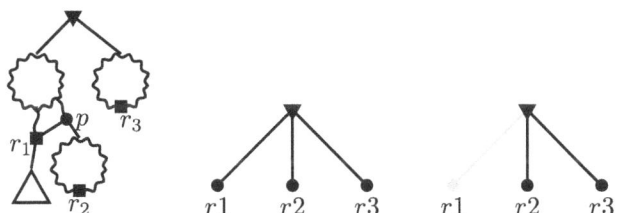

Fig. 8. (Left) Network with three reticulations r_1, r_2, r_3. Triangles are subtrees (no reticulations). (Centre) Dependency tree of network on left. (Right) Subdependency tree of left, with removed vertices greyed-out. Reticulation r_1 could not be reduced without the cherry reduction of r_2 first, so this subdependency does not produce any reticulation-trimmed subnetwork.

5.3 6-vertex dependency trees

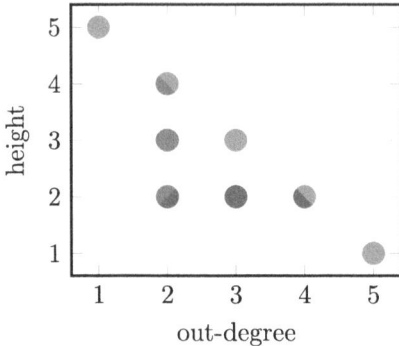

Fig. 9. Each point represents a cluster of distinct topologies with the same height and out-degree. When the cluster size is > 1, each tree is represented equally in the corresponding results shown in Fig. 4. The colours correlate with the colours used on the plots, with grey representing the experiment on the right of Fig. 4. As a matter of interest, the most dense clusters are (2,3) (5 trees) and (2,4) (4 trees) and the clusters with each only 1 tree are (2,2), (1,5), and (5,1).

5.4 Ranking heuristic on real data

num. reticulations		0		
cherry distance	0	15	30	45
time (ms)	0	0	0	0
% accuracy	100	100	100	100
% of worst solution	0	0	0	0

num. reticulations		1		
cherry distance	0	15	30	45
time (ms)	2	1	1	1
% accuracy	100	100	100	100
% of worst solution	0	0	0	0

num. reticulations		2		
cherry distance	0	15	30	45
time (ms)	6	7	6	5
% accuracy	100	87	99	100
% of worst solution (abs. value)	0	0.13 (2)	0.07 (2)	0

num. reticulations		3		
cherry distance	0	15	30	45
time (ms)	31	26	23	20
% accuracy	100	100	100	100
% of worst solution	0	0	0	0

num. reticulations		4		
cherry distance	0	15	30	45
time (ms)	100	85	76	65
% accuracy	100	99	100	100
% of worst solution (abs. value)	0	0.13 (2)	0	0

num. reticulations		5		
cherry distance	0	15	30	45
time (ms)	166	229	172	110
% accuracy	100	95	94	95
% of worst solution (abs. value)	0	0.13(2)	0.8 (24)	0.2 (10)

All tests are on 27-leaf Malus data set, randomly modified by the specified cherry distance. For all tests, $t = 0.5$.

5.5 rNNI

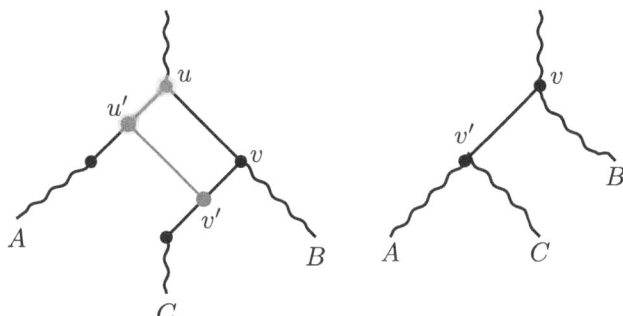

Fig. 10. An rNNI move on a level-1 network is demonstrated, as specified in [11], where it is shown to always preserve level. (Left) The original network is in black. An edge uv is selected such that it is not in a 3-cycle (a biconnected component made of three vertices), the vertices u and v are not in unique nontrivial biconnected components, and that neither u nor v are reticulations. The sibling edge to uv is bisected with new vertex u', and so is one child edge of v with new vertex v'. An edge $v'u'$ is added. The edge uu' is removed, requiring its endpoints be identified with a neighbour to preserve the proper network definition precluding a vertex v with $v^- = 1$ and $v^+ = 1$. Additions are represented in blue and removals are highlighted in red. It is not necessarily the case that $A \cap B \cap C = \emptyset$. (Right) The resulting network after the rNNI move is made.

5.6 rNNI on simulated networks

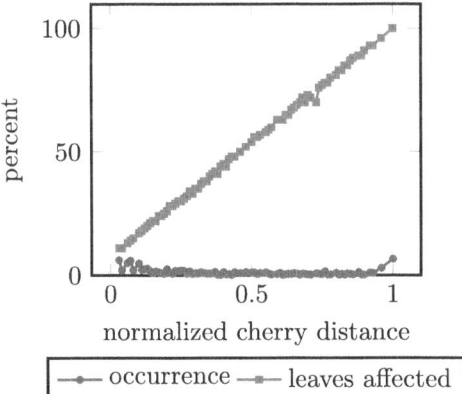

Fig. 11. Results on a test of $n = 1000$ simulated networks, a pair who differ by one rNNI move, each with 27 leaves and an equal number of replicates on each number of reticulations 0–5.

References

1. Huson, D.H., Bryant, D.: Application of phylogenetic networks in evolutionary studies. Molecular Biol. Evol. **23**, 254–267 (2005)
2. Penny, D., Hendy, M.: The use of tree comparison metrics. Syst. Zool. **34**(1), 75–82 (1985)
3. Allen, B.L., Steel, M.: Subtree transfer operations and their induced metrics on evolutionary trees. Ann. Comb. **5**, 1–15 (2001)
4. Robinson, D.F.: Comparison of labeled trees with valency three. J. Comb. Theory, Ser. B **11**(2), 105–119 (1971)
5. Moore, G.W., Goodman, M., Barnabas, J.: An iterative approach from the standpoint of the additive hypothesis to the dendrogram problem posed by molecular data sets. J. Theor. Biol. **38**(3), 423–457 (1973)
6. Bordewich, M., Semple, C.: Computing the hybridization number of two phylogenetic trees is fixed-parameter tractable. IEEE/ACM Trans. Comput. Biol. Bioinf. **4**(3), 458–466 (2007)
7. Bordewich, M., Linz, S., Semple, C.: Lost in space? Generalising subtree prune and regraft to spaces of phylogenetic networks. J. Theor. Biol. **423**, 1–12 (2017)
8. Francis, A., Huber, K.T., Moulton, V., Wu, T.: Bounds for phylogenetic network space metrics. J. Math. Biol. **76**, 1229–1248 (2018)
9. Huber, K.T., Moulton, V., Wu, T.: Transforming phylogenetic networks: moving beyond tree space. J. Theor. Biol. **404**, 30–39 (2016)
10. Gambette, P., Van Iersel, L., Jones, M., Lafond, M., Pardi, F., Scornavacca, C.: Rearrangement moves on rooted phylogenetic networks. PLoS Comput. Biol. **13**(8), e1005611 (2017)
11. Huber, K.T., Linz, S., Moulton, V., Wu, T.: Spaces of phylogenetic networks from generalized nearest-neighbor interchange operations. J. Math. Biol. **72**(3), 699–725 (2016)
12. Robinson, D.F., Foulds, L.R.: Comparison of phylogenetic trees. Math. Biosci. **53**(1–2), 131–147 (1981)
13. Day, W.H.: Optimal algorithms for comparing trees with labeled leaves. J. Classif. **2**, 7–28 (1985)
14. Baroni, M., Semple, C., Steel, M.: A framework for representing reticulate evolution. Ann. Comb. **8**, 391–408 (2005)
15. Cardona, G., Rosselló, F., Valiente, G.: Tripartitions do not always discriminate phylogenetic networks. Math. Biosci. **211**(2), 356–370 (2008)
16. Cardona, G., Llabrés, M., Rosselló, F., Valiente, G.: Metrics for phylogenetic networks I: generalizations of the Robinson-Foulds metric. IEEE/ACM Trans. Comput. Biol. Bioinf. **6**(1), 46–61 (2008)
17. Lu, B., Zhang, L., Leong, H.W.: A program to compute the soft Robinson-Foulds distance between phylogenetic networks. BMC Genomics **18**, 1–10 (2017)
18. Humphries, P.J., Linz, S., Semple, C.: Cherry picking: a characterization of the temporal hybridization number for a set of phylogenies. Bull. Math. Biol. **75**(10), 1879–1890 (2013)
19. Linz, S., Semple, C.: Attaching leaves and picking cherries to characterise the hybridisation number for a set of phylogenies. Adv. Appl. Math. **105**, 102–129 (2019)

20. Janssen, R., Murakami, Y.: Linear time algorithm for tree-child network containment. In: Martín-Vide, C., Vega-Rodríguez, M.A., Wheeler, T. (eds.) Algorithms for Computational Biology: 7th International Conference, AlCoB 2020, Missoula, MT, USA, April 13–15, 2020, Proceedings, pp. 93–107. Springer International Publishing, Cham (2020). https://doi.org/10.1007/978-3-030-42266-0_8
21. Erdős, P.L., Semple, C., Steel, M.: A class of phylogenetic networks reconstructable from ancestral profiles. Math. Biosci. **313**, 33–40 (2019)
22. Janssen, R., Murakami, Y.: On cherry-picking and network containment. Theoret. Comput. Sci. **856**, 121–150 (2021)
23. Landry, K., Teodocio, A., Lafond, M., Tremblay-Savard, O.: Defining phylogenetic network distances using cherry operations. IEEE/ACM Trans. Comput. Biol. Bioinf. **20**(3), 1654–1666 (2023)
24. Landry, K., Tremblay-Savard, O., Lafond, M.: A fixed-parameter tractable algorithm for finding agreement cherry-reduced subnetworks in level-1 orchard networks. J. Comput. Biol. **31**(4), 360–379 (2024)
25. Kong, S., Pons, J.C., Kubatko, L., Wicke, K.: Classes of explicit phylogenetic networks and their biological and mathematical significance. J. Math. Biol. **84**(6), 47 (2022)
26. Solis-Lemus, C., Coen, A., Ane, C.: On the identifiability of phylogenetic networks under a pseudolikelihood model. arXiv preprint arXiv:2010.01758 (2020)
27. Solís-Lemus, C., Ané, C.: Inferring phylogenetic networks with maximum pseudolikelihood under incomplete lineage sorting. PLoS Genet. **12**(3), e1005896 (2016)
28. Solís-Lemus, C., Bastide, P., Ané, C.: PhyloNetworks: a package for phylogenetic networks. Mol. Biol. Evol. **34**(12), 3292–3298 (2017)
29. Allman, E.S., Baños, H., Rhodes, J.A.: NANUQ: a method for inferring species networks from gene trees under the coalescent model. Algorithms Mol. Biol. **14**, 1–25 (2019)
30. Landry, K., Tremblay-Savard, O., Lafond, M.: Finding agreement cherry-reduced subnetworks in level-1 networks. In: Comparative Genomics Jahn, K., Vinař, T., eds., pp. 179–195, Springer Nature Switzerland, Cham (2023). https://doi.org/10.1007/978-3-031-36911-7_12
31. Landry, K., Tremblay-Savard, O., Lafond, M.: Finding agreement cherry-reduced subnetworks in level-1 networks. arXiv preprint arXiv:2305.00033 (2023)
32. Cardona, G., Rosselló, F., Valiente, G.: Extended newick: it is time for a standard representation of phylogenetic networks. BMC Bioinformatics **9**, 1–8 (2008)
33. Liu, B., et al.: Phylogenomic conflict analyses in the apple genus malus s.l. reveal widespread hybridization and allopolyploidy driving diversification, with insights into the complex biogeographic history in the northern hemisphere. J. Integr. Plant Biol. **64**(5), 1020–1043 (2022)

Sequence Analysis

Detecting and Mapping Local Model Violations During Biomolecular Sequence Analysis: a *RE*sampling and *V*isual *EvAL*uation Approach

Meijun Gao[1] and Kevin J. Liu[1,2,3(✉)]

[1] Department of Computer Science and Engineering, Michigan State University, East Lansing, MI, USA
kjl@msu.edu
[2] Ecology, Evolution, and Behavior Program, Michigan State University, East Lansing, MI, USA
[3] Genetics and Genome Sciences Program, Michigan State University, East Lansing, MI, USA

Abstract. In phylogenetics and phylogenomics, a common assumption made during biomolecular sequence analysis is that a single "global" evolutionary model sufficiently describes the evolution of all sites within a given dataset. However, numerous evolutionary processes can violate this assumption and introduce local model mis-specification. While both traditional statistical methods and machine learning approaches have been employed to address this issue, they are typically constrained by specific modeling assumptions and often are specialized for one or a few closely related tasks, limiting their applicability.

In this study, we introduce REVEAL ("REsampling and Visual EvALuation"), a general-purpose statistical framework for detecting and mapping local model mis-specification during biomolecular sequence analysis. REVEAL does not impose additional modeling assumptions beyond those used during global model-based sequence analysis. REVEAL leverages sequence-aware statistical resampling techniques to extract a local support matrix along the input sequences, enabling the identification of potential local model violations. Performance benchmarking using simulation experiments demonstrates that REVEAL achieves robust type I and type II error, with as high as 90% precision and 85% recall across a range of experimental conditions that have different sources of local model mis-specification. We also employ REVEAL to analyze genomic sequence data for mouse and mosquito, and REVEAL detects local model violations that align with findings from previously published studies.

Keywords: Biomolecular sequence analysis · Statistical resampling · Model mis-specification · Phylogenetic estimation

---

Supplementary Information The online version contains supplementary material available at https://doi.org/10.1007/978-3-031-94928-9_13.

1 Introduction

Statistical phylogenetic methods typically rely on parametric evolutionary models to reconstruct and analyze evolutionary histories. These models are designed to account for the complex evolutionary processes that shape genetic data over time. For instance, maximum likelihood estimation methods require a model of sequence evolution to describe changes in biomolecular sequences over time. However, a given parametric model can fail to adequately describe underlying evolutionary processes and bias statistical estimation – an outcome referred to as a model violation. Model violations generally fall into two categories: "global" and "local" model violations.

Global model violations occur when the evolutionary model fails to adequately represent evolutionary processes that apply to all sites in the locus or loci under study. For example, substitution model violations can occur in traditional phylogenetic analyses, and computational methods have been proposed to detect and address these model violations [1, 4].

Local model violations occur when a given evolutionary model may be suitable for some sites and/or regions in a biomolecular sequence dataset but not other sites/regions. Complex evolutionary processes such as recombination, natural selection, incomplete lineage sorting, and horizontal gene transfer can lead to deviations from the global modeling assumption that all sites under study evolved in an independent and identically distributed (i.i.d.) manner [6]. Local model violations can manifest as various types of heterogeneity within sequences, including evolutionary rate heterogeneity across sites and lineages [16], base frequency heterogeneity, and local topological heterogeneity across sites. Past studies have utilized parametric model-based methods to examine biomolecular sequence patterns associated with specific evolutionary processes. For example, local genealogical variation due to genetic drift and incomplete lineage sorting (ILS) [5], recombination and recombination hotspots [14], and the complex interplay between substitution, recombination, and gene conversion [9] have been investigated in genetic and genomic sequence data. Other studies have applied machine learning methods for similar purposes. For example, supervised machine learning and deep neural networks have been used to identify genomic regions that evolved via introgression [21], map recombination breakpoints and recombination hotspots [2, 17], and detect genomic signatures of selective sweeps [26]. While parametric model-based and machine learning-based methods can be effective in certain contexts, they are typically designed for specific evolutionary processes that require a priori modeling assumptions, are specialized to particular biomolecular sequence analysis tasks, and often require labeled training data, which may not always be available or accurately annotated.

In this study, we aim to detect local model violations without making any assumptions about sequence evolution (beyond those made in a global model-based analysis) or imposing restrictions on the types of local model violations. The problem formulation under study and our algorithmic solution for the problem point to an automated, data-driven alternative to the traditional approach of formulating *a priori* modeling assumptions for specific biomolecular sequence analysis tasks and then performing iterative model refinement.

2 Methods

A primary contribution of this study is a new general-purpose statistical method for detecting and mapping local model mis-specification during biomolecular sequence analysis. A key requirement is that the new method requires no additional modeling assumptions beyond those used for global data analysis (i.e., the traditional simplifying assumption that a single "global" statistical model adequately captures the underlying processes that generated all parts of the input dataset). Stated another way, the new method does not utilize any additional parametric models for any subsets of the dataset and/or the entire dataset (beyond the single model used for global inference and learning on the entire dataset).

We now define the computational problem under study. The input consists of a multiple sequence alignment A with N aligned sequences and K sites, a global model θ, and a global model-based estimation method f_θ. The output is a classification of each site of $a_i \in A$ for $1 \leq i \leq K$ to one of z model classes. The most basic task for detecting local model violation utilizes $z = 2$ classes, where a "background" class corresponds to the θ model and a "locally variable" class corresponds to local model(s) (which violate the assumption that θ suffices as a single global model for all sites in A). As a proof of concept, our study's experiments focus on a particular model and task: finite-site models of nucleotide substitution (i.e., the GTR model [22] and nested models of nucleotide substitution) and their use in maximum likelihood estimation (MLE) of phylogenetic trees. We note that the specific computational problem under study, our algorithmic solution to the computational problem, and study design can readily be adapted to other statistical models and inference/learning tasks. (We expand on this point in the Discussion and Conclusions sections).

To address this problem, we introduce REVEAL – a "REsampling and Visual EvALuation" framework to detect and map local model violations during biomolecular sequence analysis. REVEAL consists of a computational pipeline with three stages. (A flowchart is provided in Fig. 1.) In stage one, local resampling is performed on the MSA A to obtain a set of local replicates Ξ, and local re-estimation is performed on each locally resampled replicate $\xi \in \Xi$ using the model-based method f_θ under the model θ. In stage two, repeatability/agreement of the set of local re-estimates is assessed using statistical calculations. The resulting site-level statistics are packaged into a 2-D matrix to facilitate the final stage of analysis. In stage three, regions in the 2-D matrix with similar site-level local statistics that suggest similar local re-estimation repeatability or lack thereof are identified. The identified regions delineate local model variation that can result in model mis-specification if not properly accounted for (e.g., as in a traditional model-based sequence analysis that assumes a single global model suffices for all data under study). The 2-D input matrix is readily visualized as image, which then naturally lends itself to unsupervised machine learning approaches for image processing. We now provide technical details for each stage of the REVEAL algorithm.

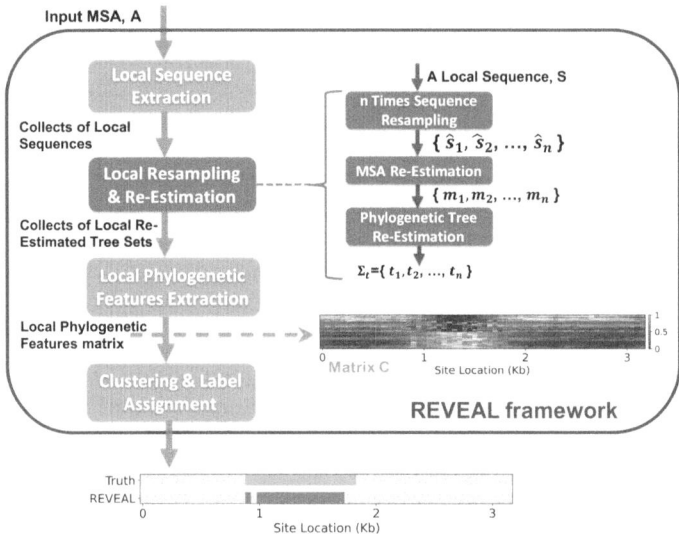

Fig. 1. Illustrated overview of REVEAL, a "REsampling and Visual EvALuation" framework for global-model-agnostic and local-model-free mapping of local model violations during biomolecular sequence analysis. REVEAL consists of a computational pipeline with three stages. (1) The first stage performs local resampling and re-estimation along an input multiple sequence alignment A. (2) The second stage quantitatively assesses agreement/disagreement among local re-estimates from the first stage. Statistics used for quantitative assessment are referred to as "local support". Reduced local support values (and reproducibility of local re-estimation) provide a key indicator of local model variation that can confound biomolecular sequence analysis if not accounted for properly. Per-site local support values are aggregated into a 2-D matrix. The illustration includes an example matrix C. (3) The 2-D matrix can be visualized as an image and also lends itself well to image processing techniques. With this insight in mind, the final stage of REVEAL uses unsupervised clustering to estimate regions with similar local model variation in the site-level 2-D matrix. The illustration includes an example with REVEAL-estimated regions shown in green (and compared against ground truth in orange).

2.1 REVEAL Algorithm

REVEAL stage 1: local resampling and re-estimation. The inputs to REVEAL consist of a multiple sequence alignment (MSA) A, where each of the N rows represents an aligned sequence corresponding to a specific taxon and each of the K columns represents a site, as well as the global model θ and estimation method f_θ for performing global analysis of the entirety of A. The first stage of REVEAL performs local resampling and re-estimation on A.

A sliding-window approach is used to perform local resampling and re-estimation. We define a local window $s_i = A[:, i - \frac{w}{2} : i + \frac{w}{2}]$ to be a subset of columns in A that are centered at the ith site and w is the window length. To systematically resample local sequences, local resampling is performed at regular intervals along A. The process begins at the first site and proceeds with a fixed step size p. The window positions are then $i = 0, p, 2p, 3p, \ldots, \lfloor \frac{K}{p} \rfloor p$, where K is the total length of the A. Consequently, the total number of extracted local sequences is $b = \lfloor \frac{K}{p} \rfloor + 1$.

Local resampling and re-estimation involves performing n non-parametric sequence resampling iterations within each window s_i to generate n resampled local replicates $\hat{s}_1, \hat{s}_2, \ldots, \hat{s}_n$. For each resampled replicate $\hat{s}$, re-estimation is performed using the estimation method f_θ under model θ that was used for global sequence analysis. For the task under study, re-estimation uses a two-phase method where unaligned sequences in the replicate are first aligned to obtain a re-estimated MSA and then a phylogenetic tree is estimated on the re-estimated MSA. The result is a set of re-estimated phylogenetic trees $\Sigma_t = \{t_1, t_2, \ldots, t_n\}$.

The non-parametric sequence resampling method used in this study is RAWR [24]. RAWR is a sequence-aware statistical resampling technique that avoids the simplifying assumption of independent and identically distributed (i.i.d.) input data – unlike standard bootstrap resampling and other widely used non-parametric resampling techniques. Here we briefly recap the RAWR resampling procedure (cf. Algorithm 1 in [24]). RAWR resampling takes the form of a random walk conducted on the input MSA A, resulting in a resampled RAWR replicate: (1) to begin, a starting site and walk direction are chosen uniformly at random, (2) sites are resampled as the walk proceeds along the initial walk direction, with walk reversals occurring with certainty at the first and last site of A and with probability γ elsewhere, (3) resampling concludes once the resampled replicate length equals the length of A, and (4) the resampled sequences are unaligned to obtain the replicate set of unaligned sequences. The experiments in our study utilize a RAWR reversal probability of $\gamma = 0$. REVEAL also uses default settings of $p = 50$, $w = 300$, and $n = 20$.

REVEAL stage 2: calculating local support values and their 2-D image matrix representation. The second stage of REVEAL calculates local support values to assess repeatability of local tree re-estimation. The local support values quantify phylogenetic branch agreement/disagreement between the local phylogenetic trees and the global phylogenetic tree T_G inferred from A using the method f_θ under model θ.

REVEAL utilizes three different classes of local support values. The first class is topological branch support $p \in \mathbb{R}^{1 \times (N-3)}$, which quantifies the occurrence frequency of branches in the global tree T_G within Σ_t. For a given branch l in the internal edge set of T_G, we define $\Sigma_{t|l}$ as the subset of Σ_t containing branch l. Consequently, the element in p corresponding to branch l in T_G is given by

$\frac{|\Sigma_{t|l}|}{|\Sigma_t|}$. The second and third classes concern re-estimated branch lengths (rather than re-estimated topologies). The second class is mean re-estimated branch length $m \in \mathbb{R}^{1\times(2\times N-3)}$, which is the average length of each branch in T_G across the re-estimated tree set Σ_t. The third class is the standard deviation of re-estimated branch lengths $d \in \mathbb{R}^{1\times(2\times N-3)}$, i.e., the standard deviation of each branch length in T_G as observed in the re-estimated tree set Σ_t.

The local support values are calculated within each window as part of REVEAL's sliding-window analysis. Let $P \in \mathbb{R}^{b\times(N-3)}$ be the matrix of topological branch support values p across all windows in sequence order; similarly, the matrices $M \in \mathbb{R}^{b\times(2\times N-3)}$ and $D \in \mathbb{R}^{b\times(2\times N-3)}$ contain the mean m and standard deviation d of locally re-estimated branch lengths across all windows in sequence order, respectively. The three matrices are combined into a single matrix $C = [P; M; D]$ that can be naturally visualized as a 2-D image.

REVEAL stage 3: mapping higher-level regions with local model variation in the lower-level 2-D support value matrix . The final step of the REVEAL framework applies clustering analysis to the 2-D image representation of the concatenated local support value matrix C. As a preprocessing step before clustering, values in the matrix C are normalized to the unit interval.

The goal is to classify each site in the input MSA, where each class corresponds to one of z different site models. In our study, we focus on $z = 2$ classes where one class corresponds to the global model θ and the other class corresponds to a local model that can cause model mis-specification if not accounted for. REVEAL uses the K-means algorithm [11] to perform unsupervised clustering on the 2-D matrix C. The output is an assignment of each window to one of z clusters, where each cluster represents a distinct site model class. The site at the center of each window is assigned the window's cluster, and cluster assignments for all other sites are based on nearest neighbor interpolation.

2.2 Simulation Study

Simulation Conditions. To evaluate the performance of the REVEAL framework, we consider a variety of evolutionary processes that can cause local model violations during biomolecular sequence analysis and we perform model-based simulations for performance benchmarking purposes. In our simulations, sequences evolve under a mixture model consisting of a background model θ_B and one or more variable region models. The background model consists of the traditional multi-species coalescent (MSC) model [12]. In contrast, the variable region models evolved under evolutionary processes that are not captured by the θ_B model and induce local model violations during θ_B model-based sequence analysis. We investigate four distinct types of variable region models – each representing a different evolutionary process – to comprehensively evaluate the performance of the REVEAL framework.

The first type of variable region model builds upon the traditional MSC model but exhibits greater evolutionary divergence than the background model.

We denote this variable-divergence variable region model as θ_H. Such local model violations can arise from various evolutionary processes. For example, regions under strong selective pressure may experience accelerated accumulation of beneficial mutations, leading to increased divergence compared to neutrally evolving background regions.

The second type of variable region model is the multi-species coalescent with recombination (MSCwR) model, which extends the traditional multi-species coalescent (MSC) model by incorporating recombination events, and we denote this model as model θ_R. This model accounts for local model violations resulting from variations in recombination rates across the genome, a common phenomenon observed in many species. Some genomic regions experience high recombination rates (recombination hotspots), while others have low or negligible recombination rates (recombination cold spots).

The third type of variable region model follows the multispecies network coalescent with recombination (MNSCwR) model, which incorporates both recombination and reticulation events. The reticulation events can capture the complex evolutionary scenarios involving gene flow between species, such as introgression, hybridization, and horizontal gene transfer. These processes often create discordant patterns from the background evolutionary model, resulting in local model violations. We denote this variable region model as model θ_I.

The last type of variable region model is the natural selection model, denoted as model θ_S. Natural selection is a well-studied driver of local model violations, particularly when it exerts differential selective pressure on different genomic regions. For instance, positive selection accelerates the fixation of advantageous mutations, leading to elevated substitution rates in specific genomic regions, creating divergence from the background evolutionary model. Conversely, purifying selection eliminates deleterious mutations, which can also contribute to reduced variation in certain regions. These selective processes result in genomic regions that deviate significantly from the expected neutral evolution, often leading to local model violations, which the θ_S model is designed to capture.

In this study, we construct the mixture model comprising one background model and either one or two variable region models. When a single variable region model is included, we examine four distinct types, each under a separate simulation model condition named after the corresponding variable region model: model conditions H, R, I, and S correspond to mixture models where the single variable region model consists of θ_H, θ_R, θ_I, and θ_S, respectively. Mixture models with two variable region models are designated by model condition M, where a pair of variable region models are randomly selected from the set of θ_H, θ_R, and θ_I models.

Simulation Procedures. Loci and sites evolving in the background region are simulated under the MSC model. First, random birth-death model trees with a tree height of 1.0 coalescent unit and $N = 10$ taxa were sampled using **r8s** version 1.7 [23]. Then, local coalescent histories and gene trees were sampled under the MSC model using **ms** [15].

To simulate local coalescent histories and gene trees under the variable region model θ_H, we follow the same procedures as the background model, except for the final local tree height h. The value of h is progressively increased to define the model conditions H.1, H.2, and H.3, each representing a higher level of local divergence.

To simulate local coalescent histories and gene trees under the variable region model θ_R, the same procedures are used as in the background model with one change: the MSCwR model is enabled by specifying the -r switch in **ms**. The finite-sites model of recombination is parameterized by a recombination rate r. To assess the impact of different recombination rates, we define the model conditions R.1, R.2, and R.3, each characterized by increasing recombination probabilities r.

For the variable region model θ_I, we begin by constructing a phylogenetic network. First, a random birth-death model tree with a height of 1.0 coalescent unit is sampled using **r8s**. Then, a single reticulation event is added by selecting a time t_M uniformly at random from the interval $(0, 1/4)$, following the method outlined in [25]. Next, **msmove** [10] is used to simulate local coalescent histories and gene trees under the MNSCwR model. To explore varying levels of introgression or gene flow, we construct the model conditions I.1, I.2, and I.3 by increasing the admixture probability β.

The variable region model θ_S is simulated using **SFS_CODE** [13], a simulation software package designed for modeling sequence evolution in populations under selection. The selection type was set to positive, indicating that mutations are beneficial with a probability of 1.0. To assess varying levels of selection intensity, we construct the model conditions S.1, S.2, and S.3, which are quantified by increasing selection coefficients γ and therefore progressively stronger selection pressure.

For the background model θ_B and all variable region models other than θ_S, local gene trees were deviated from ultrametricity using the method of [18] with a deviation factor of $c = 2.0$. Sequences were then simulated under a finite-sites nucleotide substitution and insertions/deletions (indels) model along these trees using INDELible v1.03 [7], with branch lengths converted following Eq. 3.1 from [12]. The substitution model follows the GTR model with base frequencies, using parameters from [9], while indels follow the medium gap length distribution from [19] with an indel rate of 0.02. For the variable region model S, sequence evolution was simulated using **SFS_CODE** [13], applying the same substitution model and indel rate but with a different gap length distribution. Full simulation commands are provided in the Supplementary Materials

Sequence regions evolving under variable region models are denoted V and those evolving under the background model are denoted as B. In our simulation procedure, regions V are randomly positioned within simulated sequences. For mixture models with a single variable region model, the total root sequence length L is set to 2000; for those with two variable region models, L is set to 4000. The root sequence length of a variable region V, denoted as L_v, follows a Gaussian distribution with a mean of 500 and a standard deviation of 100, and the variable region position is randomly selected within the range of 100 to $L - L_v - 100$.

For each model condition, the simulation and experimental procedures are repeated to obtain 20 independent replicates. Model condition parameters and summary statistics for the simulated datasets are presented in Table 1. Model parameters for the background model θ_B are set to $h = 1.0$, $r = 0$, $\beta = 0$, and $\gamma = 0$. Only parameters that differ from the background model are reported in Table 1.

Performance evaluation criteria. REVEAL's performance was assessed in terms of both type I and type II error using precision and recall. To construct the confusion matrix for calculating precision and recall, we compared the REVEAL's site class prediction against ground truth. The confusion matrix consists of four elements. True positives (TP) consist of sites within true variable regions V that are correctly identified. False positives (FP) consist of sites within true background regions B that are incorrectly classified as part of V. True negatives

Table 1. Model condition parameters and summary statistics for simulated datasets. Each θ_H-based model condition is named H.1 through H.3, reflecting a generally increasing order of evolutionary divergence and local model violation intensity, as explained in the text. The other model conditions are named similarly. The number of taxa was set to $N = 10$ for all model conditions. Average normalized Hamming distance ("ANHD"), gappiness ("Gap."), and MSA length are reported as an average for background and variable regions in each model condition. Results are reported across 20 experimental replicates per model condition.

Model condition	Parameters setting	Global MSA			MSA in region V		
		Length	ANHD	Gap.	Length	ANHD	Gap.
H.1	$h_1 = 2.0$	2938.6	0.521	0.314	828.8	0.622	0.416
H.2	$h_2 = 4.0$	3281.0	0.537	0.387	1194.3	0.696	0.599
H.3	$h_3 = 8.0$	3428.6	0.526	0.409	927.3	0.731	0.748
R.1	$r_1 = 0.01$	2818.6	0.518	0.286	702.4	0.510	0.288
R.2	$r_2 = 0.05$	2826.8	0.494	0.284	794.1	0.503	0.333
R.3	$r_3 = 0.1$	2939.5	0.510	0.307	818.5	0.500	0.347
I.1	$\beta_1 = 0.4$	2797.4	0.504	0.280	743.2	0.510	0.279
I.2	$\beta_2 = 0.5$	2804.7	0.507	0.284	757.2	0.500	0.269
I.3	$\beta_3 = 0.6$	2835.3	0.514	0.291	778.2	0.498	0.279
S.1	$\gamma_1 = 20$	3151.3	0.405	0.357	875.5	0.510	0.446
S.2	$\gamma_2 = 50$	3220.4	0.409	0.367	977.0	0.528	0.465
S.3	$\gamma_3 = 100$	3227.8	0.397	0.369	913.4	0.535	0.490
M.1	$m_1, m_2 \in \{h_1, r_1, \beta_1\}$	5677.8	0.515	0.293	1517.2	0.540	0.337
M.2	$m_1, m_2 \in \{h_2, r_2, \beta_2\}$	6132.4	0.522	0.336	1920.2	0.576	0.455
M.3	$m_1, m_2 \in \{h_3, r_3, \beta_3\}$	6739.3	0.524	0.385	2584.0	0.556	0.510

(TN) consist of sites within true background regions B that are correctly identified. False negatives (FN) consist of sites within true variable regions V that are mistakenly classified as part of B. Precision is then calculated as $\frac{TP}{TP+FP}$, and recall is defined as $\frac{TP}{TP+FN}$.

Computational runtime and peak memory usage was also reported for REVEAL. All experiments were conducted on the MSU Institute for Cyber-Enabled Research (ICER) High-Performance Computing Center (HPCC). We utilized HPCC computing nodes that were equipped with Intel Xeon Gold 6148 CPUs running at 2.40 GHz and between 5 and 10 GiB of memory.

2.3 Empirical Study

We also performed REVEAL analyses of two empirical datasets. The first dataset consists of genomic sequence data for wild-derived strains of house mouse (*Mus musculus*) and *M. spretus*. The clade is an emerging model of adaptive interspecific introgression [20] and genomic maps of adaptive introgression have been reported in past studies [20]. We downloaded whole genome sequences and genome-wide SNP data for classical and wild-derived mouse strains from the Mouse Genomes Project, where the house mouse genome version GRCm39 served as the reference genome. We used bcftools to filter out non-biallelic variants and retain only those with a missing genotype call rate of less than 10%. Haplotype phasing was then performed using SHAPEIT [3]. From the SHAPEIT output, we extracted SNP haplotype alignments along with corresponding genomic coordinate information. We focused exclusively on wild-derived inbred strains, as they better reflect natural genomic variation compared to classical mouse strains. The dataset includes wild-derived strains from five mouse species and/or subspecies: *M. spretus*, *M. musculus musculus*, *M. musculus castaneus*, *M. musculus molossinus*, and *Mus musculus domesticus* (Supplementary Table S1). Summary statistics for mouse dataset are presented in Supplementary Table S2. Detailed information about the mouse samples can be found in the Supplementary Appendix.

The second dataset came from [8]'s study of adaptive introgression in mosquito. See Supplementary Appendix for more details about the dataset and REVEAL analyses of the dataset.

3 Results

3.1 Simulation Study

Figure 2 shows precision and recall of REVEAL's site-level classification across all model conditions with a single variable region model for different values of n (the number of local resampling and re-estimation iterations for the local sequence). REVEAL's overall performance is high across the model conditions, with some minor variability. These results highlight the robustness and effectiveness of the framework under different evolutionary processes that can cause local model violations. Precision consistently ranges between 0.9 and 1.0, with the recall maintaining values above 0.85. Furthermore, neither precision nor recall shows a distinct upward or downward trend as the parameter of the variable region model increases under the same type of model conditions, with a minor exception noted for recall in model conditions H and S. In Fig. 2, we also compare the estimation performance under different values of n (i.e., the number of local resampling and re-estimation iterations for local sequences). Under the same type of model conditions, as the value of n increases, the estimation performance tends to stabilize or improve slightly on most model conditions.

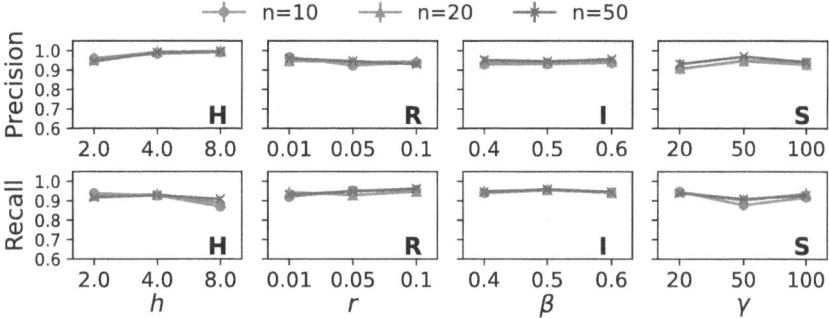

Fig. 2. Estimation performance of the REVEAL framework across model conditions with a single variable region model. All parameters for the REVEAL use default values except for n (the number of local resampling replicates). The mean and standard error of the Precision and Recall for the REVEAL estimation performance are shown for each model condition across 20 experimental replicates per model condition. The estimation performance is compared across three different amounts of resampling replication (n) used for the REVEAL framework.

The runtime of the REVEAL algorithm is primarily affected by the length of the input MSA, the step size for local sequence extraction p, and the number of local resampling replicates n. Figure 3 illustrates the runtime and memory usage of the REVEAL framework across various model conditions with different values of n. As expected, the runtime increases as the value of n rises across all model conditions. It reveals a trade-off between computational cost and performance stability as n increases. However, while the runtime grows substantially at $n = 50$, the improvement in performance is smaller. This indicates that setting n to 20 in practice suffices for achieving robust estimation performance without excessive runtime overhead. In contrast to runtime, memory usage remains consistent across all model conditions and values of n, fluctuating between 70 MB and 90 MB.

We also present precision and recall results and computational resource usage of the REVEAL framework for model conditions with two variable region models in Fig. 4. Precision remains consistently high across all model conditions – exceeding 0.9 – and shows a slight increase with greater evolutionary divergence

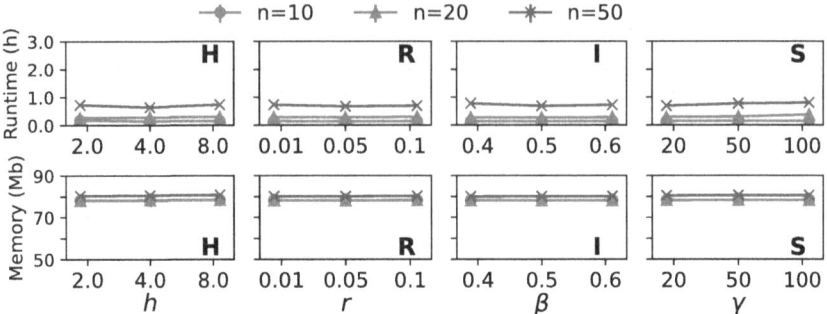

Fig. 3. Runtime and memory usage of the REVEAL Framework across model conditions with a single variable region model. All parameters for REVEAL use default values except for n (the number of local resampling replicates). For each model condition, the mean and standard error of the runtime and memory of the REVEAL framework across 20 experimental replicates per model condition are reported.

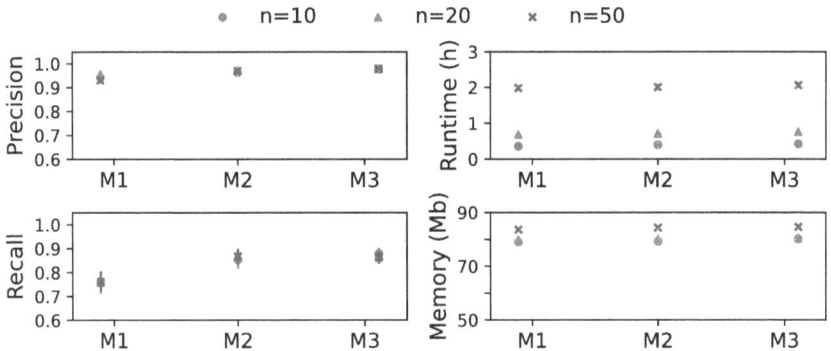

Fig. 4. Estimation performance, runtime, and memory usage of the REVEAL framework across model conditions with two variable region models. All parameters for REVEAL use default values except for n (the number of local resampling replicates). The mean and standard error of the REVEAL framework's precision, recall, runtime, and memory are reported across 20 experimental replicates for each model condition.

within the variable region models. While recall drops below 0.8 for model condition M.1, it improves as evolutionary divergence increases, reaching values above 0.85 for model conditions M.2 and M.3. This trend highlights REVEAL's ability to recover detection performance as sequence variability grows. Despite the added complexity of detecting and mapping two variable region models, REVEAL continues to demonstrate reliable detection performance, maintaining high precision and recall. Additionally, model conditions with two variable region models require a longer runtime compared to those with a single variable region model, primarily due to the increased length of the input multiple sequence alignment (MSA). However, memory usage remains stable across different model conditions.

3.2 Empirical Study

Figure 5 presents results for REVEAL's analysis of the mouse dataset. The local support value matrix C produced by REVEAL analysis is shown for selected regions of the house mouse genome. As a point of comparison, the figure also includes a visualization of introgression patterns identified by PhyloNet-HMM (adapted from [20]) – a statistical method that is specifically designed for mapping introgression patterns in genomes. PhyloNet-HMM performs inference and learning under a bespoke model that combines a multi-species coalescent model, a finite-sites substitution model, and a hidden Markov model. In the local phylogenetic matrix C of the REVEAL framework, we observe a marked decrease in local support values in and around the PhyloNet-HMM-inferred introgression regions, as compared to neighboring regions. We present the complete local phylogenetic matrix C for all mouse chromosomes in the Supplementary Materials.

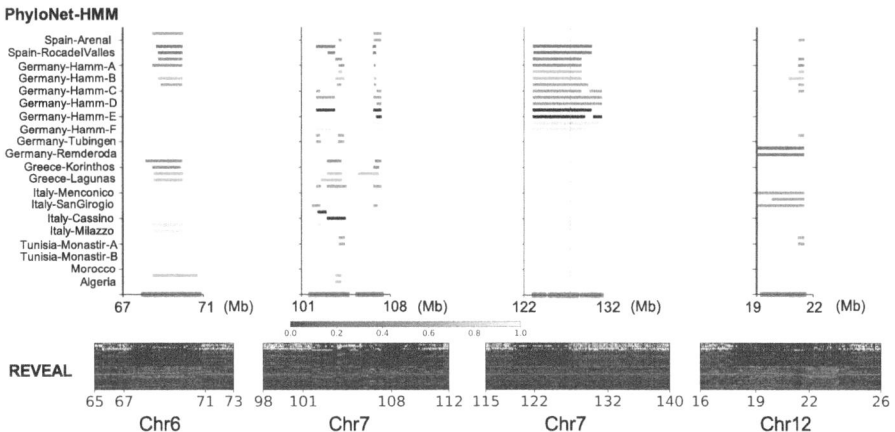

Fig. 5. Empirical study: a comparison of introgression patterns identified by the PhyloNet-HMM method in [20] and REVEAL's local support value matrix C. The top section of each panel illustrates selected introgression regions identified across 20 *M. m. domesticus* samples using PhyloNet-HMM (reproduced from Fig. 4 of [20]). Red squares along the x-axis indicate the locations of genes within these introgression regions. The bottom section of each panel shows the local support value matrix C from REVEAL's analysis, focusing on the regions where the PhyloNet-HMM model detected introgression.

4 Discussion

A key advantage of the REVEAL framework is that it operates without imposing any assumptions or restrictions on the evolutionary processes that cause model violations in genomic sequences. Unlike traditional approaches that are designed to detect one specific evolutionary process, it can detect a diverse range of local model violations without being constrained by predefined evolutionary models. Furthermore, it does not depend on supervised learning techniques and does not require manually labeled data, making it a more flexible and scalable solution for local model violation detection.

The simulation study validates the robustness and effectiveness of the REVEAL framework across a range of simulation and experimental conditions. REVEAL successfully detected local model violations caused by divergence heterogeneity, recombination, introgression, and natural selection. Even in more complex cases with multiple locally variable regions, REVEAL's precision remained consistently above 0.9, although recall experienced a slight decline. Importantly, as the intensity of local model violations increases, detection and site mapping performance improves. These results confirm that REVEAL can generalize across diverse evolutionary scenarios without requiring predefined assumptions.

We fully acknowledge that a fully parametric model-based analysis is expected to outperform a model-agnostic method like REVEAL, under specific assumptions including: (1) the correct model or a very accurate model can be assumed to be available a priori, and (2) one or a few closely-related sequence analysis tasks is/are under study. However, these assumptions are quite strong and constraining. If the assumed local model(s) are incorrect or inadequate, then model mis-specification can impair detection and mapping of local model variation. If different tasks are performed on a dataset in a study, then purpose-built models and model-based methods must be developed and applied for each task.

Rather, REVEAL points to a different and practical alternative. In lieu of requiring strong a priori modeling assumptions, REVEAL provides an automated, data-driven, global-model-agnostic, and local-model-free approach to detect and pinpoint local model variation as part of a biomolecular sequence analysis. The results of a REVEAL analysis can then (a) inform practitioners that the original global analysis is insufficient and requires modeling improvement, and (b) precisely map regions in the dataset that require locally variable models, resulting in improved model-based sequence analysis.

5 Conclusions

In this study, we introduce REVEAL, a general-purpose framework for detecting and mapping local model violations during biomolecular sequence analysis. Unlike existing approaches, REVEAL is readily adapted to different biomolecular sequence analysis tasks and requires no additional modeling assumptions beyond those required for traditional global sequence analysis. Our simulation experiments validate the robustness and effectiveness of REVEAL across various evolutionary scenarios and causes of local model violations. Furthermore, we apply REVEAL to two empirical datasets and identify widespread local model violation regions across chromosomes. These results corroborate findings in past empirical studies that were inferred using parametric model-based algorithms for narrowly specialized inference and learning tasks.

While this study provides an initial proof of concept, future research can unlock further algorithmic enhancements. For example, REVEAL's algorithmic formulation is generalizable to other biomolecular sequence analysis tasks such as genome rearrangement mapping and structural variant analysis.

This work has been supported by the NSF (CCF-1565719, CCF-1714417, DEB-1737898, and IOS-1740874 to KJL). Computational experiments and analyses were performed on the MSU High Performance Computing Center.

References

1. Shepherd, A.D., Klaere, S.: How well does your phylogenetic model fit your data? Systematic Biol. **68**(1), 157–167 (2019)
2. Adrion, J.R., Galloway, J.G., Kern, A.D.: Predicting the landscape of recombination using deep learning. Mol. Biol. Evol. **37**(6), 1790–1808 (2020)
3. Browning, B.L., Tian, X., Zhou, Y., Browning, S.R.: Fast two-stage phasing of large-scale sequence data. Am. J. Hum. Genet. **108**(10), 1880–1890 (2021)
4. Burgstaller-Muehlbacher, S., Crotty, S.M., Schmidt, H.A., Reden, F., Drucks, T., von Haeseler, A.: ModelRevelator: fast phylogenetic model estimation via deep learning. Mol. Phylogenet. Evol. **188**, 107905 (2023)
5. Dutheil, J.Y., Ganapathy, G., Hobolth, A., Mailund, T., Uyenoyama, M.K., Schierup, M.H.: Ancestral population genomics: the coalescent hidden markov model approach. Genetics **183**(1), 259–274 (2009)
6. Felsenstein, J.: Confidence limits on phylogenies: an approach using the bootstrap. Evolution **39**(4), 783–791 (1985)
7. Fletcher, W., Yang, Z.: INDELible: a flexible simulator of biological sequence evolution. Mol. Biol. Evol. **26**(8), 1879–1888 (2009)
8. Fontaine, M.C., et al.: Extensive introgression in a malaria vector species complex revealed by phylogenomics. Science **347**(6217), 1258524 (2015)
9. Gao, M., Liu, K.J.: Statistical analysis of GC-biased gene conversion and recombination hotspots in eukaryotic genomes: a phylogenetic hidden Markov model-based approach. In: Proceedings of the 12th ACM Conference on Bioinformatics, Computational Biology, and Health Informatics, pp. 1–24 (2021)
10. Garrigan, D., Geneva, A.: MSMOVE: a modified version of Hudson's coalescent simulator MS allowing for finer control and tracking of migrant genealogies (2014)
11. Hartigan, J.A., Wong, M.A.: Algorithm AS 136: a k-means clustering algorithm. J. Royal Stat. Soc. Ser. C (Appl. Stat.) **28**(1), 100–108 (1979)
12. Hein, J., Schierup, M., Wiuf, C.: Gene genealogies, variation and evolution: a primer in coalescent theory. Oxford University Press, USA (2004)
13. Hernandez, R.D.: A flexible forward simulator for populations subject to selection and demography. Bioinformatics **24**(23), 2786–2787 (2008)
14. Hobolth, A., Christensen, O.F., Mailund, T., Schierup, M.H.: Genomic relationships and speciation times of human, chimpanzee, and gorilla inferred from a coalescent hidden Markov model. PLoS Genet. **3**(2), e7 (2007)

15. Hudson, R.R.: Generating samples under a Wright-Fisher neutral model of genetic variation. Bioinformatics **18**(2), 337–338 (2002)
16. Jayaswal, V., Wong, T.K., Robinson, J., Poladian, L., Jermiin, L.S.: Mixture models of nucleotide sequence evolution that account for heterogeneity in the substitution process across sites and across lineages. Syst. Biol. **63**(5), 726–742 (2014)
17. Li, Y., Chen, S., Rapakoulia, T., Kuwahara, H., Yip, K.Y., Gao, X.: Deep learning identifies and quantifies recombination hotspot determinants. Bioinformatics **38**(10), 2683–2691 (2022)
18. Liu, K., Raghavan, S., Nelesen, S., Linder, C.R., Warnow, T.: Rapid and accurate large-scale coestimation of sequence alignments and phylogenetic trees. Science **324**(5934), 1561–1564 (2009)
19. Liu, K., et al.: SATe-II: very fast and accurate simultaneous estimation of multiple sequence alignments and phylogenetic trees. Syst. Biol. **61**(1), 90 (2012)
20. Liu, K.J., Steinberg, E., Yozzo, A., Song, Y., Kohn, M.H., Nakhleh, L.: Interspecific introgressive origin of genomic diversity in the house mouse. Proc. Natl. Acad. Sci. **112**(1), 196–201 (2015)
21. Ray, D.D., Flagel, L., Schrider, D.R.: IntroUNET: Identifying introgressed alleles via semantic segmentation. PLoS Genet. **20**(2), e1010657 (2024)
22. Rodriguez, F., Oliver, J., Marin, A., Medina, J.: The general stochastic model of nucleotide substitution. J. Theor. Biol. **142**, 485–501 (1990)
23. Sanderson, M.J.: r8s: inferring absolute rates of molecular evolution and divergence times in the absence of a molecular clock. Bioinformatics **19**(2), 301–302 (2003)
24. Wang, W., Hejasebazzi, A., Zheng, J., Liu, K.J.: Build a better bootstrap and the RAWR shall beat a random path to your door: phylogenetic support estimation revisited. Bioinformatics **37**(Supplement_1), i111–i119 (2021)
25. Wuyun, Q., VanKuren, N.W., Kronforst, M., Mullen, S.P., Liu, K.J.: Scalable statistical introgression mapping using approximate coalescent-based inference. In: Proceedings of the 10th ACM International Conference on Bioinformatics, Computational Biology and Health Informatics, pp. 504–513 (2019)
26. Zhao, H., Souilljee, M., Pavlidis, P., Alachiotis, N.: Genome-wide scans for selective sweeps using convolutional neural networks. Bioinformatics **39**(Supplement_1), i194–i203 (2023)

A Simple Way to Find Related Sequences with Position-Specific Probabilities

Martin C. Frith[1,2,3]

[1] Department of Computational Biology and Medical Sciences, University of Tokyo, Chiba 277-8568, Japan
[2] Artificial Intelligence Research Center, AIST, Tokyo 135-0064, Japan
[3] AIST-Waseda University Computational Bio Big-Data Open Innovation Laboratory, Tokyo 169-8555, Japan
mcfrith@edu.k.u-tokyo.ac.jp

Abstract. One way to understand biology is by finding genetic sequences that are related to each other. Often, a family of related sequences has position-varying probabilities of substitutions, insertions, and deletions: we can use these to find distant and subtle relationships. Current software tools for this task either do not use all probability evidence (e.g. PSI-BLAST, MMseqs2), or have excessive complexity and minor biases (e.g. HMMER). This complexity inhibits fertile development of alternative tools. This study describes a simplest reasonable way to find related sequences with position-specific probabilities, using all probability evidence. The algorithms likely use the fewest operations that such algorithms possibly could.

A full version of this paper is available at bioRxiv: https://doi.org/10.1101/2025.03.14.643233

Keywords: Homology · Alignment · Probability · Sequence · Profile

1 Introduction

This study was motivated by seeking subtly-related DNA sequences. We recently found gene-regulating DNA conserved between humans, molluscs, arthropods, and even corals [4]. These DNA regions all control gene expression in embryonic development: a control system conserved since early Precambrian animals.

These sequences have position-varying rates of substitutions, insertions, and deletions, for example, a conserved "CCAAT box" near the right of Fig. 1. This suggests we could find more-subtly related sequences by using these position-specific probabilities. This is a classic approach: a set of position-specific rates is termed a "profile".

The currently popular profile search methods are of two types. The first is exemplified by PSI-BLAST: it uses position-specific rates of substitution, but its insertion and deletion parameters are position-nonspecific and ad hoc. Also,

```
gggtggtctc-tctac-tttggtgagctgttgctaagcagctaataatag---tcatgtttgctgagtaatttctgccct-ccgcgagccaatcg  human PTCH1
gggtggtctc-tctac-tttggtgagctgttgctaagcagctaataatag---tcgtgtttgctgagtaatttctgcct--tcctcggccaatga  chimaera ptch1
gggtggtcct-ggccac-tttggtgagctgttgcttagcagctaataatagt--agatgtttgctaagtaatttgcttcttcccttttagccaatgg chimaera ptch2
gtgtggtctg-tcttc-tttggcgagccgtttcctagcagcaaatgacggg--ccgtgttactcgagcaaattgtggag--cccgcggccaatca  lancelet
gtgtggtcta-cctcttttttctggtgacgttactaagcaactaatgagaac--tcgtgttacatgagtcatttggccg---tgagtgaccaatca  acorn worm
gtgtggtcttctccac-tttagcgaaccgttgcttggcaactaatgcaaggttcgctgttacgagaggagctagaaag---ccacgagccaatcg  horseshoe crab
gtgtggtcttctatac-tttagtaagccgttgcttggcaactaatacaaggctcgctattacgagaggagctagaaag---ccacgagccaatcg  horseshoe crab
gggtggtctt-tctac-tttggtgagccgttgcctggcaactaatgtgcagc-tgatgttacttgagcaatctgcgcta--taattgaccaatca  chiton
gggtggtcta-tctac-ttgggtgagctgttgctaagcaactaatgttgggc-tggtgttacacgagcaatcctactga--taattgaccaatca  sea hare
gggtggtctg--ctac-ctgggggagctgttgccaggcaactaatgttgggc-tggtgttacacgagcaatcctactga--taattgaccaatca  abalone
gggtggtctt-tctac-tttggtgagctgttgctaggcaactaatgtcgggc-tgcagttacatgagctgcctgtgggc--ttaagaaccaatca  shamisen shell
```

Fig. 1. Alignment between promoter DNA sequences of *Ptch* (patched) genes in some animals [4].

it considers just one alignment between potentially related regions. It would be more powerful to combine the evidence from alternative ways of aligning them.

The other type of method uses position-specific probabilities of substitution, insertion, and deletion, and combines alternative alignments. The de facto standard tool seems to be HMMER, which is complex and quirky [1]. The perceived need for this complexity might explain why few of the many sequence comparison tools use this full-probability approach. The present study shows that this complexity is not necessary.

All these tools produce a similarity score, indicating the strength of evidence for related sequences. It is important to know the probability of a similarity score occurring by chance, in a random sequence. Unfortunately, this is hard to determine. There is a solution for gapless local alignment, in the limit of infinitely long sequences [5]. Gapped alignment relies on conjecture and simulations, which limit PSI-BLAST to its simple profile search. HMMER uses conjectures that apply to its sophisticated profile search. The scope of HMMER's conjectures is unclear: their practical application does not always give accurate results [1], and they seem to contradict the gapless solution described by Karlin and Altschul [5].

This study shows a minimal, bias-free way to find sequence parts related to a profile, using position-specific substitution, deletion, and insertion rates, and combining evidence from alternative alignments. It defines E-values as: the expected number of distinct sequence regions with score $\geq s$, if we compare a profile to a random sequence. It extends a recent position-nonspecific method [3]: the non-obvious aspect of this extension is how precisely to define alignment probabilities.

2 Methods

We wish to find related parts between a profile of length m, and a sequence of length n ($Q_0, Q_1, \ldots, Q_{n-1}$). The profile has these letter and gap probabilities:

$\theta_i(y)$ $(0 \leq i < m)$ probability of letter type y at position i
$\psi(y)$ background probability of letter type y
δ_i $(0 \leq i < m)$ probability of starting a deletion
ϵ_i $(0 < i \leq m)$ probability of extending a deletion
α_i $(0 \leq i \leq m)$ probability of starting an insertion
β_i $(0 \leq i \leq m)$ probability of extending an insertion

The complete, precise definition of the profile's probabilities is shown in Fig. 2.

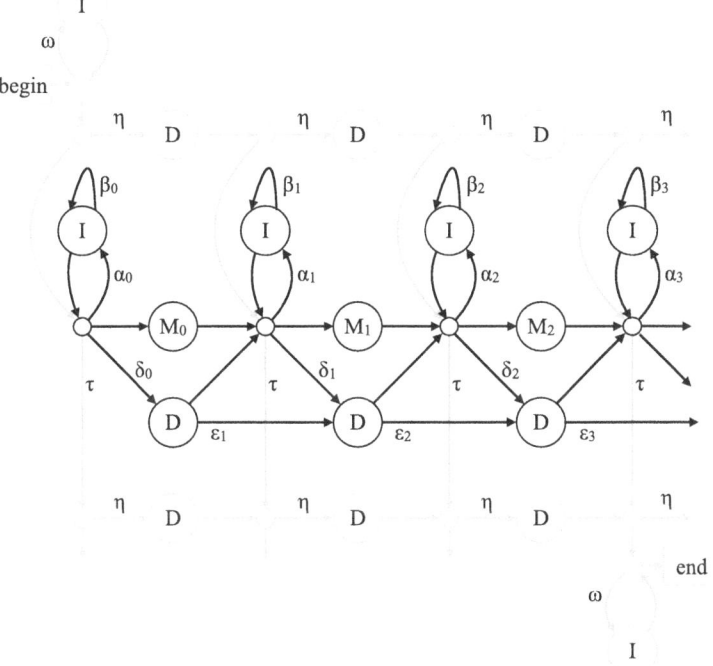

Fig. 2. A scheme for assigning probability to: a sequence with one segment aligned to a profile. The black part handles alignment to the profile; the grey parts handle unaligned flanks. In this example, the profile has length $m = 3$. Any path from **begin** to **end** represents an alignment: its probability is the product of the arrow probabilities (for example: ω, η) and letter probabilities. Each pass through an **I** matches the next letter in the sequence, with probability $\psi(y)$ for letter type y. Passing through M_i matches the next letter in the sequence, with probability $\theta_i(y)$.

To combine evidence from alternative alignments, we can define a similarity score as a sum of probability ratios:

$$\text{score} = \log \sum_{\text{alignments}} \frac{\text{prob(alignment)}}{\text{prob(length-0 alignment)}}. \tag{1}$$

Here, prob(alignment) is the product of probabilities of insertions, deletions, aligned letters, and unaligned letters.

There are several choices for which alignments to include in the sum. Yu and Hwa suggested an "end-anchored" approach: include all alignments that start anywhere and end at some fixed coordinates (i, j) in the profile and the sequence [6]. Another "mid-anchored" approach includes all alignments that pass through some fixed coordinates (i, j) [3].

The following algorithm calculates the end-anchored sum (w), over all alignments that end at (i, j). So the score is: $\log w$. The mid-anchored sum is the product of the end- and start-anchored sums.

$$\begin{aligned}
& X_{-1\ j-1} \leftarrow 0 \qquad Y_{-1\ j} \leftarrow 0 & & 0 \leq j \leq n \\
& X_{i-1\ -1} \leftarrow 0 \qquad Z_{i\ -1} \leftarrow 0 & & 0 \leq i \leq m \\
& \left. \begin{aligned} & w \leftarrow X_{i-1\ j-1} + Y_{i-1\ j} + Z_{i\ j-1} + 1 \\ & X_{ij} \leftarrow S'_i(Q_j) w \\ & Y_{ij} \leftarrow d'_i w + e'_i Y_{i-1\ j} \\ & Z_{ij} \leftarrow a'_i w + b'_i Z_{i\ j-1} \end{aligned} \right\} & & \begin{aligned} 0 \leq i \leq m \\ 0 \leq j \leq n \end{aligned}
\end{aligned}$$

The algorithm parameters in terms of the probabilities are:

$$\begin{aligned}
a'_i &= \alpha_i(1 - \beta_i) & (0 \leq i \leq m) \tag{2} \\
b'_i &= \beta_i & (0 \leq i \leq m) \tag{3} \\
d'_i &= \delta_i(1 - \epsilon_{i+1}) & (0 \leq i < m) \tag{4} \\
e'_i &= \epsilon_i(1 - \epsilon_{i+1})/(1 - \epsilon_i) & (0 < i < m) \tag{5} \\
S'_i(y) &= (1 - \alpha_i - \delta_i)\theta_i(y)/\psi(y) & (0 \leq i < m) \tag{6}
\end{aligned}$$

This assumes that $\tau \approx 0$ and $\omega \approx 1 \approx \eta$. These choices for ω, η, and τ mean there is no alignment length bias [2].

If we use end-, start-, or mid-anchored scores, with no length bias, we can estimate the E-value for a score s to occur in a random sequence of length n, with letter frequencies $\psi(y)$. Based on a conjecture of Yu and Hwa [6]:

$$E = Kn/\exp(s). \tag{7}$$

K is an unknown parameter, which is typically different for different profiles. It can be estimated by fitting to similarity scores of random sequences [3].

Disclosure of Interests. The author has no competing interests to declare that are relevant to the content of this article.

References

1. Eddy, S.R.: A probabilistic model of local sequence alignment that simplifies statistical significance estimation. PLoS Comput. Biol. **4**(5), e1000069 (2008)
2. Frith, M.C.: How sequence alignment scores correspond to probability models. Bioinformatics **36**(2), 408–415 (2020)
3. Frith, M.C.: A simple method for finding related sequences by adding probabilities of alternative alignments. Genome Res. **34**(8), 1165–1173 (2024)
4. Frith, M.C., Ni, S.: DNA conserved in diverse animals since the Precambrian controls genes for embryonic development. Mol. Biol. Evol., msad275 (2023)
5. Karlin, S., Altschul, S.F.: Methods for assessing the statistical significance of molecular sequence features by using general scoring schemes. Proc. Natl. Acad. Sci. **87**(6), 2264–2268 (1990)
6. Yu, Y.K., Hwa, T.: Statistical significance of probabilistic sequence alignment and related local hidden markov models. J. Comput. Biol. **8**(3), 249–282 (2001)

Position Specific Scoring Is All You Need? Revisiting Protein Sequence Classification Tasks

Sarwan Ali[1(✉)], Taslim Murad[2], Prakash Chourasia[2], Haris Mansoor[3], Imdad Ullah Khan[3], Pin-Yu Chen[4], and Murray Patterson[2]

[1] Columbia University, New York, USA
sa4559@columbia.edu
[2] Georgia State University, Atlanta, USA
{tmurad2,pchourasia1}@student.gsu.edu, mpatterson30@gsu.edu
[3] Lahore University of Management Sciences, Lahore, Pakistan
{16060061,imdad.khan}@lums.edu.pk
[4] IBM T. J. Watson Research Center, Yorktown Heights, USA
pin-yu.chen@ibm.com

Abstract. Understanding the structural and functional characteristics of proteins are crucial for developing preventative and curative strategies that impact fields from drug discovery to policy development. An important and popular technique for examining how amino acids make up these characteristics of the protein sequences with position-specific scoring (PSS). While the string kernel is crucial in natural language processing (NLP), it is unclear if string kernels can extract biologically meaningful information from protein sequences, despite the fact that they have been shown to be effective in the general sequence analysis tasks. In this work, we propose a weighted PSS kernel matrix (or W-PSSKM), that combines a PSS representation of protein sequences, which encodes the frequency information of each amino acid in a sequence, with the notion of the string kernel. This results in a novel kernel function that outperforms many other approaches for protein sequence classification. We perform extensive experimentation to evaluate the proposed method. Our findings demonstrate that the W-PSSKM significantly outperforms existing baselines and state-of-the-art methods in terms of predictive performance.

Keywords: Sequence Classification · Representation Learning · Embedding Generation · Position Weight Matrix

1 Introduction

Protein sequence classification is a fundamental problem in bioinformatics, with applications in protein structure or function prediction, viral host specificity, and drug design [10,21,26]. Various feature-engineering-based methods to assist protein classification by mapping sequences to numerical form are proposed, for example, [13] creates one-hot encoding vectors. Other similar techniques [1,17] work by taking the k-mer frequencies and position distribution information into account. Machine learning classifier uses these embeddings. However, the final prediction performance of such a pipeline may be affected by the particular properties associated with each embedding. For example, [13] faces the curse of dimensionality challenge and lacks model information regarding the relative positions of amino acids. Similarly, [1,17] are computationally expensive and can exhibit sparsity issues.

Several neural networks and transformer-based methods have been popularly used such as WDGRL [22], AutoEncoder [27], and Evolutionary Scale Modeling or ESM-2 [15]. Additionally, several pre-trained models for protein classification, such as Protein Bert. [3], Seqvec [8], UDSMProt [24], TAPE [20] etc. are proposed.

Another widely followed approach to protein sequence classification is machine learning algorithms that make use of kernel functions, particularly support vector machines (SVMs), due to their ability to handle highly dimensional data and their robustness to noise [28]. The kernel function—a key component of approaches such as SVM—defines the similarity between pairs of sequences. For the classification of protein sequences, kernel-based methods are typically favored over representation-based methods for the following reasons:

- Protein sequences are complex, and contain structures (e.g., secondary or tertiary structure) that may not be captured by the straightforward representation-based embeddings. To handle this complexity, kernel-based techniques convert the sequences into a higher dimensional latent feature space. This feature space can be constructed by taking into account such complicated patterns or structures to classify the sequences [2].
- Managing numerous forms of information: Protein sequences contain a variety of data, including information about the secondary structure, evolutionary relationships, and the makeup composition of amino acids. By utilizing multiple kernel functions, different sorts of information can be captured using kernel-based methods.
- Better performance: It has been demonstrated that for protein sequence classification, kernel-based algorithms outperform representation-based methods in vasrios research works [14,25]. This is because kernel-based approaches can successfully handle the complexity of the sequences and collect more information, as mentioned above.

Although several efforts have been made to propose string kernel methods in the literature [2], these methods are general-purpose, i.e., they are not designed to consider the specific nature of protein sequences. To bridge this gap, we propose a novel kernel function for protein sequence classification called the weighted position-specific scoring kernel matrix (W-PSSKM).

Our contributions to this paper are the following:

1. We propose a kernel function, called W-PSSKM, to efficiently design a kernel matrix specifically for protein sequences.
2. Using kernel PCA, we design feature embeddings that enable the use of our kernel matrix with a wide variety of non-kernel classifiers (along with kernel classifiers, such as SVM) for supervised analysis of protein sequences.
3. By demonstrating the performance of our kernel function on different real-world protein sequence datasets, we show that the W-PSSKM achieves high predictive

accuracy and outperforms recent baselines and state-of-the-art methods from the literature.

The rest of the paper is organized as follows: Sect. 2 provides related work followed by Sect. 3 with the proposed approach. Section 4 describes the data sources, baseline methods, and evaluation techniques. Section 5 presents the results of the study. Finally, we conclude in Sect. 6 summarizing the findings, our contributions, and possible future directions for research.

2 Related Work

Biological sequence study is a popular topic in research, like protein analysis [4] is essential for inferring its functional and structural properties, which helps in understanding diseases and building prevention mechanisms like drug discovery, etc. Various feature embedding-based methods are put forward to gain a deeper understanding of the biological sequences like [13] proposed a one-hot encoding technique to classify spike protein sequences. PWKmer [17] method uses position distribution information and k-mers frequencies to do a phylogenetic analysis of HIV-1 viruses. However, these methods are computationally expensive and can face the curse of dimensionality challenge.

Another popular approach includes methods employing neural networks to generate numerical representations like WDGRL [22], AutoEncoder [27], and Evolutionary Scale Modeling or ESM-2 [15] etc. WDGRL is an unsupervised technique that uses a neural network to extract numerical embeddings from the sequences. AutoEncoder follows the encoder-decoder architecture and the encoder network yields the feature embeddings for any given sequence. However, they require training data. Although ESM-2 uses advanced language models to predict protein functions, it may not emphasize evolutionary conservation to the same extent. A set of pre-trained models to deal with protein classification are also introduced like Protein Bert [3], Seqvec [8], UDSMProt [24], TAPE [20] etc. In Protein Bert an NN model is trained using protein sequences and this pre-trained model can be employed to get embeddings for new sequences. Likewise, SeqVec provides a pre-trained deep language model for generating protein sequence embeddings. In UDSMProt a universal deep sequence model is put forward which is pre-trained on unlabeled protein sequences from Swiss-Prot and further fine-tuned for the classification of proteins. The Tasks Assessing Protein Embeddings (TAPE) [20] framework introduces five semi-supervised learning tasks relevant to protein biology, each with specific training, validation, and test splits to ensure meaningful biological generalization. However, all these methods have heavy computational costs.

Several kernel-based analysis techniques are proposed, like gapped k-mer (Gkm) string kernel [5] enables the usage of string inputs (biological sequences) for training SVMs. It determines the similarity between pairs of sequences using gapped k-mers, which eradicates the sparsity challenge associated with k-mers. However, the interpretation of gkmSVMs can be challenging. GkmExplain [23] is an extension of Gkm which claims to be more efficient in performance. The string kernel [16] is a kernel function that is based on the alignment of substrings in sequences but it's space inefficient.

Likewise, in recent years, there has been growing interest in kernel functions that are based on the position-specific scoring matrix (PSSM) representation of protein sequences. The PSSM is a widely-used representation of protein sequences that encodes information about the frequency and conservation of each amino acid in a sequence relative to a set of multiple sequence alignments [9]. The PSSM-based kernel functions have several advantages over other kernel functions, including their ability to capture more complex relationships between sequences.

3 Proposed Approach

In this section, we first discuss the algorithm and overall pipeline used to generate the proposed weighted kernel matrix (W-PSSKM).

3.1 Proposed Algorithmn

The generation of the Kernel matrix for protein sequences is the aim of this research work. The suggested method's algorithmic pseudocode is provided in Algorithms 1 and 2. Moreover, Fig. 1 depicts the overall pipeline for our proposed method. There are three main steps involved in generating W-PSSKM:

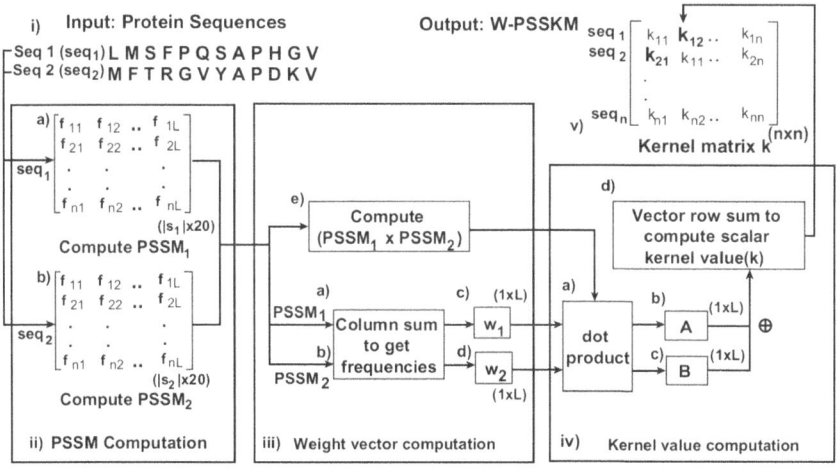

Fig. 1. Flow chart for the process of constructing a W-PSSKM.

Step 1: For the pair of given sequences (seq_1 and seq_2) and set of unique Amino Acids L (where $L = 20$ for protein sequences), we first compute the position-specific scoring matrix (PSSM) using Algorithm 2, which is called as a function in lines 6 and 7 of Algorithm 1 (also shown in Fig. 1-ii). In Algorithm 2, we scan the given protein sequence and increment the respective row (position of amino acid in given sequence) and column (position of amino acid among 20 unique amino acids i.e. "ACDEFGHIKLMN-PQRSTVWY") value of position specific scoring matrix (performed using AAIndex function). Note that we use data padding in the scenario where sequences do not have fixed lengths. We noted that this step did not change the behavior of the supervised analysis in the case where we have unaligned sequence data.

Step 2: After getting the position-specific scoring matrix (lines 6 and 7 of Algorithms 1) for two given sequences (i.e. $PSSM_1$ and $PSSM_2$), these matrices are used (column sum) to get the frequency count of each amino acid ($freq_1$, $freq_2$) for both sequences (see lines 8 and 9 in Algorithms 1). Then, the Weight vectors (i.e. w_1 and w_2) are computed by normalizing the frequencies as shown in lines 10 and 11 in Algorithm 1.

Step 3: Finally, the kernel value is computed by first taking the element-wise product of $PSSM_1$ and $PSSM_2$ matrices and then multiplying with respective weight vectors (i.e. w_1 and w_2). This will give us the vectors A and B (see lines 12 and 13) in Algorithm 1. The A and B vectors are added together and their sum is taken to get the resultant kernel Eq. 1. The complete flow of kernel value computation between two sequences is given in Fig. 1 (in the appendix), Note that line 16 in Algorithm 1 assigns the value to the lower half of the Kernel matrix since it is a symmetric matrix.

$$k(x_i, x_j) = \mathbb{1}^T \{(PSSM_i \odot PSSM_j)w_i \\ +(PSSM_j \odot PSSM_i)w_j\} \quad (1)$$

where $\mathbb{1}$ is a column vector of ones used to perform summation. Suppose,

$$A = (PSSM_1 \odot PSSM_2)w_1 = [a_1, a_2, ... a_s]^T \quad (2)$$

$$B = (PSSM_2 \odot PSSM_1)w_2 = [b_1, b_2, ... b_s]^T \quad (3)$$

Then,

$$k(x_i, x_j) = \mathbb{1}^T(A + B) = \sum_{i=1}^{s}(a_i + b_i) \quad (4)$$

After generating the kernel matrix as shown in Fig. 1 (in the appendix) and Algorithm 1, we use kernel PCA [11] to design the feature embeddings for protein sequences using the principal components. This enables us to use non-kernel classifiers (e.g. KNN) and compare them with the kernel classifiers (e.g. SVM) to perform supervised analysis on protein sequences. For SVM, we use W-PSSKM as input while for all other classifiers, we use kernel-PCA-based embeddings as input to perform supervised analysis.

In practice, a large data set leads to a large K, and storing K may become a problem. Kernel-PCA can help in this regard to convert K into a low-dimensional subspace. Since our data is highly non-linear, kernel PCA can find the non-linear manifold. Using the W-PSSKM kernel, the originally linear operations of PCA are performed in a reproducing kernel Hilbert space. It does so by mapping the data into a higher-dimensional space but then turns out to lie in a lower-dimensional subspace of it. So Kernel-PCA increases the dimensionality to be able to decrease it. Using Kernel-PCA, we compute the top principal components from K and use them as embeddings for supervised analysis as input for any linear and nonlinear classifiers.

Algorithm 1. Weighted PSSM Kernel Matrix (W-PSSKM)

Input: Set of Protein Sequences ($sequences$)
Output: Kernel Matrix (K)
1: **for** $ind_1 \in 1 : |sequences|$ **do**
2:     **for** $ind_2 \in ind_1 : |sequences|$ **do**
3:         **if** $ind_1 \leq ind_2$ **then**   ▷ Only upper triangle
4:             $seq_1 \leftarrow sequences[ind_1]$
5:             $seq_2 \leftarrow sequences[ind_2]$
6:             $PSSM_1 \leftarrow \textsc{ComputePSSM}(seq_1)$
7:             $PSSM_2 \leftarrow \textsc{ComputePSSM}(seq_2)$
8:             $freq_1 \leftarrow \textsc{ColumnSum}(PSSM_1)$
9:             $freq_2 \leftarrow \textsc{ColumnSum}(PSSM_2)$
10:            $\boldsymbol{w_1} \leftarrow \frac{freq_1}{\text{sum}(freq_1)}$ ▷ Compute weight vec.
11:            $\boldsymbol{w_2} \leftarrow \frac{freq_2}{\text{sum}(freq_2)}$ ▷ Compute weight vec.
12:            $A \leftarrow (PSSM_1 \odot PSSM_2)\boldsymbol{w_1}$
13:            $B \leftarrow (PSSM_2 \odot PSSM_1)\boldsymbol{w_2}$
14:            $\boldsymbol{V} \leftarrow A + B$
15:            K$[ind_1, ind_2] \leftarrow \textsc{Sum}(\boldsymbol{V}) = \sum_{i=1}^{s}(a_i + b_i)$
16:            K$[ind_2, ind_1] \leftarrow$ K$[ind_1, ind_2]$ ▷ Symmetry
17:         **end if**
18:     **end for**
19: **end for**
20: **return** K

Algorithm 2. Compute PSSM Matrix

1: **function** $\textsc{ComputePSSM}(sequence)$
2:     $s \leftarrow \text{len}(sequence)$ ▷ Compute length of sequence
3:     $L = 20$ ▷ Unique Amino Acid Count
4:     $PSSM \leftarrow \text{zeros}(s, L)$ ▷ Initialize PSSM matrix
5:     **for** $i \leftarrow 1$ **to** s **do**
6:         $intAA \leftarrow \textsc{AAIndex}(sequence[i])$
7:         $PSSM[i, intAA] + +$
8:     **end for**
9:     **return** $PSSM$
10: **end function**

4 Experimental Evaluation

We use the Coronavirus Host protein sequence dataset for the experimentation, whose summary is given in Table 1. We use several baseline and state-of-the-art (SOTA) methods to compare performance with the proposed W-PSSKM kernel. These baseline and SOTA methods are summarized in Table 2.

All experiments are conducted using an Intel(R) Xeon(R) CPU E7-4850 v4 @ 2.10 GHz having Ubuntu 64-bit OS (16.04.7 LTS Xenial Xerus) with 3023 GB memory. Moreover, we used 70–30% train-test data split with 10% data from the training set used as a validation set for hyperparameters tuning. We repeat the experiments 5 times (on random splits) and report average and standard deviation (std.) results.

Table 1. Dataset Statistics for three datasets used in performing the evaluation.

Name	Seq.	Classes	Sequence Statistics			Reference	Description
			Max	Min	Mean		
Coronavirus Host	5558	21	1584	9	1272.36	ViPR [19], GISAID [6]	The spike sequences belonging to various clades of the Coronaviridae family accompanied by the infected host label e.g. Humans, Bats, Chickens, etc.

Table 2. The summary of all the baseline methods which are used to perform the evaluation.

Category	Method	Description
Feature Engineering based embedding	Spike2Vec	It creates the numerical feature embeddings of a sequence by taking the frequencies of its k-mers into account
	PWM2Vec	This approach creates the numerical form of the biological sequences by assigning weights to the amino acids in a k-mer based on the location of the amino acids in the k-mer's position weight matrix (PWM)
	Spaced k-mer	For a sequence, it creates the feature vector by counting the number of its spaced k-mers.
Neural Network based embedding	WDGRL	This unsupervised technique uses a neural network to extract numerical embeddings from the sequences.
	Autoencoder	It follows an auto-encoder architecture-based neural network to extract the embeddings. The output of the encoder contains the numerical embeddings for a given spike sequence.
	ESM-2	ESM-2 is a model trained to predict masked amino acids in protein sequences, learning from billions of such predictions to capture evolutionary information.
Kernel Function	String Kernel	String kernel uses a number of matched and mismatched k-mers to calculate the similarity between two sequences.
End-To-End Deep Learning	LSTM	2 LSTM layer with 200 Units, Leaky Layer with alpha=0.05, dropout with 0.2 with Sigmoid Activation Function and ADAM Optimizer
	CNN	2 Conv. layers with 128 Filters, Kernel size 5, a Leaky Layer with alpha=0.05, a max pooling layer of size 2, a Sigmoid Activation Function along with ADAM Optimizer
	GRU	1 GRU layer with 200 Units, a Leaky Layer with alpha=0.05, dropout with 0.2 as well as Sigmoid Activation Function with ADAM Optimizer
Pretrained Language Models	SeqVec	A pre-trained language model (using ELMO) that takes protein sequences as input and generates vector representation as output.
	Protein Bert	It is a deep language pre-trained model based on the transformer specifically designed for proteins.
	TAPE	LLM model with a self-supervised pretraining method for molecular sequence embedding generation.

Classification is performed using Support Vector Machine (SVM), Naive Bayes (NB), Multi-Layer Perceptron (MLP), K-Nearest Neighbors (KNN), Random Forest

(RF), Logistic Regression (LR), and Decision Tree (DT) ML models. The evaluation metrics used for assessing the performance of these ML models are average accuracy, precision, recall, F1 (weighted), F1 (macro), Receiver Operator Characteristic Curve Area Under the Curve (ROC AUC), and training runtime. Since we have multi-class classifications, the one-vs-rest approach is used for ROC AUC computation.

5 Results and Discussion

In this section, we present the classification results for the proposed W-PSSKM and compare them with the results of baselines and SOTA methods on multiple datasets.

The average classification results (of 5 runs) for Coronavirus Host data are reported in Table 3. We can observe that the proposed W-PSSKM outperforms baselines and SOTA methods for all but one evaluation metric (training runtime). On comparing the average accuracy, when compared to feature engineering-based techniques (Spike2Vec, PWM2Vec, Spaced k-mers), W-PSSKM shows up to 10.1% improvement in comparison to the second best (Spike2Vec with random forest and logistic regression). In comparison to NN-based models (WDGRL and Autoencoder), the W-PSSKM achieved up to 15.5% improvement to the second best (Autoencoder with random forest).

In Table 3, compared to the string kernel method, which is designed while focusing on NLP problems in general, W-PSSKM achieves up to 28.7% improvement than the second best (string kernel with random forest). Moreover, while comparing to pre-trained language models for protein sequences (SeqVec and Protein Bert), the W-PSSKM achieves up to 12.2% improvement compared to the second best (SeqVec with random forest classifier). Overall, we can see that the proposed W-PSSKM significantly outperforms different types of baselines and SOTA methods from the literature.

The standard deviation (std) results (of 5 runs) for Coronavirus Host data are reported in Table 4. With the exception of the training runtime, the std values for all evaluation metrics are significantly lower for all evaluation metrics, baselines, SOTA, and the proposed W-PSSKM method. For training runtime, the WDGRL method with the Naive Bayes classifier takes the least time due to the fact that its embedding dimension is the lowest among others. Compared to the end-to-end deep learning models (LSTM, GRU, and CNN), we can observe that the proposed approach significantly outperforms these models for all evaluation metrics. The main reason for the lower performance of deep learning models is that it is well established from previous works that deep learning (DL) methods do not work efficiently as compared to simple tree-based methods for tabular data [7, 12, 18].

5.1 Class-Wise Comparison

An example of a pair of proteins belonging to the same class and, to a different class is shown in Fig. 2 for the Coronavirus Host dataset. Figure 2(a) and (b) represent the k-mers spectrum for the environment label. As they belong to the same class, we expect pairwise distance to be small and a large kernel value. The Euclidean distance metric cannot capture the similarity effectively as compared to the W-PSSKM-based distance measure. Similarly, Fig. 2(c) and (d) represent k-mers for different classes, and we

Table 3. Classification results (averaged over 5 runs) for different evaluation metrics for **Coronavirus Host Dataset**. The best values for each method are underlined, while the overall best values are shown in bold.

Embeddings	Algo.	Acc. ↑	Prec. ↑	Recall ↑	F1 (Weig.) ↑	F1 (Macro) ↑	ROC AUC ↑	Train Time (sec.) ↓
Spike2Vec	SVM	0.848	<u>0.852</u>	0.848	0.842	0.739	<u>0.883</u>	191.066
	NB	0.661	0.768	0.661	0.661	0.522	0.764	10.220
	MLP	0.815	0.837	0.815	0.814	0.640	0.835	46.624
	KNN	0.782	0.794	0.782	0.781	0.686	0.832	82.112
	RF	<u>0.853</u>	0.848	<u>0.853</u>	0.845	0.717	0.864	15.915
	LR	<u>0.853</u>	<u>0.852</u>	<u>0.853</u>	<u>0.846</u>	<u>0.757</u>	0.879	60.620
	DT	0.829	0.827	0.829	0.825	0.696	0.855	<u>4.261</u>
PWM2Vec	SVM	0.799	0.806	0.799	0.801	0.648	0.859	44.793
	NB	0.381	0.584	0.381	0.358	0.400	0.683	<u>2.494</u>
	MLP	0.782	0.792	0.782	0.778	0.693	0.848	21.191
	KNN	0.786	0.782	0.786	0.779	0.679	0.838	12.933
	RF	<u>0.836</u>	<u>0.839</u>	<u>0.836</u>	<u>0.828</u>	<u>0.739</u>	<u>0.862</u>	7.690
	LR	0.809	0.815	0.809	0.800	0.728	0.852	274.917
	DT	0.801	0.802	0.801	0.797	0.633	0.829	4.537
Spaced k-mers	SVM	0.830	0.836	0.830	0.825	0.645	0.832	4708.264
	NB	0.711	0.792	0.711	0.707	0.621	0.809	291.798
	MLP	0.829	0.842	0.829	0.823	0.586	0.774	1655.708
	KNN	0.780	0.783	0.780	0.775	0.589	0.790	2457.727
	RF	0.842	0.850	0.842	0.835	0.632	0.824	542.910
	LR	<u>0.844</u>	<u>0.851</u>	<u>0.844</u>	<u>0.837</u>	<u>0.691</u>	<u>0.833</u>	187.966
	DT	0.830	0.839	0.830	0.826	0.640	0.827	<u>56.868</u>
WDGRL	SVM	0.329	0.108	0.329	0.163	0.029	<u>0.500</u>	2.859
	NB	0.004	0.095	0.004	0.007	0.002	0.496	**0.008**
	MLP	0.328	0.136	0.328	0.170	0.032	0.499	5.905
	KNN	0.235	0.198	0.235	0.211	<u>0.058</u>	0.499	0.081
	RF	0.261	0.196	0.261	<u>0.216</u>	0.051	0.499	1.288
	LR	<u>0.332</u>	0.149	<u>0.332</u>	0.177	0.034	<u>0.500</u>	0.365
	DT	0.237	<u>0.202</u>	0.237	0.211	0.054	0.498	0.026
Auto-Encoder	SVM	0.602	0.588	0.602	0.590	0.519	0.759	2575.95
	NB	0.261	0.520	0.261	0.303	0.294	0.673	21.747
	MLP	0.486	0.459	0.486	0.458	0.216	0.594	29.933
	KNN	0.763	0.764	0.763	0.755	0.547	0.784	<u>18.511</u>
	RF	<u>0.800</u>	<u>0.796</u>	<u>0.800</u>	<u>0.791</u>	0.648	<u>0.815</u>	57.905
	LR	0.717	0.750	0.717	0.702	0.564	0.812	11072.67
	DT	0.772	0.767	0.772	0.765	0.571	0.808	121.362
String Kernel	SVM	0.601	0.673	0.601	0.602	0.325	0.624	5.198
	NB	0.230	0.665	0.230	0.295	0.162	0.625	<u>0.131</u>
	MLP	0.647	0.696	0.647	0.641	0.302	0.628	42.322
	KNN	0.613	0.623	0.613	0.612	0.310	0.629	0.434
	RF	<u>0.668</u>	0.692	<u>0.668</u>	<u>0.663</u>	<u>0.360</u>	<u>0.658</u>	4.541
	LR	0.554	<u>0.724</u>	0.554	0.505	0.193	0.568	5.096
	DT	0.646	0.674	0.646	0.643	0.345	0.653	1.561
Neural Network	LSTM	0.325	0.103	<u>0.325</u>	0.154	0.021	0.502	21634.34
	CNN	<u>0.442</u>	0.101	0.112	0.086	<u>0.071</u>	<u>0.537</u>	17856.40
	GRU	0.321	<u>0.139</u>	0.321	<u>0.168</u>	0.032	0.505	126585.01
SeqVec	SVM	0.711	0.745	0.711	0.698	0.497	0.747	0.751
	NB	0.503	0.636	0.503	0.554	0.413	0.648	<u>0.012</u>
	MLP	0.718	0.748	0.718	0.708	0.407	0.706	10.191
	KNN	0.815	0.806	0.815	0.809	0.588	0.800	0.418
	RF	<u>0.833</u>	<u>0.824</u>	<u>0.833</u>	<u>0.828</u>	<u>0.678</u>	<u>0.839</u>	1.753
	LR	0.673	0.683	0.673	0.654	0.332	0.660	1.177
	DT	0.778	0.786	0.778	0.781	0.618	0.825	0.160
Protein Bert		0.799	0.806	0.799	0.789	0.715	0.841	15742.95
ESM-2	SVM	0.981	0.978	0.981	0.977	0.744	**0.898**	8.463
	NB	0.759	0.950	0.759	0.757	0.573	0.810	8.595
	MLP	0.978	0.975	0.978	0.974	0.706	0.875	62.442
	KNN	0.977	0.975	0.977	0.973	0.707	0.865	<u>2.202</u>
	RF	**0.983**	0.978	**0.983**	**0.979**	**0.783**	0.893	17.740
	LR	0.982	**0.979**	0.982	0.978	0.777	0.889	107.879
	DT	0.982	**0.979**	0.982	0.978	0.742	0.890	3.733
TAPE	SVM	0.818	0.823	0.818	0.811	0.711	<u>0.854</u>	3.201
	NB	0.482	0.587	0.482	0.442	0.400	0.712	0.494
	MLP	0.812	0.819	0.812	0.802	0.665	0.828	3.737
	KNN	0.793	0.797	0.793	0.789	0.633	0.818	<u>0.150</u>
	RF	<u>0.830</u>	<u>0.834</u>	<u>0.830</u>	<u>0.823</u>	<u>0.725</u>	0.846	13.656
	LR	0.779	0.797	0.779	0.764	0.628	0.794	11.325
	DT	0.785	0.786	0.785	0.782	0.578	0.798	4.675
W-PSSKM (ours)	SVM	0.952	0.950	0.952	0.950	0.704	0.886	0.894
	NB	0.634	0.710	0.634	0.634	0.426	0.725	<u>0.101</u>
	MLP	0.874	0.878	0.874	0.873	0.535	0.778	15.964
	KNN	0.939	0.933	0.939	0.934	0.555	0.773	0.345
	RF	<u>0.955</u>	<u>0.953</u>	<u>0.955</u>	<u>0.951</u>	0.710	0.833	4.933
	LR	0.950	0.946	0.950	0.946	<u>0.775</u>	0.847	13.007
	DT	0.933	0.938	0.933	0.934	0.608	0.836	1.369

Table 4. Standard Deviation values of 5 runs for Classification results on the proposed and SOTA methods for **Coronavirus Host dataset**.

Embed. Method	ML Algo.	Acc.	Prec.	Recall	F1 weigh.	F1 Macro	ROC-AUC	Train. runtime (sec.)
Spike2Vec	SVM	0.02187	0.03118	0.02187	0.02506	0.01717	0.01059	0.52562
	NB	0.00954	0.03743	0.00954	0.01682	0.01232	0.00833	0.00818
	MLP	0.00295	0.00383	0.00295	0.00393	0.00025	0.00284	6.35600
	KNN	0.01892	0.02670	0.01892	0.02128	0.02197	0.01347	0.01097
	RF	0.00898	0.00720	0.00898	0.00712	0.00227	0.00176	0.04175
	LR	0.00802	0.01287	0.00802	0.01179	0.00698	0.00363	0.03516
	DT	0.01608	0.01437	0.01608	0.01533	0.01894	0.00981	0.00880
PWM2Vec	SVM	0.01459	0.01735	0.01459	0.01737	0.01599	0.01051	0.42070
	NB	0.01808	0.02434	0.01808	0.02163	0.01882	0.00905	0.00863
	MLP	0.01898	0.02002	0.01898	0.01990	0.01293	0.00793	7.39342
	KNN	0.01098	0.01342	0.01098	0.01097	0.00904	0.00441	0.04144
	RF	0.02330	0.01223	0.02330	0.02348	0.01934	0.01331	0.04276
	LR	0.01896	0.01959	0.01896	0.02155	0.01807	0.01092	0.00840
	DT	0.00974	0.01461	0.00974	0.01171	0.00781	0.00558	0.01814
String Kernel	SVM	0.00892	0.00545	0.00892	0.00737	0.00150	0.01176	0.08293
	NB	0.03446	0.04765	0.03446	0.03270	0.02739	0.01621	0.00054
	MLP	0.02197	0.06306	0.02197	0.02880	0.02930	0.01528	3.28975
	KNN	0.01546	0.01811	0.01546	0.01752	0.01364	0.00600	0.00257
	RF	0.02143	0.03719	0.02143	0.02293	0.02613	0.01234	0.07206
	LR	0.00898	0.03013	0.00898	0.01760	0.01914	0.00551	0.00171
	DT	0.02911	0.03146	0.02911	0.03035	0.03428	0.01948	0.00829
WDGRL	SVM	0.008378	0.005078	0.008378	0.006888	0.00141	0.002417	0.034498
	NB	0.008720	0.055455	0.00872	0.022475	0.022506	0.007747	0.000352
	MLP	0.016103	0.010655	0.016103	0.018511	0.014246	0.007622	2.437331
	KNN	0.010047	0.009635	0.010047	0.010275	0.011106	0.006126	0.002751
	RF	0.013395	0.019497	0.013395	0.015266	0.018384	0.008316	0.035652
	LR	0.007675	0.094971	0.007675	0.008903	0.005274	0.00175	0.001188
	DT	0.009280	0.008941	0.00928	0.009266	0.007579	0.004472	0.004392
Spaced Kernel	SVM	0.00908	0.00555	0.00908	0.00751	0.00152	0.01198	0.08447
	NB	0.03510	0.04853	0.03510	0.03330	0.02790	0.01651	0.00055
	MLP	0.02238	0.06423	0.02238	0.02934	0.02984	0.01557	3.35067
	KNN	0.01575	0.01845	0.01575	0.01785	0.01389	0.00611	0.00262
	RF	0.02183	0.03788	0.02183	0.02336	0.02662	0.01257	0.07340
	LR	0.00915	0.03069	0.00915	0.01793	0.01949	0.00561	0.00174
	DT	0.02965	0.03204	0.02965	0.03091	0.03492	0.01984	0.00845
Autoencoder	SVM	0.00956	0.00974	0.00956	0.01059	0.00127	0.00094	0.35010
	NB	0.05871	0.04630	0.05871	0.05414	0.02578	0.01233	0.03049
	MLP	0.00846	0.01142	0.00846	0.00776	0.01468	0.00882	1.00276
	KNN	0.00631	0.00803	0.00631	0.00807	0.00699	0.00493	0.01380
	RF	0.00338	0.00548	0.00338	0.00381	0.01029	0.00786	0.72100
	LR	0.00982	0.00982	0.00982	0.01072	0.00128	0.00093	0.19975
	DT	0.01025	0.00968	0.01025	0.00968	0.02163	0.00821	0.09998
SeqVec	SVM	0.00729	0.00924	0.00729	0.00924	0.00102	0.00031	0.29267
	NB	0.14408	0.06203	0.14408	0.11723	0.02721	0.01650	0.01298
	MLP	0.01185	0.01247	0.01185	0.01129	0.02103	0.00999	0.68591
	KNN	0.01281	0.01485	0.01281	0.01456	0.02329	0.01131	0.05557
	RF	0.01050	0.01599	0.01050	0.01294	0.01490	0.00771	0.36725
	LR	0.00795	0.00984	0.00795	0.01007	0.00126	0.00053	0.22741
	DT	0.01119	0.01183	0.01119	0.01277	0.02537	0.00989	0.12363
ESM-2	SVM	0.002996	0.004791	0.002996	0.003504	0.004206	0.002321	63.325091
	NB	0.010739	0.00679	0.010739	0.012877	0.005521	0.002957	1.637343
	MLP	0.005551	0.005859	0.005551	0.004859	0.006128	0.002757	37.433558
	KNN	0.004605	0.00998	0.004605	0.005275	0.015125	0.006012	1.023186
	RF	0.00622	0.018651	0.00622	0.007063	0.006244	0.003455	2.860571
	LR	0.005202	0.006391	0.005202	0.005831	0.003528	0.001892	24.725468
	DT	0.011207	0.010541	0.011207	0.010715	0.007195	0.003521	3.567783
TAPE	SVM	0.005941	0.010001	0.005941	0.005494	0.060968	0.023419	0.137464
	NB	0.009549	0.010933	0.009549	0.01013	0.044898	0.020246	0.023094
	MLP	0.016546	0.010195	0.016546	0.01605	0.04501	0.026421	0.567482
	KNN	0.012213	0.015524	0.012213	0.011534	0.050628	0.029621	0.009386
	RF	0.008764	0.012207	0.008764	0.008546	0.054582	0.030222	0.465312
	LR	0.006207	0.005478	0.006207	0.00697	0.062016	0.032901	0.530472
	DT	0.009587	0.008818	0.009587	0.009863	0.050643	0.027967	0.358515
Weighted PSSKM Kernel	SVM	0.00459	0.00418	0.00459	0.00435	0.04749	0.01768	0.07872
	NB	0.02022	0.01953	0.02022	0.01775	0.05769	0.02999	0.02715
	MLP	0.00591	0.00586	0.00591	0.00583	0.03420	0.02921	3.88783
	KNN	0.00544	0.00658	0.00544	0.00567	0.06500	0.03523	0.02241
	RF	0.00283	0.00290	0.00283	0.00329	0.03443	0.02141	0.18349
	LR	0.00263	0.00331	0.00263	0.00323	0.05910	0.02767	1.81866
	DT	0.00510	0.00403	0.00510	0.00444	0.03242	0.02751	0.05445

expect the Euclidean measure to return large values and kernel to smaller values. However, W-PSSKM distance can capture these differences more effectively than simple Euclidean-based distance metrics.

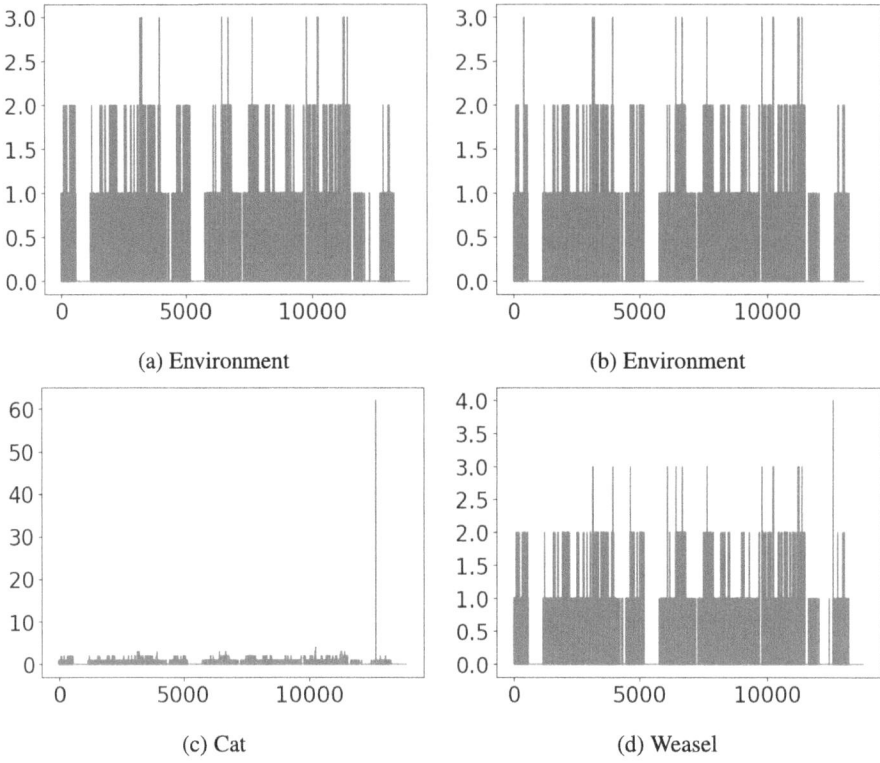

Fig. 2. K-mers spectrum of two pairs of classes. (a) and (b) belongs to the same class, while (c) and (d) belong to different classes for **Coronavirus Host dataset**. The Gaussian kernel distance for (a) and (b) is almost 0 while for the W-PSSKM model is **3.23** (larger distance is better). The Gaussian kernel for (c) and (d) is **0.48** while for the W-PSSKM model is **0.39** (smaller distance is better).

5.2 Inter-class Embedding Interaction

We utilize heat maps to analyze further whether our proposed kernel can better identify different classes. These maps are generated by first taking the average of the similarity values to compute a single value for each pair of classes and then computing the pairwise cosine similarity of different class's embeddings with one another. The heat map is further normalized between [0-1] to the identity pattern. The heatmaps for the baseline Spike2Vec and its comparison with the W-PSSKM embeddings are reported in Fig. 3. We can observe that in the case of the Spike2Vec heatmap, the embeddings for the label are similar, for all classes. This eventually means it is difficult to distinguish between different classes (as seen from Spike2Vec results in Table 3) due to high pairwise similarities among their vectors. On the other hand, we can observe that the pairwise similarity between different class embeddings is distinguishable for W-PSSKM embeddings. This essentially means that the embeddings that belong to similar classes are highly similar to each other. In contrast, the embeddings for different classes are very different, indicating that W-PSSKM can accurately identify similar classes and different classes. This can also be verified by observing the higher predictive accuracy for the proposed method-based embeddings in Table 3.

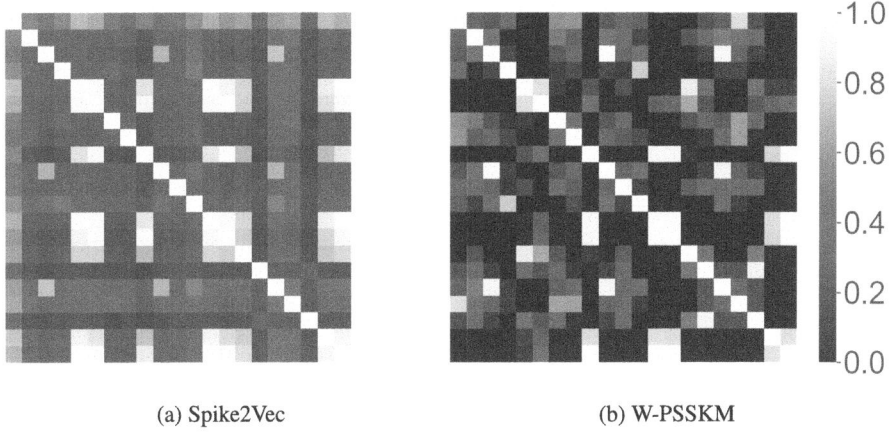

(a) Spike2Vec   (b) W-PSSKM

Fig. 3. Heatmap comparison for classes in **Coronavirus Host**.

6 Conclusion

In this paper, we propose a weighted position weight matrix-based kernel function to design a kernel matrix that we can use to perform supervised analysis of protein sequences. Using kernel PCA, we explored the classification performance of non-kernel classifiers along with the kernel SVM and showed that we could achieve higher predictive performance using classifiers such as KNN and random forest. In the future, we will evaluate the performance of a W-PSSKM kernel for nucleotide sequences. Applying this method to other domains (e.g., music or video) would also be an interesting future direction.

7 Limitations

Despite the significant improvements and promising results achieved by the W-PSSKM in protein sequence classification, there are several limitations to consider. For example, storing the $n \times n$ dimensional kernel matrix in memory could be an issue if the value of n is high. This overhead may limit the scalability of the method when applied to very large databases. Although detailed results are shown for different protein sequence datasets, it is not clear if this method could generalize well on other biological datasets, such as nucleotide sequences and SMILES strings.

References

1. Ali, S., Patterson, M.: Spike2Vec: an efficient and scalable embedding approach for COVID-19 spike sequences. In: IEEE International Conference on Big Data (Big Data), pp. 1533–1540 (2021)
2. Ali, S., Sahoo, B., Khan, M.A., Zelikovsky, A., Khan, I.U., Patterson, M.: Efficient approximate kernel based spike sequence classification. IEEE/ACM Trans. Comput. Biol. Bioinform. (2022)
3. Brandes, N., Ofer, D., Peleg, et al.: ProteinBERT: a universal deep-learning model of protein sequence and function. Bioinformatics **38**(8), 2102–2110 (2022). https://doi.org/10.1093/bioinformatics/btac020
4. Buchan, D.W., Jones, D.T.: The PSIPRED protein analysis workbench: 20 years on. Nucleic Acids Res. **47**(W1), W402–W407 (2019)
5. Ghandi, M., Lee, D., et al.: Enhanced regulatory sequence prediction using gapped k-mer features. PLoS Comput. Biol. **10**(7), e1003711 (2014)
6. GISAID Website (2022). https://www.gisaid.org/. Accessed 17 Feb 2023
7. Grinsztajn, L., Oyallon, E., Varoquaux, G.: Why do tree-based models still outperform deep learning on tabular data? arXiv preprint arXiv:2207.08815 (2022)
8. Heinzinger, M., Elnaggar, A., Wang, Y., et al.: Modeling aspects of the language of life through transfer-learning protein sequences. BMC Bioinform. **20**(1), 1–17 (2019)
9. Henikoff, S., Henikoff, J.G.: Amino acid substitution matrices from protein blocks. Proc. Natl. Acad. Sci. **89**(22), 10915–10919 (1992)
10. Hirokawa, T., Boon-Chieng, S., Mitaku, S.: SOSUI: classification and secondary structure prediction system for membrane proteins. Bioinformatics (Oxford, England) **14**(4), 378–379 (1998)
11. Hoffmann, H.: Kernel PCA for novelty detection. Pattern Recogn. **40**(3), 863–874 (2007)
12. Joseph, M., Raj, H.: GATE: gated additive tree ensemble for tabular classification and regression. arXiv preprint arXiv:2207.08548 (2022)
13. Kuzmin, K., et al.: Machine learning methods accurately predict host specificity of coronaviruses based on spike sequences alone. Biochem. Biophys. Res. Commun. **533**(3), 553–558 (2020)
14. Leslie, C., Eskin, E., et al.: Mismatch string kernels for SVM protein classification. In: Advances in Neural Information Processing Systems, pp. 1441–1448 (2003)
15. Lin, Z., et al.: Language models of protein sequences at the scale of evolution enable accurate structure prediction. BioRxiv **2022**, 500902 (2022)
16. Lodhi, H., Saunders, C., Shawe-Taylor, J., et al.: Text classification using string kernels. J. Mach. Learn. Res. **2**, 419–444 (2002)
17. Ma, Y., Yu, Z., Tang, R., Xie, X., Han, G., Anh, V.V.: Phylogenetic analysis of HIV-1 genomes based on the position-weighted k-mers method. Entropy **22**(2), 255 (2020)
18. Malinin, A., Prokhorenkova, L., Ustimenko, A.: Uncertainty in gradient boosting via ensembles. In: International Conference on Learning Representations (ICLR) (2021)
19. Pickett, B.E., et al.: VIPR: an open bioinformatics database and analysis resource for virology research. Nucleic Acids Res. **40**(D1), D593–D598 (2012)
20. Rao, R., et al.: Evaluating protein transfer learning with tape. Adv. Neural Inf. Process. Syst. **32** (2019)
21. Rognan, D.: Chemogenomic approaches to rational drug design. Br. J. Pharmacol. **152**(1), 38–52 (2007)
22. Shen, J., Qu, Y., Zhang, W., Yu, Y.: Wasserstein distance guided representation learning for domain adaptation. In: AAAI Conference on Artificial Intelligence (2018)

23. Shrikumar, A., Prakash, E., Kundaje, A.: GkmExplain: fast and accurate interpretation of nonlinear gapped k-mer SVMs. Bioinformatics **35**(14), i173–i182 (2019)
24. Strodthoff, N., Wagner, P., et al.: UDSMProt: universal deep sequence models for protein classification. Bioinformatics **36**(8), 2401–2409 (2020)
25. Toussaint, N.C., Widmer, C., Kohlbacher, O., Rätsch, G.: Exploiting physico-chemical properties in string kernels. BMC Bioinform. **11**(8), 1–9 (2010)
26. Whisstock, J.C., Lesk, A.M.: Prediction of protein function from protein sequence and structure. Q. Rev. Biophys. **36**(3), 307–340 (2003)
27. Xie, J., Girshick, R., Farhadi, A.: Unsupervised deep embedding for clustering analysis. In: International Conference on Machine Learning, pp. 478–487 (2016)
28. Xing, Z., Pei, J., Keogh, E.: A brief survey on sequence classification. ACM SIGKDD Explor. Newsl. **12**(1), 40–48 (2010)

Epidemiology

Residual Immunity and Seasonality of an Epidemic

Siyu Chen[1,2](✉) and David Sankoff[3]

[1] Princeton University, Princeton, NJ 08544, USA
[2] Cornell University, Ithaca, NY 14850, USA
siyu.chen@cornell.edu
[3] University of Ottawa, Ottawa, ON K1N 6N5, Canada
sankoff@uottawa.ca

Abstract. We present a dynamical model of the onset and severity of cyclical epidemic disease taking account only of seasonal boosts of antibody during the infectious season and residual immunity remaining from one season to the next. We also compile data from public health sources on the annual number of cases of influenza A and peak infectivity month over a quarter century. In these data, we discover that there is a negative correlation between the change in number of cases from one year to the next and the shift of peak infectivity month between the two seasons, although this does not extend to a prediction of epidemic timing or case number based on the previous season's statistics. Simulating the mathematical model, we discover that there is also a negative correlation between the change in titer from one season to the next and the shift of peak infectivity month between the two seasons, suggesting that the empirical results can be explained by our minimal boost-and-wane model. In addition, the model predicts that suppressing the epidemic for one season, or witnessing a strong surge for one season, both have lasting effects for a number of successive seasons.

Keywords: seasonal epidemic · dynamic model · antibody titer · immunity

1 Introduction

We propose a minimal model of the immunity to a seasonal epidemic disease, exemplified by influenza, taking account only of the exposure-induced boosts during the infectious season and the waning of residual immunity remaining from one season to the next. This offers a direct approach to resolving the contentious issue of how much the timing and severity of one season's epidemic affects the onset and severity of the next season [1]. Classical models deriving from Kermack and McKendrick [2] focus on the individuals in a population, whether they are susceptible to infection, exposed to the pathogen, symptomatic, hospitalized, in ICU, recovered or deceased. Populations are subdivided into compartments by age, sex, socioeconomic level and/or geographical location. These models calculate the number of individuals in different compartments and account for their movements from one group to another driven by the immunity level (e.g., [3]), concentrating with few exceptions [4] on the details of a single season. Departing from these models, a direct accounting for the population antibody against the implicated pathogen should track the risk trajectory of the epidemic, avoiding the numerous parameters required in compartmental models, and facilitate the study of the time-course of the epidemic across many seasons.

Our model features an exponential waning process during and between annual seasons, imposed on the temporal distribution of antibody boost, reflecting exposures during one season. The latter distribution, especially the timing of peak infectivity, interacting with the waning function, is all that is necessary to reproduce, in mathematically tractable form, the mechanical cycle of immunity characteristic of seasonal infectious disease.

We can naturally iterate the cyclical boost-and-wane process to simulate immunity trajectories over many years and thus quantify the relationship between residual immunity and the time elapsed between annual infectivity peaks, namely a relation between end-of-season titer and the length of the interval between annual infectivity peaks.

Determining the onset of epidemic outbreak, whether of an emerging, newly recognized, virulent pathogen [5] or of a recurrent seasonal infection [6–8], is a vexing problem both for epidemiological research and for population health planning. This has been stressed for a variety of recurrent viral epidemics worldwide [9–11]. Although the onset and severity of an epidemic are largely unpredictable, there was widespread prediction that the lowering of population immunity against influenza [12] and RSV [13] as a side-effect of non-pharmaceutical interventions during the COVID-19 pandemic (masking, testing, lockdown, isolation,...) would lead to early onset and increased severity of epidemics of these diseases after the pandemic.

To provide an initial assessment of these ideas, we considered a quarter-century of epidemic influenza A statistics from several public health repositories [14–17]. We compared these statistics and the COVID-19-era predictions with the behaviour of our boost-and-wane model. The result is that the influence of one season's timing and severity on the next season's timing and severity is barely perceptible, if at all, in the statistical record, and is not predicted by our model. Moreover in the three post-COVID-19 years, the influenza A seasons in the US and much of the world only returned to levels commensurate with, or less than pre-COVID-19 activity, although the Canada did undergo stronger rebound levels two years later, as did the UK.

Despite the equivocal evidence of the influence of one season on the next, simulations of our model does reveal an unexpected connection between the timing and severity of successive seasons; namely that the change of immune level from one season to the next is substantially and negatively correlated with the span of time between the peak activity of the two consecutive season. This surprising trend is also apparent in the US data and is strongly supported in the Canadian case.

Our model can incorporate a probabilistic determination of peak infectivity, and is thus naturally adaptable to hypotheses about the link between peak month and severity predictions and the pre-season titers. This feature enables a series of experiments to show that after a major disruption, such as a complete suppression or substantial elevation, of one season's epidemic activity, the re-establishment of pre-existing levels of infectivity is not immediate but requires several years to achieve.

2 Background

The World Health Organization [14], Health Canada [15], The Center for Disease Control in the US [16], the UK Health Security Agency [17], among many others, publish weekly reports on influenza strains, including variants of influenza A, dating back to the late 1990s. These data on strains, tests and positivity rates, do not of course include total population antibody levels, but we can use total number of cases detected each week/month during an infectious season as a proxy for these levels, while the peak month can be identified from the visual displays. These data are plotted in Fig. 1 for Canada and the US.

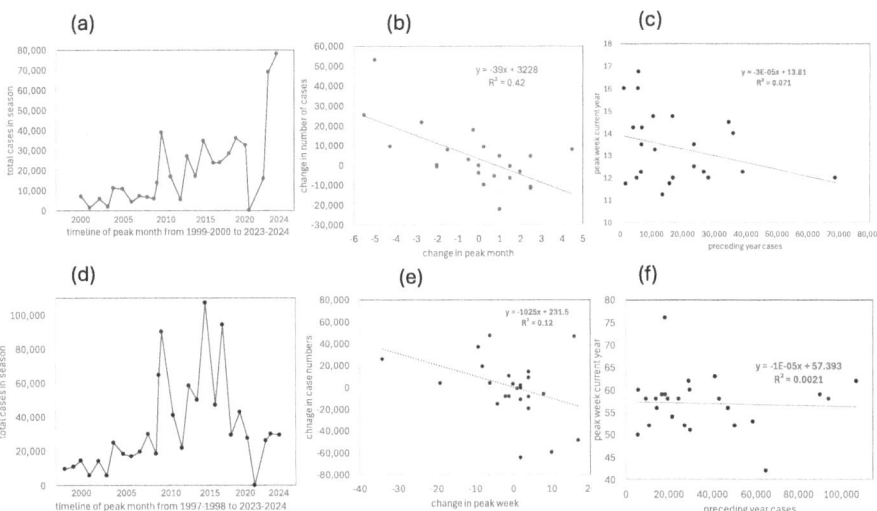

Fig. 1. Comparison of influenza epidemic history in Canada (a–c) and the US (d–f). (a) and (d): trajectory of influenza A epidemics over a quarter of a century, x-axis denotes the peak month of the season while y-axis represents the total number of cases in that season. (b) and (e): change in number of cases as a function of change in peak month, and (c) and (f): peak month as a function of number of cases in preceding season.

The timeline of the Canadian and US influenza A experience depicted in Fig. 1 shows a clear and dramatic increase in number of cases in the years following the 2009 pandemic. And another major, but not immediate, increase after the COVID-19 pandemic is apparent in the Canadian data and worldwide [14], though not the US. Panels (b) and (e) in Fig. 1 show that the change in number of cases from one year to the next is negatively correlated with the shift in peak month, while the panels (c) and (f) do not reveal any strong influence of the peak month of one year on the number of cases in the next year.

3 The Model

3.1 An Infectious Cycle for a Single Season

The average antibody titer of a population will increase because of new exposures but decrease due to antibody waning, as sketched in Fig. 2. Our model contains a minimum of elements, namely a probability distribution $f_\mu(t)$ reflecting the stringency of infection, e.g., the rate at which susceptible individuals in a population acquire an infectious disease, and an antibody decay rate ω.

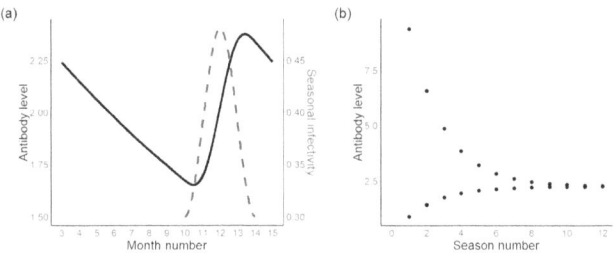

Fig. 2. (a) An antibody trajectory over a hypothetical infectious season, March until March the next year, with infectivity curve overlayed. The antibody level starts at 2.24 in March and ends at 2.24 in March of the next year. The dashed brown curve shows the distribution of stringency of infection during the season starting in October ending in February. The peak month of infectivity, predefined at December, corresponds to the steepest increase of antibody level. (b) Antibody trajectories converge to one fixed point at 2.24 over several seasons from two initial levels, 9.37 and 0.88. y-axis is the antibody level at the end of each season, i.e., March of the year.

Then the change of antibody titers between time point $t = T_0$, before an infectious season, and $t = T_1$, after the season, can be calculated by the relation:

$$X_{T_1} = X_{T_0} e^{-\omega(T_1-T_0)} + \int_{T_0}^{T_1} \beta f_\mu(t) e^{-\omega(T_1-t)} dt, \tag{1}$$

where β is the amplitude or severity of the infectious season.

We illustrate with $f_\mu(t)$ as a raised cosine distribution [18], though any distribution f with support on an interval within T_0 and T_1, even a Dirac function, would produce results similar to what we will show. The raised cosine function has an amplitude and vertical shift of $\frac{a}{2\pi}$, a phase shift of μ and a period of $\frac{2\pi}{a}$, parameterized as:

$$f_\mu(t) = \frac{a}{2\pi}(1 + \cos a(t-\mu)), \text{if } t \in [t_o, t_e] \; 0, \text{ otherwise} \tag{2}$$

where an infectious season starts from date $t_o = \mu - \frac{\pi}{a}$ and ends on date $t_e = \mu + \frac{\pi}{a}$, where μ is the peak month of the severity of the current infectious season.

From the definition in Eq. (2), Eq. (1) becomes

$$X_{T_1} = X_{T_0} e^{-\omega(T_1 - T_0)} + \beta e^{-\omega T_1} \int_{t_o}^{t_e} \frac{a}{2\pi}(1 + \cos a(t-\mu))e^{\omega t} dt, \tag{3}$$

Then, using the definition of t_o and t_e, the integral in Eq. (3) has closed form

$$e^{\mu w} \frac{\sinh(\pi\theta)}{\pi\theta(1+\theta^2)}, \tag{4}$$

where $\theta = \frac{\omega}{a}$. Equation (3) can then be rewritten

$$X_{T_1} = X_{T_0} e^{-\omega(T_1 - T_0)} + \beta e^{-\omega(T_1 - \mu)} \frac{\sinh(\pi\theta)}{\pi\theta(1+\theta^2)}. \tag{5}$$

3.2 A Fixed Point of an Iterated Model for Multiple Seasons

Using a given initial titer X_{T_0} to calculate X_{T_1} by Eqs. (1) or (5), and using the calculated X_{T_1} to calculate X_{T_2}, and continuing the same way for $X_{T_3}, X_{T_4}, \cdots$, we have defined a discrete dynamic system, at times $T_0 < T_1 < \cdots$. We can assume $T_i - T_{i-1} > \Delta > 0$, for some Δ and for all $i \geq 1$. Given two initial titers X_{T_0} and Y_{T_0}, we can see that

$$\|X_{T_i} - Y_{T_i}\| = \|X_{T_{i-1}} - Y_{T_{i-1}}\| e^{-\omega(T_i - T_{i-1})} \tag{6}$$

since the integral is constant, for $i = 1, 2, \cdots$. Now $e^{-\omega\Delta} < 1$, so this process is contractive, with Lipschitz constant $e^{-\omega\Delta}$.

It has a fixed point

$$\frac{\beta e^{-\omega\delta} \frac{\sinh(\pi\theta)}{\pi\theta(1+\theta^2)}}{1 - e^{-\omega\Delta}}, \tag{7}$$

in the simplest case, where $\beta = 1$, $\delta = T_i - \mu$ and $\Delta = T_i - T_{i-1}$ for all i.

For example, we may examine the default parameters $\beta = 1$, $\omega = \frac{1}{24}$, $\mu = 12$, $a^{-1} = \frac{2}{\pi}$, $\Delta = T_i - T_{i-1} = 12$, $\delta = T_i - \mu = 3$, to determine the fixed point $X = 2.24$. This may be compared to the numerical results in Fig. 2(b) tracking two trajectories of the model over 20 seasons.

We note that the fixed point X is particularly sensitive to the parameter μ, with derivative of form

$$\frac{dX}{d\mu} = -C\omega e^{-\omega\mu}, \tag{8}$$

where C is a positive constant. This result is important for understanding the tendencies in Fig. 1(b) and (e), and the results in Sect. 4.1. Although this result is based on the raised cosine model, similar results can be found for other f. In addition, in all the calculations involving fixed points, the factor β does not play an important role, since it multiplies X in a single year and in the fixed point solution in the same way.

4 Long-Term Trends

4.1 Random Peaks

As a benchmark experiment, we concatenated successive instances of the model of a single cycle described in Sect. 3 to carry out a simulation of 20 recurrent infectious seasons interspersed with quiescent periods for the rest of each cycle (i.e., year). Initialized with a random titer at date T_1 corresponding to the end of a typical infectious period, the peak of next infectious month m was chosen from a uniform distribution over a wide range, September (month 9) to April (month 16), and the first iteration was performed at T_2 set to be $m + \frac{\pi}{a}$, with output X_{T_2}. The output X_{T_2} at time T_2 from the first cycle, is then used as the starting titer X_{T_3} at time T_3 of the second cycle. The peak of next infectious season is again randomly chosen from September to April. This calculation is repeated for the third and subsequent cycles. Figure 3(a) shows five typical trajectories of titers over the 20-year seasons. The randomness in the choice of peaks prevented the process from simply degenerating into a fixed point limit cycle, but maintained a stable pattern of random variation indefinitely.

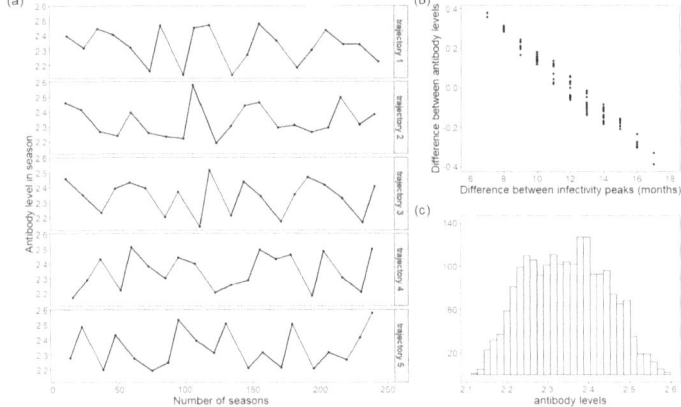

Fig. 3. (a) Sample trajectories of 20 seasons produced by iterating the model, five replicates. (b) Association of titer change with length of inter-season times. (c) Long-term distribution of titers of the iterated model.

How can we characterize this pattern formally? There is a contractive model as in Eq. (1) corresponding to each value of μ. However, as the model is changing with most iterations of the process, the approach to the fixed point(s) is disrupted at each step.

There are r different values of μ, corresponding to the months from September to April ($r = 8$). If we imposed the unrealistic conditions that the process repeated indefinitely along a fixed trajectory among all the r values of μ then there would exist a fixed orbit of r points, to which the successive titers in the trajectory would converge over time. However, once we allow complete randomness in the successive values of μ among the r possibilities, as is necessary in our epidemic model, then no such fixed orbit exists. Instead, we can only say that the distribution of titer values converges to a fixed distribution with support $[s_1, s_r]$, where s_1 and s_r represent the minimum and maximum fixed points of the r models, as in Fig. 3(c).

We plotted the change in titer $\Delta_X = X_{T_i} - X_{T_{i-1}}$ as a function of $\Delta_\mu = \mu_i - \mu_{i-1}$, the shift in peak infections between the $i-1$-st and i-th cycle. The results in Fig. 3(b) show a tight linear relation between the two quantities. The slope is -0.71 titer units/12 months differential, or 0.059 units/month. This compares to $\log \omega = 0.042$/month. The explanation for this result lies in Eq. (8). The way the titer decreases as a function of μ ensures that an early season followed by a late season will result in a low titer, while a late season followed by an early season will have a relatively high titer. Two seasons that occur during the same month will results in an intermediate titer.

4.2 Training for Seasonality

The negative correlation between the change of seasonal titer and the time elapsed between successive peaks in Fig. 3(b) must be seen as a function only of the waning time between two seasons, since there is no mechanism within the model to affect one year's peak μ as a direct function of the previous year's output titer $X_{T_{i-1}}$. Such a mechanism, however, is widely thought to be of importance to recurrent epidemics, and so we investigate it here.

To train the model as a predictor of seasonality, we made use of the results of the experiment described in Sect. 4.1. Based on each year's $X_{T_{i-1}}$, a random choice of next year's peak month μ_i was effected. This choice involved two steps: The first was the deterministic choice of one of four bins, $B_1, \cdots, B_4$, with B_1 containing the highest values of $X_{T_{i-1}}$, and B_4 containing the lowest values of $X_{T_{i-1}}$. The second step was a uniform random choice among three consecutive months. The months in each bin overlapped those in the adjacent bins, with B_1 containing January through March, B_2 containing December through February, B_3 containing November through January and B_4 containing October through December.

A 200-season experiment could then be initiated by a random choice of X_{T_0}. There was no further addition of noise or other intervention over the length of the experiment. Because setting a new peak month determines a different fixed point for X_{T_i}, each iteration of the model gives an different output from the previous season's. Figure 4(a) shows one aspect of the outcome, comparing the titers of each pair of successive seasons. The extent of the scatter is largely the effect of the large overlapping bins from which to choose the peak month.

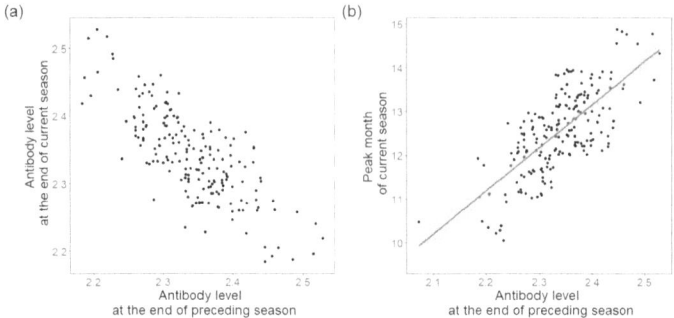

Fig. 4. (a) Effect of training on titers of successive seasons. (b) Effect of training on choice of peak month, with regression line.

4.3 Using the Model to Predict the Peak of the Next Epidemic Season

We regress the peak month chosen against the input titer for the data in Sect. 4.2. This produces the following result:

$$\text{peak month} = 9.9 \times \text{titer} - 10.5 + \epsilon \tag{9}$$

where the normal error ϵ has a standard deviation of 0.45. This result can then be used as a predictive tool. Given an end-of-season titer X_{T_i}, Eq. (9) can predict the likely peak month and the shape of the titer distribution during the season. For example, using the parameters in Fig. 4(b), an end-of-season titer of 4.7 predicts an early January peak, but distributed with a standard deviation of about two weeks.

4.4 Lasting Effects of a Missing Season

In Sect. 2, we presented evidence that the dramatically diminished influenza epidemic resulting from the COVID-19 pandemic-inspired non-pharmaceutical interventions in 2020–21 and early 2022 was not followed by an early onset and severe influenza season in 2022–2023, contrary to many predictions [12], based on the putative lack of newly acquired immunity during the pandemic.

From the modeling viewpoint, this aspect of the seasonality-titer relationship, can be expressed by simply introducing a single missing (zero-amplitude) season, and following the trajectory of the process in subsequent years. To accomplish this we started with the model in Sect. 4.2 above. In one experiment over fifty years, we introduced an '$\beta = 0$' year every eighth year. In a second experiment we introduced an '$\beta = 0$' year every four years. We then compared the trajectory from these two experiments with that from the original, unmodified, experiment, where $\beta = 1$ for every season.

The results show that for the experiment with an 8-year cycle, the effect of a reduced titer, colored orange in Fig. 5(a), persisted over five to seven years,

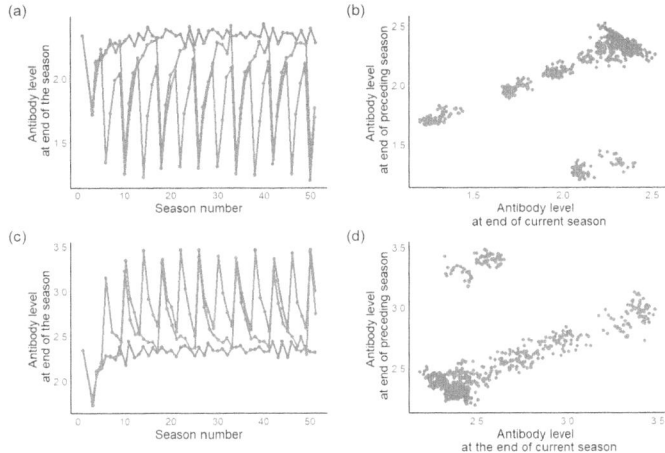

Fig. 5. (a) End-of-season titer trajectories with an 8-year pattern of a missing season (orange), a 4-year pattern (purple) and with no missing seasons (green). (b) Change of titer from one season to the next. Displaced clusters represent the immediate effect of the missing boost. (c) End-of-season titer trajectories with an 8-year pattern of severe seasons (orange), a 4-year pattern (purple) and with no severe seasons (green). (d) Change of titer from one season to the next. Displaced clusters represent the immediate effect of the amplified boost. (Color figure online)

before catching up to the sequence of unmodified seasons. On the other hand, with a four-year cycle, colored purple in Fig. 5(a), the output titer never quite caught up. Figure 5(b) shows how the titers of the two missing-seasons trajectories gradually increase towards the unmodified (no missing seasons) pattern.

4.5 Lasting Effects of a Severe Season

In a way analogous to the method in the Sect. 4.4, we can model the after-effects of a severe season, by setting $\beta = 2$ for one season out of four, or one season out of eight, while $\beta = 1$ for all the remaining seasons. The results of this experiment mirror those of the 'missing season' experiment, but in the opposite direction. The orange lines in Fig. 5(c) show how the titers in the 8-year pattern eventually settle down to rates comparable to the unmodified sequence, while the 4-year pattern shown by purple lines in Fig. 5(c) retains elevated titers throughout. Figure 5(d) shows how the titers of the two severe-seasons trajectories gradually decrease towards the unmodified (no severe seasons) pattern.

Both experiments, one invoking a missing season and the other creating a severe season, suggest a gradual return to normal patterns over few years, rather than a disproportionate reaction in the immediately following season. The evidence in Fig. 1 is not unequivocal, but corroborates this, at least partially.

5 Discussion and Conclusions

The core of our experimental model takes into account titer increases during a few months-long season of an infectious disease with fixed infectivity peak μ, plus a continuous process of waning, expressed by a negative exponential with parameter ω. These two processes can reproduce the cycle of boosting and waning characteristic of recurrent seasonal infectious disease. Iterating this model, however, is not suitable for generating long term trajectories of an epidemic. Mathematically, it is a contractive process that would quickly converge towards a sequence of identical seasons. Replacing the assumption of fixed peak times with a random choice among several months, however, modified the contractive tendency, so that simulations could explore the variation in titer as a function of the peak location parameter μ. This revealed a strong association between shift of peak month and change of titer from one season to the next, so that Fig. 3(b) expresses a mathematically-based explanation of the negative correlated pattern discovered in national statistics (Fig. 1(b) and (d)).

Introducing an element of causality into this association, we trained the model so that a lower titer in the previous year would lead to an early onset of the epidemic in the current year, via a slight bias in the random selection of the peak month, while a larger titer would delay the season. The pattern that emerged from this training then allowed us to establish a prediction rule (9) so that from the titer at the end of one season, we can predict the timing of the next one, in terms of a probability distribution of the timing of the peak month.

Inspired by predictions that the dramatic drop of infections by non-COVID-19 respiratory viruses like influenza and RSV in the 2020–2021 seasons would be followed by a strong resurgence in the following year, we adapted our model by setting the amplitude (severity) parameter β to zero for one year out of four or one year out of eight and observed how such perturbations affected subsequent years. These experiments showed that after each of the "missing" years, the trajectories of the post-season titers inevitably recovered towards the usual pattern, largely in the case of the four-year cycle, but completely in the case of the eight-year cycle.

To study the effects of a year with increased severity, we doubled the parameter β for one year out of four or one year out of eight and observed how such perturbations affected subsequent years, similar to the experiment with missing years. These experiments showed that after each of the severe years, the trajectories of the post-season titers inevitably reverted towards the usual pattern, largely in the case of the four-year cycle, but completely in the case of the eight-year cycle. This timing is of course dependent on our choice of parameters and protocols: ω, β, the training protocol, and predicting the distribution of μ. Nevertheless, it illustrates the kind of investigation possible with our titer-based modeling.

In further work, we could consider population heterogeneity by coupling several models representing antibody dynamics in different subgroups. In contrast to the movement of individuals in classical compartmental models, our model would not involve the transfer of titers between subpopulations. The effect of one compartment affecting another would modeled in the choice of μ and/or β in one compartment depending on the recent state of the process in a 'neighbouring' compartment, such as age group, spatial proximity, health workers versus general public or interacting occupational groups.

In this work, we have not assumed anything about viral strains. Apparent waning over several seasons may reflect mutational drift or selection in the antigen, rather than immunological processes per se. This does not distract from the pertinence of our model as a basis for analyzing the cyclical behaviour of boosting and waning. Many other factors may enter into the timing of the induction of an infectious season, such as climate, demographic changes, antigenic shift in the pathogen, changes in transmission patterns [9], and others. Whether or not post-season residual titer levels remain a likely driver of onset and peak times and severity remains an open question, on the basis of our work on influenza A.

Acknowledgements. We thank Yves Bourgault for key pointers leading to our discussion of dynamical systems. S.C. was funded by a Postdoctoral Research Fellowship in the High Meadows Environmental Institute, Princeton University. D.S. is funded by NSERC Discovery Grant RGPIN 5212-2022, the Canada Research Chair in Mathematical Genomics, and a University of Ottawa Distinguished University Professorship.

References

1. Lee, S.S., Viboud, C., Petersen, E.: Understanding the rebound of influenza in the post COVID-19 pandemic period holds important clues for epidemiology and control. Int. J. Infect. Dis. **122**, 1002–1004 (2022). https://doi.org/10.1016/j.ijid.2022.08.002. Epub 2022 Aug 4. PMID: 35932966; PMCID: PMC9349026
2. Kermack, W.O., McKendrick, A.G.: A contribution to the mathematical theory of epidemics. Proc. Royal Soc. Lond. Seri. A **115**(772), 700–721 (1927)
3. White, L.J., Medley, G.F.: Microparasite population dynamics and continuous immunity. Proc. Royal Soc. Lond. Ser. B Biol. Sci. **265**(1409), 1977–1983 (1998)
4. Zhao, X., Ning, Y., Chen, M.I., Cook, A.R.: Individual and population trajectories of influenza antibody titres over multiple seasons in a tropical country. Am. J. Epidemiol. **187**(1), 135–43 (2018)
5. Chen, S., Flegg, J.A., White, L.J., Aguas, R.: Levels of SARS-CoV-2 population exposure are considerably higher than suggested by seroprevalence surveys. PLoS Comput. Biol. **17**(9), e1009436 (2021)
6. Rhodes, C.J., Hollingsworth, T.D.: Variational data assimilation with epidemic models. J. Theor. Biol. **258**(4), 591–602 (2009)
7. Vattiato, G., Lustig, A., Maclaren, O.J., Plank, M.J.: Modelling the dynamics of infection, waning of immunity and re-infection with the Omicron variant of SARS-CoV-2 in Aotearoa New Zealand. Epidemics **41**, 100657 (2022)
8. Won, M., Marques-Pita, M., Louro, C., Gonçalves-Sá, J.: Early and real-time detection of seasonal influenza onset. PLoS Comput. Biol. **13**(2), e1005330 (2017)
9. Dalziel, B.D., Bjørnstad, O.N., van Panhuis, W.G., Burke, D.S., Metcalf, C.J., Grenfell, B.T.: Persistent chaos of measles epidemics in the pre-vaccination United States caused by a small change in seasonal transmission patterns. PLoS Comput. Biol. **12**(2), e1004655 (2016)

10. Baker, R.E., Park, S.W., Yang, W., Vecchi, G.A., Metcalf, C.J.E., Grenfell, B.T.: The impact of COVID-19 nonpharmaceutical interventions on the future dynamics of endemic infections. Proc. Natl. Acad. Sci. USA **117**(48), 30547–30553 (2020)
11. Messacar, K., Baker, R.E., Park, S.W., Nguyen-Tran, H., Cataldi, J.R., Grenfell, B.: Preparing for uncertainty: endemic paediatric viral illnesses after COVID-19 pandemic disruption. The Lancet **400**(10364), 1663–5 (2022)
12. Krauland, M.G., Galloway, D.D., Raviotta, J.M., Zimmerman, R.K., Roberts, M.S.: Impact of low rates of influenza on next-season influenza infections. Am. J. Prev. Med. **62**(4), 503–510 (2002)
13. Hamid, S., et al.: Seasonality of respiratory syncytial virus - United States, 2017-2023. MMWR Morb. Mortal Wkly. Rep. **72**(14), 355–361 (2023). https://doi.org/10.15585/mmwr.mm7214a1. PMID: 37022977; PMCID: PMC10078848
14. Flunet. https://www.who.int/tools/flunet. Accessed 8 Aug 2024
15. Public Health Agency of Canada, FluWatch. https://www.canada.ca/en/public-health/services/diseases/flu-influenza/influenza-surveillance.html. Accessed 8 Aug 2024
16. Center for Disease Control and Prevention. Fluview. https://www.cdc.gov/flu/weekly/index.htm. Accessed 8 Aug 2024
17. UK Health Security Agency. https://www.gov.uk/government/statistics/surveillance-of-influenza-and-other-seasonal-respiratory-viruses-in-the-uk-winter-2023-to-2024. Accessed 8 Aug 2024
18. Chattamvelli, R., Ramalingam, S.: Continuous distributions in engineering and the applied sciences–Part I (2022)

Adapting the Cov2clusters Tool for Clustering MPOXV Whole Genome Sequences

Eric CH Chen[1,2], Tara Newman[2,3], John Tyson[1,2], Anthea Lam[2], Michael Chan[2], Agatha Jassem[1,2], Natalie Prystajecky[1,2], Shannon Russell[1,2(✉)], and James Zlosnik[1,2(✉)]

[1] Department of Pathology and Laboratory Medicine, University of British Columbia, Vancouver V6T 2B5, Canada
[2] Public Health Laboratory, BC Centre for Disease Control, Vancouver V5Z 4R4, Canada
{shannon.russell,jzlosnik}@bccdc.ca
[3] National Microbiology Laboratory, Winnipeg R3E 3M4, Canada

Abstract. Motivation: The multinational outbreak of human mpox virus (MPOXV) in the summer of 2022 highlighted the need for improved tools to assist public health officials in tracking and responding to new local outbreak clusters. Phylogenetic characterization of MPOXV can support local case investigations by shedding light on whether the virus may have been acquired locally, belongs to endemically circulating strains, or represents a new introduction. In this work, we adapt clustering tools developed for SARS-CoV-2 surveillance to track local MPOXV outbreaks.

Results: We present an adapted version of cov2clusters, originally developed for monitoring SARS-CoV-2 cases in British Columbia. The tool offers stable cluster codes between trees and has been improved with optimizations in execution time and memory management. We also demonstrate the advantages of adapting previously developed tools, validated against the pathogen they were originally designed for, to monitor a new pathogen. This approach can conserve resources that would otherwise be spent on developing new tools and facilitate faster deployment in public health settings.

Keywords: MPOXV · genome clustering · public health · comparative genomics

1 Introduction

Mpox (MPX), formerly known as monkeypox, is a zoonotic disease caused by a double-stranded DNA Orthopoxvirus. Despite the original name of "monkeypox", the natural reservoirs of MPX virus (MPXV) have not been clearly determined [22]. The presence of MPXV has been identified in non-human primates and small rodents including squirrels, prairie dogs, and giant pouched rats [3]. Currently, two main clades exist with Clade I historically in Central Africa and Clade II in West Africa [6]. In May 2022, a multinational outbreak emerged, primarily involving Clade IIb MPXV. As of November 30th 2024, this outbreak has affected 127 countries and resulted in over 117,000 confirmed cases [33], with most cases identifying as gay, bisexual, same-gender-loving, and other men who have sex with men [14,34]. Human-to-human transmission during the 2022 outbreak mainly occurs through direct physical contact [22]. Respiratory transmission and infection from fomite exposure is also possible but less common [14,25].

While the number of cases has reduced significantly from the initial 2022 outbreak, small-scale resurgence has continued across the world [33]. In addition, the emergence of Clade Ib strains around the world resulted in the World Health Organization declaring the MPX outbreak "... a public health emergency of international concern" on August 14th, 2024 [35]. Public health authorities around the world continue to require tools to track and monitor the existing outbreaks, the spread of the virus, and the emerging variants of concern.

This virus remains poorly understood; the scarcity of epidemiological data on MPOXV cases, likely compounded by the intimate nature of human-to-human transmission, has made understanding local transmission networks and source attribution challenging [14,34]. Questions about whether there is a natural reservoir, how it interacts with the host's immune system, and other related aspects remain ongoing research topics [6]. Additionally its relatively large and stable genome size (~197kb), with very little genomic variation, further complicates efforts to monitor the spread of the MPOXV. Thus far, multi-omics and clinical studies [2,13,20,32] have been slowly chipping away at the problem. Some have also leveraged insights gained from HIV [2,17,20] to guide MPOXV research. In parallel, public health agencies have increasingly employed genomics to augment the resolution of limited epidemiological data, thereby enabling the generation of hypotheses regarding the sources and patterns of transmission.

In this study, our goal is to adapt the tool cov2clusters [28] for MPOXV clustering. This tool utilizes a phylogenetic tree and combines it with sample collection dates to infer sample clustering [28]. The tool also provides stable cluster codes that can be maintained as new samples are added to the tree [28], so it remains an integral tool of ongoing SARS-CoV-2 monitoring operation in British Columbia (BC).

2 Methods and Materials

2.1 Data

The 265 BC MPOXV assemblies in this study come from laboratory-confirmed samples tested at or submitted to the BC Centre for Disease Control (BCCDC) Public Health Laboratory from regional health authorities, between June 2022 and January 2025. These samples passed sample quality control and genome assembly quality assessment by Nextclade [1]. The samples were sequenced with short-read Whole Genome Sequencing (WGS) using amplicon tiling scheme, and assembled using an ARTIC pipeline [31] modified for MPOXV (https://github.com/BCCDC-PHL/mpxv-artic-nf).

In brief, the reads were trimmed by trim_galore [15], variants were called by FreeBayes [11], mapped by bwa-mem [16], processed by samtools [8], and the consensus assembly was constructed by bcftools [8]. Assemblies with greater than 90% coverage on reference are retained for the analysis. These sequences are available on GISAID (hMpxV/Canada/BC-PHL or hMpxV/Canada/BC-BCCDC flag) [9].

A total of 3,385 non-BC MPOXV sequences were downloaded from GenBank using Nextstrain's MPOXV pipeline, with data updated to January 6, 2025. The reference sequence (NC_063383.1) and annotations were downloaded from NCBI. We selected this reference sequence to maintain consistency with the Nextstrain pipeline [12].

2.2 Phylogenetic Pipeline and Clustering

We used Nextstrain's mpox pipeline (https://github.com/nextstrain/mpox) for the phylogenetic analysis of both BC and non-BC samples. Specifically, we utilized the pipeline to construct phylogenetic trees, refine them with time data, and produce trees and visualizations [1,12]. In particular, the tree-building step in Nextstrain employs IQ-TREE [21], which summarizes valuable parameters, such as mutation rates, that are used as inputs in simulations.

FAVITES-Lite [29] is used to simulate MPOXV spread using the standard Susceptible-Exposed-Infected-Removed (SEIR) model [27] for transmission networks and the Newman-Watts-Strogatz (NWS) model [23] for contact networks. Parameters for the SEIR (Susceptible, Exposed, Infectious, and Recovered) model are based on published epidemiological data [20,32,36], while genomic mutation parameters are estimated from Nextstrain's tree-building process. Notably, the mutation rates are further calibrated to align with the official Nextstrain MPOXV mutation rate for Clade IIb samples. Simulated sequences from each FAVITES-Lite run [29] are then passed into Nextstrain [1,12] for phylogenetic analysis and clustering comparison (Fig. 1).

Clustering is performed using the modified cov2clusters tool [28]. We chose this tool because it was originally developed to predict stable genomic clusters of SARS-CoV-2 cases and is currently deployed at BC CDC for routine surveillance. The tool produces stable genomic clusters over time and integrates both genomic data and metadata, such as sample collection dates, to cluster samples. Specifically, it models the probability that any two samples could belong to the same cluster using the logit model described by Eqs. 1 and 2 [28].

$$P_{ij} = \frac{1}{1 + e^{-Z_{ij}}} \quad (1)$$

$$Z_{ij} = \text{coe}_0 + \text{coe}_a \cdot \text{BranchDistance}_{ij} + \text{coe}_b \cdot \text{DateDistance}_{ij} \quad (2)$$

In brief, coe_0 is a positive number, where the higher it is, the more likely a pair of samples will be considered in the same cluster. coe_a is a negative number that penalizes how different genomically the two samples are. coe_b is also a negative number; it penalizes how far apart in time the two samples were collected.

To adapt this tool for MPOXV, we need to readjust the values of coe_0, coe_a, and coe_b to fit the observed cases. We began by exploring the boundaries of the plausible solution space through testing on BC-only samples. This ensures that the clusters formed are more refined than lineage assignments, providing additional information for public health authorities. Next, we conducted a more detailed exploration of the solution space using simulated data.

The simulation setup (Fig. 1) compares outputs from cov2clusters [28] against clusters assigned by Gephi's [5] modularity function on the simulated transmission network with the parameter *r6c2*. This parameter splits the large transmission network into blocks, similar to different MPOXV lineage assignments, which are delineated by genomic differences and time on the Nextstrain tree [12]. During development, we also updated the internal functions to use *data.table* [4], resulting in improvements in memory usage (less garbage collection pressure) and computation speed (Table 1).

Table 1. Quick benchmark on BC and global data. Performed on a MacBook Pro M1 Max with 64 GB RAM over 30 iterations using R.

Method Data	Samples Count	Median Time	memory_alloc	gc per second	num gc
original-BC+Global	3493	10.3 s	4.75 GB	8.04	2485
original-BC only	247	84.7 ms	12.06 MB	6.87	17
modified-BC+Global	3493	3.99 s	6.78 GB	2.61	317
modified-BC Only	247	28.46 ms	30.88 MB	6.52	6

Finally, we used three metrics to characterize the clustering performance of the adapted cov2clusters parameters: the adjusted Rand index [30], split-join distance [7], and the number of unclustered samples. The adjusted Rand index is used to compare how similar cov2clusters' clustering is to a manually curated snapshot. The split-join distance describes the granularity of cov2clusters [28], and the number of unclustered samples measures the stringency of the parameters.

To explore the effects of masking APOBEC3 mutations, we used squirrel (https://github.com/aineniamh/squirrel) [24] to identify likely mutation positions, and then masked these positions via the Nextstrain pipeline as a new tree was generated [12].

3 Results

3.1 Simulation and Parameter Tuning

We created simulated MPOXV outbreaks using FAVITES-Lite [29] (Fig. 1 and Fig. 2) that exhibit branching patterns similar to production MPOXV trees of

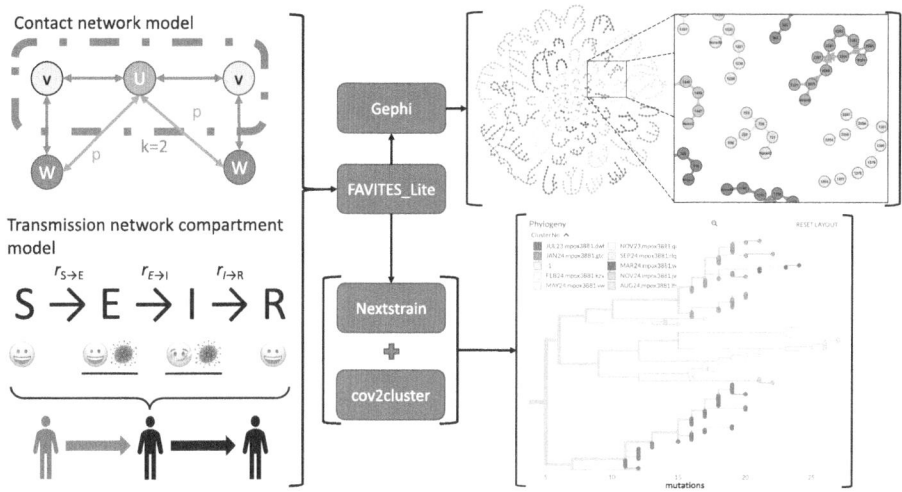

Fig. 1. Simulation and validation setup. The contact network, NWS, connects each node with its k nearest neighbour, and for each edge (n, v), adds a new random edge (u, w) with probability p. In this study, we use $k = 2$ and $p = 0.1$ and $n = 2500$. The transmission network compartment model, SEIR (Suscepti, has its transition rates estimated from published epidemiological data [20,32,36]. These models are used by FAVITES-Lite [29] to simulate the outbreak. Gephi [5] is used to generate the "truth" cluster. Nextstrain [12] is used to build the time-calibrated phylogenetic tree, and cov2clusters [28] for sample clustering.

Clade IIb built on BC and global datasets (Fig. 3). Specifically, two characteristics that we aimed to replicate in the simulation are: first, small clusters that fade out quickly over time, such as lineages B.1.4, B.1.7, B.1.8, and B.1.12; and second, lineages that not only persist through time, but also form the base of new descendant lineages, such as lineage B.1.20 and C.1, which are the parent lineages of F and E, respectively [18]. Another encouraging, though unsurprising, result is visualizing the simulated "true clusters" clustering successfully on the phylogenetic tree generated by Nextstrain (Fig. 2). These results indicate that we are able to simulate plausible MPOXV outbreaks.

Depending on the desired clustering pattern, we can adjust the parameters to achieve either more granular clustering or more relaxed clustering (Fig. 4). In the absence of a "ground truth" dataset, the criteria for balancing our parameter selection are as follows: fewer outliers and more granular clustering than Gephi's modular classes, when they differ.

For example, while mpox3988 recovered the Gephi clustering most accurately, there may be use cases where more granular clustering is preferable, even at the expense of having more outliers in the clustering output. It may be prudent to take a more conservative approach when linking cases, reducing the chance of false positives in labeling local outbreaks. Additionally, a comparison of mpox3988 clustering with manually annotated clusters from BC-only data—

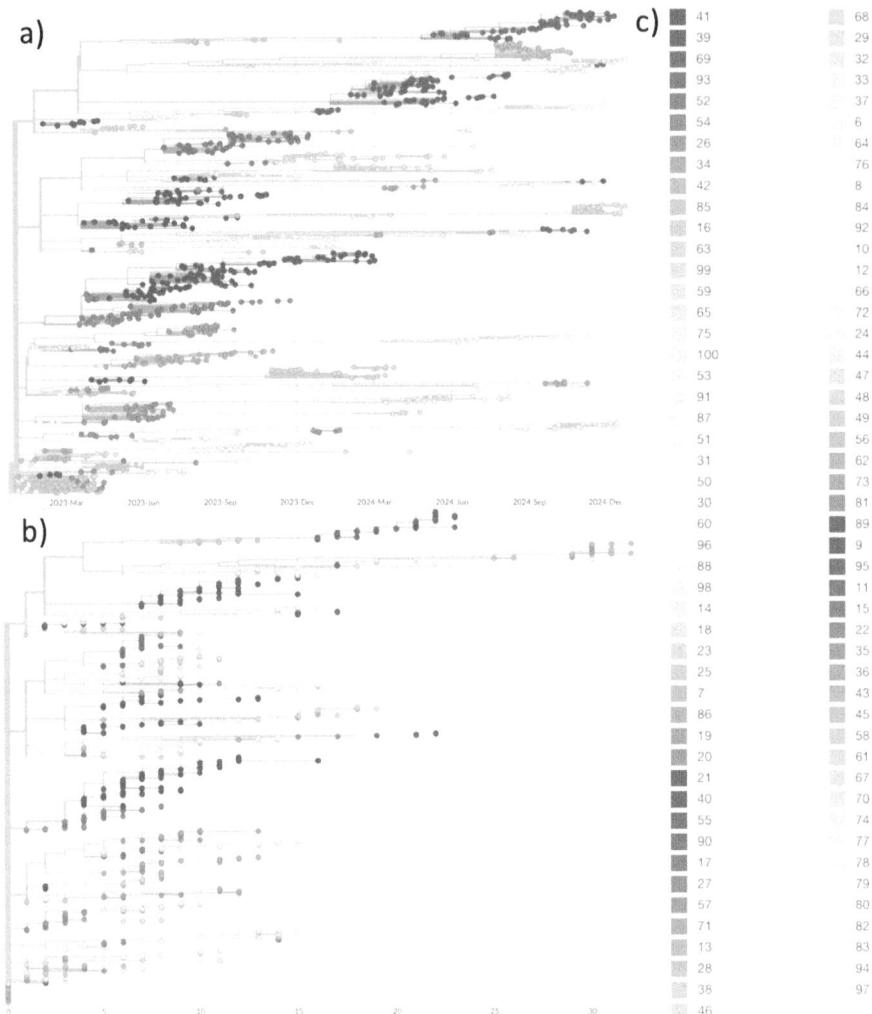

Fig. 2. Simulation example. The parameters are adjusted to mimic the MPOXV tree (Fig. 3). a) View with "Date" as the x-axis. b) View with the number of mutations as the x-axis. c) Legend showing the "true" cluster assignment from the simulation and Gephi's modularity function.

using a 3-SNP genomic distance threshold based on SARS-CoV-2 guidelines—yields an adjusted Rand Index of 0.76. This indicates that the parameters are producing results similar to manually annotated clustering. With additional epidemiological data, we may incorporate these as additional terms in the logical model or use them in downstream post-clustering analysis to better align with local public health needs.

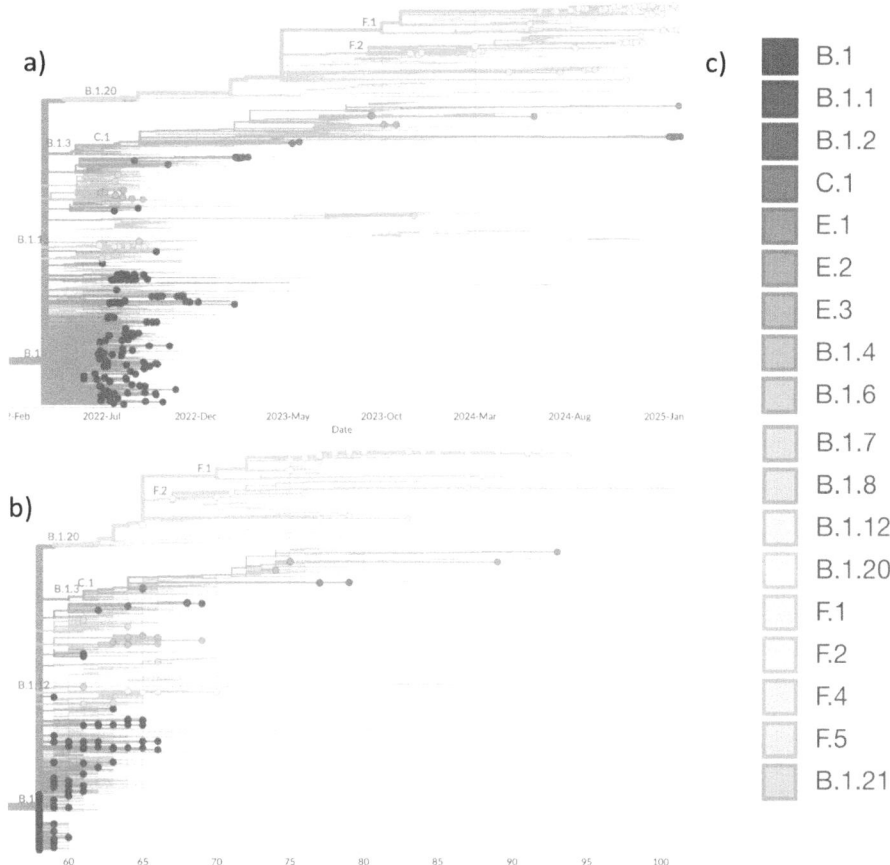

Fig. 3. MPOXV Clade IIb tree produced from BC samples and global GenBank dataset. BC samples highlighted with bubbles and global samples are hidden (no bubbles). a) View with "Date" as the x-axis. b) View with the number of mutations as the x-axis. c) Detailed MPOXV lineage legend

3.2 Clustering on BC and Global Data

While the focus of the tool redevelopment is primarily to understand clustering among BC cases, contextualizing BC cases with available global sequence data can help pinpoint genetic outliers and identify sources of new introductions.

Comparing clusters derived from BC-only data with those based on BC-plus-global data can provide valuable insights. For instance, certain samples that appear as outliers in the BC-only dataset may become part of a cluster when global data is included, as shown in Fig. 5.

In this figure, we focus on the B.1.20 lineage of the phylogenetic tree. The outliers, located at the bottom of the tree, are easily identified. Based on their position, it is plausible to link these samples to introductions from international

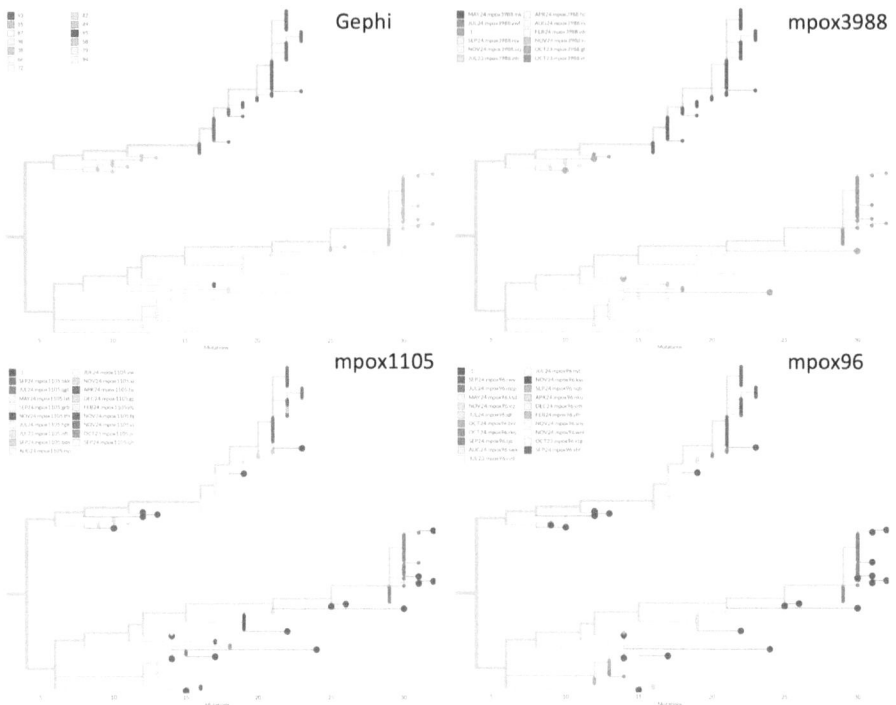

Fig. 4. Example parameter fitting. Three different parameters generating different clusters are compared with Gephi's modular class (top left). The parameters of each set are summarized in Table 2. Notably, the clusters produced are not all separated on divergent branches.

Table 2. Top simulation results. The first set of three rows shows the best results over five replicates. The second set of four rows includes other interesting results that are still biologically plausible and may even be preferred for epidemiological purposes. The Split Join Distance, which is the average number of split and join operations needed to transform one clustering result to another; the lower the better. The Adjusted Rand Index (adjRandIndex) ranges between 1 and −1, with higher values indicating more similar and zero being random. The "Unclust" measures the number of unclustered samples; the lower the better.

ParamSet	coe_0	coe_a	coe_b	SplitJoinDistance_Avg	adjRandIndex_Avg	Unclust_Avg
mpox3988	35	−2.00E+06	−0.15	317	0.527	22
mpox3371	54.5	−3.17E+06	−0.2	317.6	0.527	22
mpox4211	46	−2.67E+06	−0.15	317.6	0.527	22
mpox1505	46	−2.50E+06	−0.3	323.4	0.506	20.6
mpox5524	51	−2.17E+06	−0.1	377.8	0.401	16.6
mpox96	20.5	−3.34E+06	−0.6	671.2	0.267	158.4
mpox1105	20	−3.17E+06	−0.3	597.6	0.343	121.2

cases, as they do not cluster with other BC samples. Moreover, two samples, indicated by red arrows, are part of clusters when the BC-plus-global dataset is used. This behaviour can be useful in guiding epidemiological analysis, providing stronger evidence when identifying travel-associated cases.

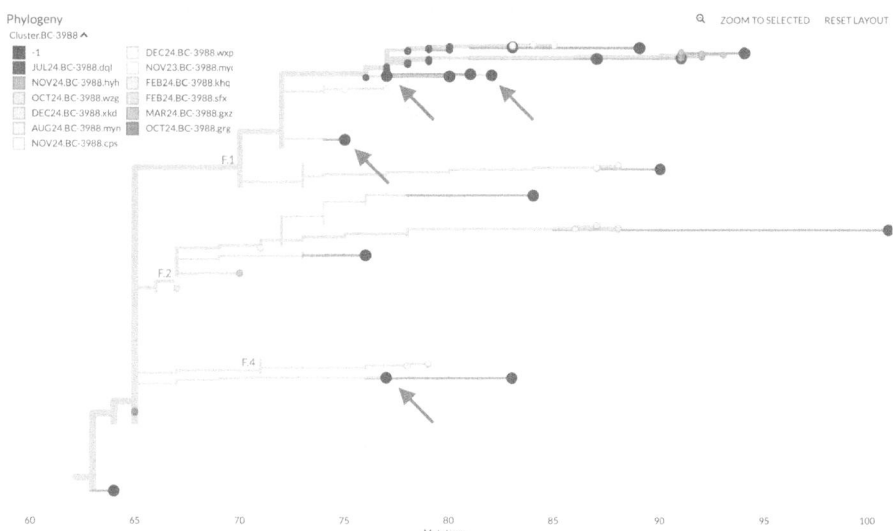

Fig. 5. MPOXV B.1.20 lineage tree with BC samples selected. Clusters given different colours and outliers are highlighted (−1) by bigger circle. Red arrow indicates four samples that changed from being assigned as an outlier to being part of a cluster when using the BC-plus-global data. (Color figure online)

3.3 Effects of Masking APOBEC3 Mutations

As part of the human immune response, human APOBEC3 enzymes are known to result in mutations in the MPOXV genome at specific dinucleotide sites [19]. Such mutations contribute to the genomic diversity seen in MPOXV, however their presence may confound the determination of true genomic differences between samples, as well as phylogenies, which in turn could affect clustering [24]. To investigate this, we evaluated the effects of masking likely APOBEC3 mutations by masking all potential APOBEC3 positions predicted by the tool squirrel [24]. It became clear that masking all such positions is too stringent. Accepted monkeypox (MPOXV) lineage calls were disrupted, with certain samples of the same lineage no longer clustering on the same branch of the phylogenetic tree (Fig. 6).

Furthermore, while the APOBEC3 family of enzymes exhibits a preference for specific DNA motifs, they can also act on other DNA combinations [19], highlighting the complexity of the mutation landscape. Since the initial outbreak in

2022, samples collected in 2023 and 2024 have shown clear phylogenetic signals, suggesting that at least some of these mutations are preserved and passed on to subsequent infections (Fig. 3). This indicates that these mutations should not all be masked outright.

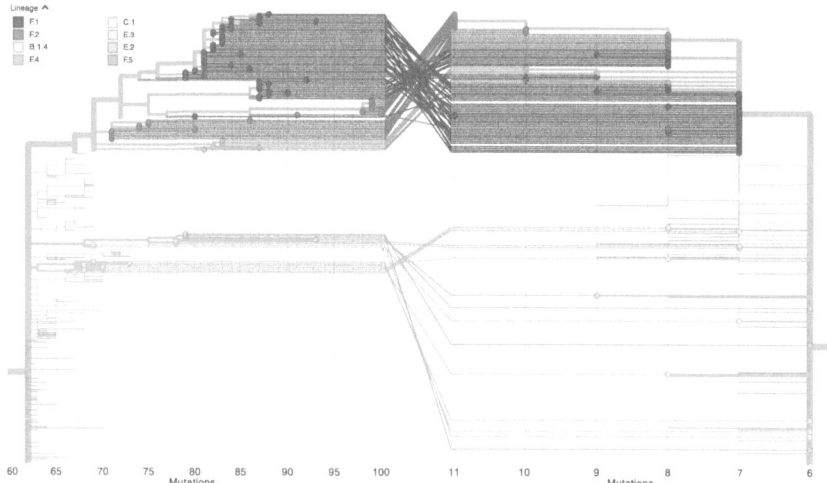

Fig. 6. Effect of APOBEC3 masking. On the left is the tree built from unmasked BC samples while on the right is the tree built from same data, but with likely APOBEC positions masked.

4 Discussion

WGS has increasingly become a part of the public health response toolkit for emerging pathogens as it synergizes well with epidemiology-based methods to better understand transmission dynamics and source attribution [10]. Tools and pipelines, such as Nextstrain [12] and ARCTIC [31] have been instrumental in obtaining and visualizing sequencing data for these pathogens. However, sample clustering remains a significant challenge. To better understand the molecular epidemiology of MPOXV cases in BC, we have redeveloped cov2clusters [28] and demonstrated that it can be adapted for MPOXV through parameter adjustments guided by genomic and incomplete epidemiological data. Although the primary limitation remains the lack of formal validation due to the absence of a "ground truth" dataset, the tool still produces plausible clustering results on simulated data (Fig. 4). This suggests that the model can still capture meaningful outbreak characteristics within the constraints of a limited dataset, as demonstrated by BC data in Fig. 5.

Another limitation is the heavy reliance on genomic differences. As currently set up, the tool would not have performed well during the initial MPOXV outbreak in May 2022, where the majority of the cases represented extremely similar sequences (Fig. 3). It is only after the initial wave, as the suspected APOBEC3 mutations begin to appear [24], that cov2clusters, or any other genomic-based clustering methods, can have a real shot at tracking and monitoring MPOXV cases as we currently do for SARS-CoV-2.

Moving forward, the flexibility offered by the logit model used in cov2clusters [28] makes it highly adaptable to different pathogens or the incorporation of different epidemiological data. Parameters can be versioned to maintain stable cluster designations as new samples are included, which is important for operational use. With the availability of new epidemiological data and biological insights from the broader research community, updates to the parameters are also possible and simple to implement.

One example is the APOBEC3 mutations [24]. Not only have these mutations been linked to the rapid accumulation of mutations in the MPOXV virus following the initial outbreak in 2022 [24], but the relationship between patients with weakened immune systems and the accumulation of APOBEC3 and non-APOBEC3 mutations remains unclear and will require further investigation [2]. While immune-deficient patients and intra-host variations have been reported [2,26], accurately quantifying the overall effects remains challenging due to the scarcity of data. Our results (Fig. 6) do not support masking all potential APOBEC3 sites, so an optimal solution, if one exists, likely involves incorporating another term to account for potential intra-host variation. Collecting multiple samples from the same patient or from different locations on the same patient is unfortunately uncommon [2,26], but overcoming this challenge will likely improve future iterations of clustering parameter adjustments.

Another example that takes advantage of the ease of including additional terms in the logit model is influenza. As an alternative to gene concatenation to summarize information from different gene trees, we may instead combine them by treating each gene tree as its own term with the associated coefficients for genomic distance in the logit model. Adding new terms is straightforward in cov2clusters by design [28], so we will only need to train or refine the appropriate parameters for influenza. This also has the added benefit of enabling the fine-tuning of the weight of genomic differences in each gene tree.

In conclusion, the results of this study highlight the potential and practicality of repurposing established analytical frameworks for rapid deployment in response to emerging diseases. The success of reusing cov2clusters to cluster MPOXV cases could encourage similar reuse of legacy frameworks or tools, maximizing efficiency and improving responses to future disease outbreaks.

Acknowledgements. We are grateful to Benjamin Sobkowiak for the helpful discussion on cov2clusters; Caroline Mburu and Aidan Nikiforuk for the discussion on converting MPOXV SEIR rates and references; and Jessica Caleta for the discussion and troubleshooting. This work is supported by Canadian Institute of Health Research (CIHR), grant number MRR-184812.

Disclosure of Interests. The authors have no competing interests to declare that are relevant to the content of this article.

References

1. Aksamentov, I., Roemer, C., Hodcroft, E.B., Neher, R.A.: Nextclade: clade assignment, mutation calling and quality control for viral genomes. J. Open Source Softw. **6**(67), 3773 (2021)
2. Akther, S., et al.: Genomic epidemiology of monkeypox virus during the 2022 outbreak in New York City. bioRxiv, p. 2024–07 (2024)
3. Alakunle, E., Moens, U., Nchinda, G., Okeke, M.I.: Monkeypox virus in Nigeria: infection biology, epidemiology, and evolution. Viruses **12**(11), 1257 (2020)
4. Barrett, T., et al.: data.table: Extension of 'data.frame' (2025). https://r-datatable.com, r package version 1.16.99. https://Rdatatable.gitlab.io/data.table, https://github.com/Rdatatable/data.table
5. Bastian, M., Heymann, S., Jacomy, M.: Gephi: an open source software for exploring and manipulating networks. In: Proceedings of the International AAAI Conference on Web and Social Media, vol. 3, pp. 361–362 (2009)
6. Brooks, J.T., et al.: How the orthodox features of orthopoxviruses led to an unorthodox mpox outbreak: what we've learned, and what we still need to understand. J. Infect. Dis. **229**(Supplement_2), S121–S131 (2024)
7. Csardi, G., Nepusz, T.: The iGraph software. Complex Syst. **1695**, 1–9 (2006)
8. Danecek, P., et al.: Twelve years of samtools and bcftools. Gigascience **10**(2), giab008 (2021)
9. Elbe, S., Buckland-Merrett, G.: Data, disease and diplomacy: GISAID's innovative contribution to global health. Global Chall. **1**(1), 33–46 (2017)
10. Ferdinand, A.S., et al.: An implementation science approach to evaluating pathogen whole genome sequencing in public health. Genome Med. **13**, 1–11 (2021)
11. Garrison, E., Marth, G.: Haplotype-based variant detection from short-read sequencing. arXiv preprint arXiv:1207.3907 (2012)
12. Hadfield, J., et al.: Nextstrain: real-time tracking of pathogen evolution. Bioinformatics **34**(23), 4121–4123 (2018)
13. Huang, Y., et al.: Multi-omics characterization of the monkeypox virus infection. Nat. Commun. **15**(1), 6778 (2024)
14. Kaler, J., Hussain, A., Flores, G., Kheiri, S., Desrosiers, D.: Monkeypox: a comprehensive review of transmission, pathogenesis, and manifestation. Cureus **14**(7) (2022)
15. Krueger, F., et al.: Felixkrueger/trimgalore: v0.6.10 - add default decompression path (2023). https://doi.org/10.5281/zenodo.7598955
16. Li, H.: Aligning sequence reads, clone sequences and assembly contigs with bwa-mem. arXiv preprint arXiv:1303.3997 (2013)
17. Li, J., Hao, Y., Wu, L., Liang, H., Ni, L., Wang, F., Wang, S., Duan, Y., Xu, Q., Xiao, J., et al.: Exploration of common pathogenesis and candidate hub genes between HIV and monkeypox co-infection using bioinformatics and machine learning. Sci. Rep. **14**(1), 26701 (2024)
18. mpox lineages: lineage-designation (2025). https://github.com/mpxv-lineages/lineage-designation. Accessed 07 Feb 2025
19. McDaniel, Y.Z., et al.: Deamination hotspots among APOBEC3 family members are defined by both target site sequence context and SSDNA secondary structure. Nucleic Acids Res. **48**(3), 1353–1371 (2020)

20. Milwid, R.M., et al.: Exploring the dynamics of the 2022 mpox outbreak in Canada. J. Med. Virol. **95**(12), e29256 (2023)
21. Minh, B.Q., et al.: Iq-tree 2: new models and efficient methods for phylogenetic inference in the genomic era. Mol. Biol. Evol. **37**(5), 1530–1534 (2020)
22. Naga, N.G., Nawar, E.A., Mobarak, A.A., Faramawy, A.G., Al-Kordy, H.M.: Monkeypox: a re-emergent virus with global health implications-a comprehensive review. Tropical Dis. Travel Med. Vaccines **11**(1), 2 (2025)
23. Newman, M.E., Watts, D.J.: Renormalization group analysis of the small-world network model. Phys. Lett. A **263**(4–6), 341–346 (1999)
24. O'Toole, Á., et al.: APOBEC3 deaminase editing in mpox virus as evidence for sustained human transmission since at least 2016. Science **382**(6670), 595–600 (2023)
25. Palich, R., et al.: Viral loads in clinical samples of men with monkeypox virus infection: a French case series. Lancet. Infect. Dis **23**(1), 74–80 (2023)
26. Rueca, M., et al.: Temporal intra-host variability of mpox virus genomes in multiple body tissues. J. Med. Virol. **95**(5), e28791 (2023)
27. Schwartz, I.B., Smith, H.L.: Infinite subharmonic bifurcation in an SEIR epidemic model. J. Math. Biol. **18**, 233–253 (1983)
28. Sobkowiak, B., et al.: Cov2clusters: genomic clustering of sars-cov-2 sequences. BMC Genom. **23**(1), 710 (2022)
29. Stadler, K., Phillips, K., Moshiri, N.: Improved user-friendliness in the design and analysis of FAVITES-lite simulations. medRxiv p. 2024–06 (2024)
30. Steinley, D.: Properties of the hubert-arable adjusted rand index. Psychol. Methods **9**(3), 386 (2004)
31. Tyson, J.R., et al.: Improvements to the artic multiplex PCR method for sars-cov-2 genome sequencing using nanopore. BioRxiv (2020)
32. Ward, T., Christie, R., Paton, R.S., Cumming, F., Overton, C.E.: Transmission dynamics of monkeypox in the united kingdom: contact tracing study. BMJ **379** (2022)
33. World Health Organization: Mpox: multi-country external situation report no. 44, published 23 December 2024 (2024). https://www.who.int/publications/m/item/multi-country-outbreak-of-mpox--external-situation-report-44---23-december-2024
34. World Health Organization: Mpox: multi-country outbreak of mpox, external situation report no. 32–30 April 2024 (2024). https://www.who.int/publications/m/item/multi-country-outbreak-of-mpox--external-situation-report-32--30-april-2024
35. World Health Organization: WHO director-general declares mpox outbreak a public health emergency of international concern (2024). https://www.who.int/news/item/14-08-2024-who-director-general-declares-mpox-outbreak-a-public-health-emergency-of-international-concern
36. Xiu, F., et al.: Impact of interventions on mpox transmission during the 2022 outbreak in Canada: a mathematical modeling study of three different cities. Int. J. Infect. Dis. 107792 (2025)

Genome Evolution

Probability-Based Sequence Comparison Finds the Oldest Ever Nuclear Mitochondrial DNA Segments in Mammalian Genomes

Muyao Huang and Martin C. Frith(✉)

The Department of Computational Biology and Medical Sciences, The Graduate School of Frontier Sciences, University of Tokyo, 5-1-5 Kashiwanoha, Kashiwa, Chiba 277-8561, Japan
mcfrith@edu.k.u-tokyo.ac.jp

Abstract. The insertion of mitochondrial genome-derived DNA sequences into the nuclear genome is a frequent event in organismal evolution, resulting in nuclear-mitochondrial DNA segments (NUMTs), which serve as a significant driving force for genome evolution. Once incorporated into the nuclear genome, some NUMTs can be conserved for extended periods, adapting to perform novel cellular functions. However, current mainstream methods for detecting NUMTs are inefficient at identifying ancient and highly degraded NUMTs, leading to their prevalence and impact being underestimated. This study focuses on identifying ancient NUMTs in mammalian genomes using enhanced high-sensitivity sequence comparison methods. A sensitive and accurate NUMT-searching pipeline was established, predicting 1,013 NUMTs in the human reference genome, 398 (39%) of which are newly detected compared to the UCSC reference human NUMTs database. Notably, 93 pre-Eutherian human NUMTs were identified, representing significantly older NUMTs than previously reported, with origins dating back at least 100 million years. The most ancient mammalian NUMTs could even date back over 160 million years, inserted into the nuclear genome of the common ancestor of therian mammals. This study provides a comprehensive exploration of the quantity and evolutionary history of mammalian NUMTs, paving the way for future research on endosymbiotic impact on the evolution of nuclear genomes. Full version is available at bioRxiv: https://doi.org/10.1101/2025.03.14.643190

Keywords: Nuclear-mitochondrial DNA segments (NUMTs) · Genome evolution · Mammals

1 Introduction

The insertion of mitochondrial genome fragments from the mitochondrion to the nucleus generates Nuclear-mitochondrial DNA segments (NUMTs), which is an ongoing and frequent process in nearly all eukaryotes, considered to have an important influence on evolution of the nuclear genome. Once inserted, NUMTs can experience various outcomes as they are not under the evolutionary constraints of the mitochondrial genome, which made most of the NUMTs lose their original functions in the mitochondrial genome [5]. Various fates complicate the identification of NUMTs and their functions, making it challenging to determine their roles as adaptive elements of the nuclear genome. Currently, the typical method for identifying NUMTs is comparing the whole mitochondrial genome to the nuclear genome using the standard BLASTN method. However, this approach may fail to detect ancient and highly degenerated NUMTs due to the less sensitive nature of DNA:DNA comparison compared to Protein:DNA comparison, which uses protein similarity scoring matrices. In this study, we propose an optimized and accurate method based on LAST combining Protein:DNA and DNA:DNA matching methods, enhancing sensitivity by learning scoring matrices specific to the given sequences [2]. Notably, our Protein:DNA matching method allows frame shifts within matches, and uses a 64×21 substitution matrix rather than a standard 20×20 matrix to improve the sensitivity [4]. Our approach is well-suited for detecting highly degenerated NUMTs. We applied this method to identify NUMTs in 16 mammalian reference genomes and trace ancient NUMTs across mammalian clades, providing estimates of their insertion times.

2 Materials and Methods

2.1 Data Collection

16 annotated published reference genomes were downloaded from NCBI, including mitochondrial and nuclear sequences from various mammalian orders. Each genome has achieved chromosome-level assembly quality in the latest assembly version (May 2024). Unplaced scaffolds were excluded to ensure accurate NUMT identification.

2.2 Optimized NUMTs-Searching Pipeline

NUMTs were detected using a novel NUMTs-searching pipeline, which combined nuclear genome-mitochondrial genome and nuclear genome-mitochondrial protein comparisons based on LAST. LAST makes efforts to find alignments for every coordinate in the nuclear genome by pre-learning the substitution and gap rates between the mitochondrial genome and the nuclear genome using `last-train` [2].

Subsequently, we repeated the search using a reversed query sequence as a negative control. The highest score obtained from the reverse search was used as a threshold to filter out alignments with score lower than it from the original search. In an important novel step, alignments that overlapped with nuclear ribosomal RNA regions were also excluded. Finally, to facilitate the subsequent analysis in building the NUMT orthology relationships network, adjacent NUMTs were assembled into larger contiguous regions, referred to as "blocks," using bedtools if the distance between them in the nuclear genome was less than 2000 bp, regardless of their strand orientation.

2.3 Identification of Ancient NUMTs in Mammalian Genomes

120 pairwise genome alignments were conducted by LAST, ensuring that only orthologous segments were identified while excluding non-homologous insertions between any two species within the 16 species [1]. A NUMT from one species was considered ancient only if it overlapped with at least two genome-to-genome alignments between that species and others. We then analyzed NUMT orthology relationships between two species based on ancient NUMTs blocks from 16 species.

3 Results

3.1 Newly Detected Human NUMTs

We annotated NUMTs in human genome hg38 and predicted a total of 1013 NUMT insertions. We used Bedtools v2.29.1 to compare our results with the UCSC human NUMTs database coordinates for accuracy assessment. UCSC NUMTs coordinates were converted to the hg38 genome assembly using the UCSC LiftOver. Among 750 UCSC NUMTs, 734 (98%) were detected in our findings (with an overlap rate exceeding 50% of the sequence length). Compared to the UCSC database, 398 out of 1013 (39%) NUMTs were newly identified in our results, demonstrating the sensitivity and accuracy of our NUMTs searching pipeline. Our hg38 NUMTs dataset formatted in BED has been uploaded to the UCSC genome browser, and adopted as their hg38 human NUMTs database.

3.2 Ancient NUMT Insertions Across Mammalian Clades

A total of 193 ancient orthologous NUMTs were identified in the human genome, with 93 of them estimated to have been inserted at least 100 million years ago (Fig. 1). To our knowledge, previous studies have not reported ancient NUMTs that predate the common ancestor of Boreoeutheria [3]. We selected the 11 most conserved NUMTs shared by humans and over 10 species as examples, grouping them into 5 groups based on their presence in inter-genome alignments. Strikingly, the most ancient NUMTs (shown in group 1) could date back at least 160 million years ago.

4 Conclusion and Discussion

It can be challenging to detect highly degraded NUMTs using conventional search strategies. To address this difficulty, our study presents a novel detection pipeline that combines sensitivity and accuracy. This study not only deepens our knowledge of NUMTs but also serves as a testament to the power of innovative computational approaches in revealing hidden layers of genomic information. These findings are expected to inspire further research into the evolutionary and functional significance of NUMTs, as well as other relics of ancient genomic processes.

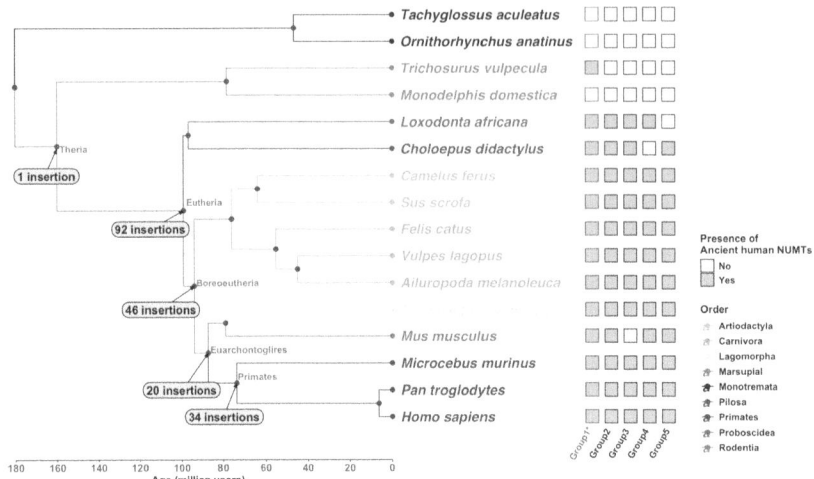

Fig. 1. The most ancient NUMTs in the human genome. The presence of the most ancient NUMT groups in human-other species alignments is shown on the right in the figure. A red border indicates potential NUMTs—highly degenerated NUMTs that could not be directly detected through NU genome:MT genome or NU genome:MT protein comparisons in the human genome but are inferred from alignments with other species.

Acknowledgements. We are grateful to Silvia Rodriguez for discussions about detecting NUMTs in various species and for her assistance in refining the pipeline. This work was supported by the Japan Science and Technology Agency [JPMJCR21N6].

Disclosure of Interests. None declared.

References

1. Frith, M.C., Kawaguchi, R.: Split-alignment of genomes finds orthologies more accurately. Genome Biol. **16** (2015). https://doi.org/10.1186/s13059-015-0670-9
2. Hamada, M., Ono, Y., Asai, K., Frith, M.C.: Training alignment parameters for arbitrary sequencers with LAST-TRAIN. Bioinformatics **33** (2017). https://doi.org/10.1093/bioinformatics/btw742
3. Uvizl, M., et al.: Comparative genome microsynteny illuminates the fast evolution of nuclear mitochondrial segments (NUMTs) in mammals. Mol. Biol. Evol. **41** (2024). https://doi.org/10.1093/molbev/msad278
4. Yao, Y., Frith, M.: Improved DNA-versus-protein homology search for protein fossils. IEEE/ACM Trans. Comput. Biol. Bioinf. (2022). https://doi.org/10.1109/tcbb.2022.3177855
5. Zhang, G.J., Dong, R., Lan, L.N., Li, S.F., Gao, W.J., Niu, H.X.: Nuclear integrants of organellar DNA contribute to genome structure and evolution in plants. Int. J. Mol. Sci. **21** (2020). https://doi.org/10.3390/ijms21030707

Unraveling Insect Immunity: A Cross-Order Comparative Genomic Analysis of Key Immune Proteins

Triveni Shelke[1], Vanika Gupta[2](✉), and Ishaan Gupta[1](✉)

[1] Department of Biochemical Engineering and Biotechnology, Indian Institute of Technology, New Delhi 110016, India
ishaan@dbeb.iitd.ac.in
[2] Department of Zoology, Delhi University, Delhi 110007, India
vgupta1@zoology.du.ac.in

Abstract. Insects have evolved diverse traits to thrive in varied ecosystems, facing constant immune challenges. This bestowed a natural ability to combat these threats, which offers valuable insights for understanding innate immunity. Although the advent of comparative genomics has improved the understanding of evolutionary dynamics in insects over the years, immunity remains relatively unexplored. This study capitalises on the extensive protein data available for 27 insect species to trace the evolutionary trajectories of key immune proteins. Our comparative genomic analysis focuses on proteins that play key roles in insect immunity. Using orthogroup analysis, we find an uneven distribution of orthologs within orthogroups among the orders *Hemiptera, Coleoptera, Hymenoptera, Lepidoptera* and *Diptera*. The unequal prevalence of insect orthologs in immune orthogroups discloses interesting gene duplications, losses and functional diversification events. Our finding reveals, that out of 27 proteins, a total of 2 proteins, PGRP-SA and diptericin show signatures of positive selection, where only diptericin had a specific site under positive selection. We find that PGRP-SA, Superoxide dismutase, Peroxidase, Cecropin A1 & C and Thioester-containing protein depict significant expansions and contractions across insect species. Our findings provide insights into how the five predominant orders of the class *Insecta* have developed varying forms of innate immunity to adapt to their geographical environments. These findings have several practical implications as they may be useful in strategizing pest management and conservation of insects.

Keywords: Insect Immunity · comparative genomics · gene family expansion · positive selection

1 Introduction

Insects represent a dominant class in the phylum *Arthropoda,* tracing their existence back to 480 mya (Misof et al. 2014). Since its origin, this class has evolved into many forms, occupying ecological niches as diverse as terrestrial, rainforests, and marine ecosystems (Misof et al. 2014; Starr & Wallace 2021). These organisms, who possess an unparalleled capacity for evolutionary advancement, exhibit a remarkable balance between speciation and extinction rates (Jouault et al. 2022, Condamine et al. 2016; Stadler & Bokma 2013). This diversity results from high reproductive capacity, short generation time, large effective population size, and other factors that have persisted for hundreds of millions of years (Bradley et al. 2009). Diverse habitats that insects occupy expose them to infections caused by bacteria, fungi, and viruses (Mondal et al. 2023). However, insects have an innate immune system that gives them robust resistance against infectious agents (Hoffmann 1995). The innate immunity is regulated by two major immune pathways Toll and Imd both of which control the expression of antimicrobial peptides (AMPs). The Toll pathway is primarily activated by fungi and Gram-positive bacteria whereas the Imd pathway responds to Gram negative bacteria (De Gregorio et al. 2002). The significance of insect immunity extends beyond individual fitness and plays a vital role in shaping population dynamics, community structure, and disease transmission within ecosystems (Eleftherianos et al. 2021).

Comparative genomics analysis has previously associated gene family expansion and contraction with traits related to detoxification (Breeschoten et al. 2022), immunity (Hou et al. 2021) and parasitism (Yin et al. 2018). Therefore, comparative genomics could yield insights into the selective pressures that have driven the evolution of immune system diversity, highlighting key adaptations to pathogens, parasites, and environmental challenges. Moreover, understanding the evolution of immune pathways in insects can inform strategies for pest management, disease control, and conservation efforts. Compellingly, insights into immune system evolution can inspire innovative approaches for enhancing beneficial insect populations (Viljakainen 2015). Recent developments in genetics and molecular biology have significantly enhanced our comprehension of the intricate immune mechanisms that enable insects to defend themselves against parasites and pathogens. Nonetheless, most of these studies are primarily concerned with the discovery and description of how resistance mechanism's function and rarely address the issue of why they have evolved in such particular ways. Initially, studies have shown that immune genes evolve faster than non-immune genes in 12 *Drosophila* species (Sackton et al. 2007). However, only ants and bees' studies could find signatures of positive and purifying selection respectively (Roux et al. 2014). Later, in *Drosophila*, signatures of positive selection were located in genes encoding pathogen receptors (Obbard et al. 2009). Such similar instances are identified in some components of antimicrobial peptides in *A. gambiae* (Lehmann et al. 2009), Imd signalling in termites (Bulmer & Crozier 2006) and RNA interference pathway in *A. aegypti* (Obbard et al. 2006). Even though studying insect immunity through evolutionary context is familiar, most have focused on closely related species. Less is known about how these pathways have been conserved across the class of *Insecta*. Comparing immune systems across multiple species can highlight both the commonalities and differences in their defence mechanisms, potentially revealing novel immune capabilities and diverse strategies for an internal defence that might be missed when studying a single species. In this study, we have explored the evolution of the innate immune system of 27 insect species across the predominant orders of *Insecta*, to trace the patterns of conservation and divergence leading to adaptation.

2 Results

2.1 Delineating Immune Orthogroups Across Insect Species

OrthoFinder (version_2.5.5) (Emms & Kelly 2019) was used for orthogroup identification in 27 insect species, of which 3 belong to the order *Coleoptera*, 12 to *Diptera*, 3 to *Hemiptera*, 5 to *Hymenoptera* and 4 to *Lepidoptera*. The annotated 27 immune-related proteins for *Drosophila melanogaster* associated with toll and imd signalling pathway, cellular immunity and antimicrobial peptides were filtered from Flybase (Gramates et al. 2022). The initial analysis involved deducing all the possible orthogroups between the 27 insect species. A total of 27,464 orthogroups were identified from the OrthoFinder results, out of which only 27 orthogroups were selected as those containing the immune proteins of interest. These 27 orthogroups comprised of orthologs and paralogs from the contributing insect species, with multiple proteins grouped together for each species. Interestingly all the 27 immune-related genes fall in separate orthogroups. Out of the 27 selected orthogroups, only 13 orthogroups had at least one orthologous gene from each species and 19 had an ortholog in at least 80% of the species. Although none of the selected groups had single-copy orthologs, inferring all proteins underwent duplication events. PGRP-SA protein orthogroup had the highest cumulative number of orthologous protein sequences across all insects (Fig. 1). It had the greatest number of copies among the *Coleoptera*, *Diptera* and *Lepidoptera* species, though peroxidase dominated *Hemiptera* and spatzle dominated *Hymenoptera* (Fig. 1). Apart from *Diptera*, all other orders did not show orthologs for immune proteins such as Dorsal immune factor, Diptericin A, Eater, Drosocin and Drosomycin (Fig. 1). Cecropin A1 and C orthologs were absent in all the other orders except *Diptera* and *Coleoptera* whereas the same was observed for nimrod except for *Diptera* and *Lepidoptera* (Fig. 1). A few instances of missing orthologs in corresponding immune orthogroups were noted for specific orders such as Defensin in *Lepidoptera*, transferrin in *Hemiptera* and Cathepsin B in *Hymenoptera*. These exceptions arose because of data available from only one insect in each order. In the case of *Lepidoptera*, only *Helicoverpa* had corresponding Nimrod orthologs, while in *Coleoptera*, *Aethina* contributed orthologs to the Cecropin A1 and C orthogroup (Fig. 1). Interestingly, among the two isoforms studied for Cathepsins, Cathepsin L1 were universally present in all insects, while Cathepsin B exhibited an unequal distribution among insect species (Fig. 1).

2.2 Key Immune Gene Families Show Significant Expansions and Contractions Across Insect Species

Immune gene family expansion and contraction were recorded using CAFE5 (version 5) (Mendes et al. 2021) which utilized OthroFinder generated orthogroup counts. The significance of gene family expansions was evaluated by comparing the observed counts to the distribution derived from all the 27,464 orthogroups identified by OrthoFinder, using the birth-death parameter lambda estimated equal to 0.9999 by CAFE5. Gene family showing expansions that fell in the top 5% ($p < 0.05$) tail of this distribution were considered significantly changed. Expansion and contraction events were noted in PGRP-SA (Fig. 2A), Spatzle, Peste, Peroxidase, Superoxide dismutase, Cecropin

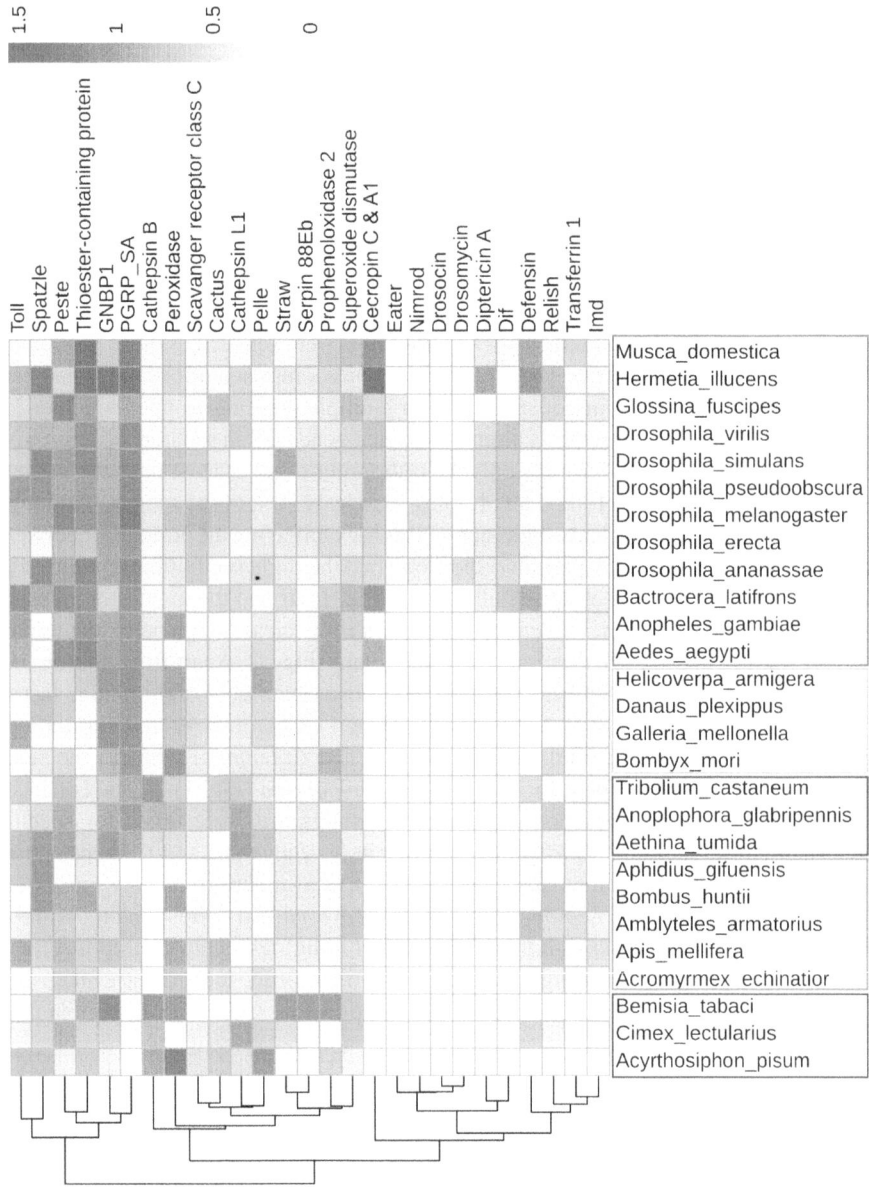

Fig. 1. Heatmap showing the contribution of each species (on right) in immune orthogroups (on top) represented by immune genes of *Drosophila melanogaster*. The species (on right) have been grouped by boxes, the pink box represents the order *Diptera*, yellow for *Lepidoptera*, blue for *Coleoptera*, green for *Hymenoptera* and red for *Hemiptera*. The colour scale represents the log10 normalized counts of orthologs in each orthogroup, the more the redness implies a greater number of orthologs of a species in that orthogroup and vice-versa. (Color figure online)

A1 & C, Cathepsin B and Thioester-containing protein (Supplementary figures: S1, S2, S3). These events in all the immune gene families were not found to be order-specific but more subjective to individual insect species. *Drosophila melanogaster* and *Apis mellifera* have the highest number of expansion events whereas *Amblyteles armatorius*, *Anopheles gambiae* and *Drosophila erecta* have the highest number of contraction events in all the immune gene families combined (Fig. 2B). Some insect species like *Bemisia tabaci*, *Bombus huntii*, *Aethina tumida*, *Tribolium castaneum*, *Dannua plexippus* and *Drosophila ananassae* did not show either expansion or contraction events in the studied gene families (Fig. 2B).

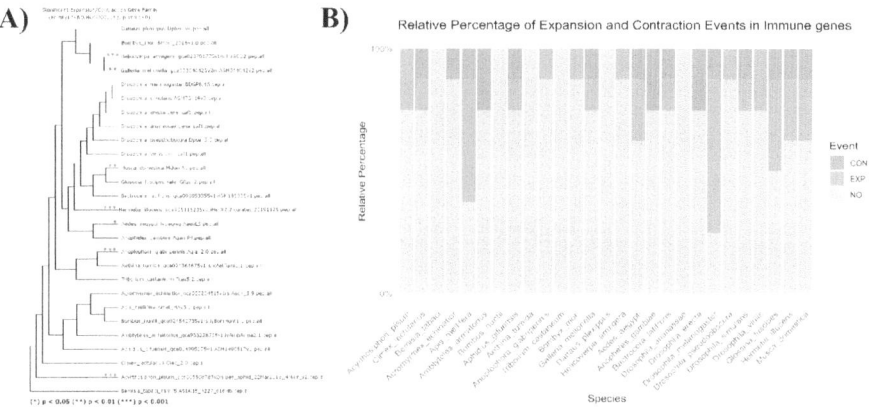

Fig. 2. A) Gene family expansion and contraction plot for PGRP-SA gene orthogroup, the tree depicts expansion (in red) and contraction (in blue) on the nodes with * implying the significance of the test (* implies p < 0.05, ** imply p < 0.01 and *** imply p < 0.001). **B)** The plot shows the relative percentage of expansion and contractions or neither event noted in overall immune orthogroups in each species, where the red fill implies expansion, blue implies contraction and grey shows none of them.

2.3 Phylogenetic Concordance Between Species Tree and Immune-Gene Based Tree

The variation or similarity of the species tree when compared to the tree made from the immune genes could provide insights into the dynamics of the evolution of immune genes versus the whole genome. To understand this, the species tree generated from OrthoFinder (version 2.5.5) (Emms & Kelly 2019) was used as a reference against the tree generated using IQtree (version 2.3.0) (Minh et al. 2020) with only the interested immune genes. The species tree generated using OrthoFinder was rooted using the internal STAG algorithm of the tool. Before generating the gene-based tree using IQtree, the multiple-copy orthologs were reduced to single-copy orthologs by using Phylopypruner *(GitHub - fethalen/phylopypruner: Tree…)*. Phylopypruner results even though provided one-to-one orthologs in species for each immune orthogroup, resulted in the loss of a few species because of highly divergent protein sequences (Supplementary Table S1). The gene-based tree was then obtained using the concatenated orthologs of all species. The IQtree generated tree was based on maximum likelihood and was unrooted, which was latter re-rooted using the mid-point rooting option from iTOL (Letunic & Bork 2024). The two trees were consistent with their divergence order, with all the species in each order clustering together (Fig. 3). According to the immune-specific tree and the species tree *Hemiptera* was found to be the older lineage branching off before the holometabolous orders. Dissimilarity was observed in the branching of species within the orders, but *Diptera* remained more or less conserved as per the species tree (Fig. 3).

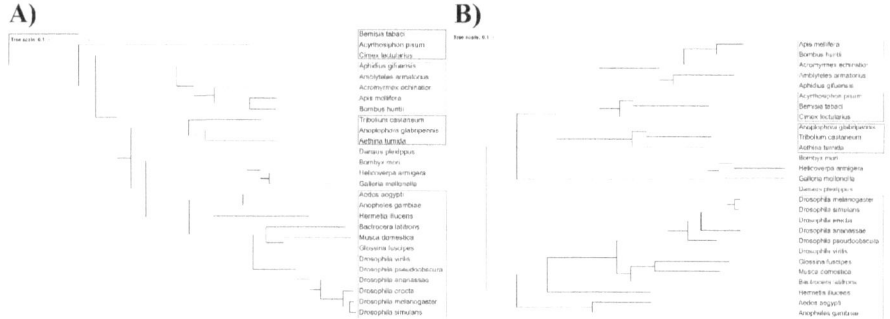

Fig. 3. A) The Overall species tree of 27 insects and **B)** The species tree of 27 insects based on the 27 immune genes. The respective Tree scales are mentioned on top of the trees. The species have been grouped using boxes that represent the orders of the species. The pink box represents the order *Diptera*, yellow for *Lepidoptera*, blue for *Coleoptera*, green for *Hymenoptera* and red for *Hemiptera*.

2.4 Purifying Selection Dominates in Immune Genes Across Insect Species

The single copy orthologs pruned by Phylopypruner *(GitHub - fethalen/phylopypruner: Tree...)* were used for positive selection analysis using codeml from the PAML (Yang 2007) package. Out of the 27 genes, only two genes that are PGRP-SA (w = 1.93780) and Diptericin A (w = 16.55279) showed signatures of positive selection. Although PGRP-SA did not have a specific site under positive selection. Diptericin A on the other hand had one site (55th position) under positive selection with posterior probability significance of 0.953* (Fig. 4). This site (in *Drosophila melanogaster*) contains alanine and located within the peptide chain subunit, indicating potential functional implications. To ascertain the functional importance and adaptive role of the identified site under positive selection, the DrosEU variant file was inspected for the presence of varying mutations in natural populations of *Drosophila*. Referring to the DrosEU variant file, it was found that this site exhibits missense variation of alanine to glycine (Ala to Gly) in natural populations of *Drosophila melanogaster*. The other 25 immune genes were found to be under purifying selection with w < 1.

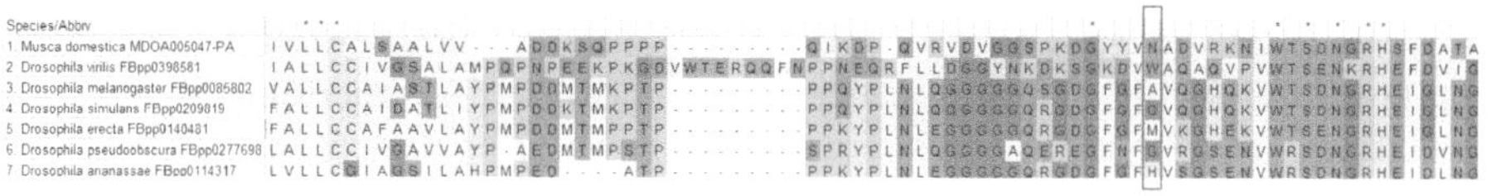

Fig. 4. The alignment of Diptericin A protein sequences of species contributed to the respective orthogroup. The red box highlights the position under positive selection.

3 Discussion

Our study provides a brief description of the evolution of immunity across varied insect species encompassing the five agriculturally and economically relevant orders of class *Insecta*. The comparative genomic analysis of 27 insect species revealed an intricate and interconnected landscape of immune gene evolution. The unequal or uneven distribution of orthogroups among the five studied insect orders underscores the non-uniform evolution of innate immunity in insects. This uneven distribution of immune gene copy number appears linked to lineage-specific ecological pressures that lead to gene loss, duplication and divergence. The ecological niches that different insects inhabit and the pathogens they encounter help shape their innate immunity (Eleftherianos 2021). The dipterans are known to be agricultural pests and blood-feeding vectors of pathogens associated with human and animal diseases (Raguso 2020). The coleopterans thrive on terrestrial environments including natural and human-modified landscapes as well as are represented in freshwater ecosystems (Alekseev et al. 2024). Many lepidopteran species are potential pests that feed on plants including agricultural crops (Hahn et al., 2015) whereas hymenopterans play a fundamental role in almost all terrestrial ecological forces (Branstetter et al. 2017). Similarly, *Hemiptera* species are major agricultural pests thriving on both the terrestrial and aquatic ecosystem (Ma et al. 2022). Habitats of specific insects are mentioned in (Table S2). Proteins like Dorsal immune factor, Diptericin A, Eater, Drosocin and Drosomycin were found to be present exclusively in *Diptera*. Apart from the eater protein, all the others belong to the Imd and Toll signalling pathways specifically responsive against bacterial and fungal infections (Tanji et al. 2007; Cabral et al. 2020). Diptericin A, Drosocin and Drosomycin all are antimicrobial peptides where Diptericin and Drosocin act against Gram-negative bacteria and Drosomycin against fungi. Diptericin A has been lost in lineages with divergent ecology such as the flower-feeding *Tephritid* lineage (Hanson et al. 2019a, 2019b). Drosocin and Drosomycin are *Drosophila*-restricted (Hanson et al. 2019a, 2019b) although they are known to be absent in *Drosophila virilis* (Seto & Tamura 2013). Nimrod, which is known to be present in certain species of Diptera (Kurucz et al. 2007), had no corresponding orthologs in other species apart from *Helicoverpa armigera* which has not been reported yet. Clusters of nimrod-like genes have been reported in *Apis* and *Anopheles* (Kurucz et al. 2007); they were not captured in the orthology analysis. Orthologs for Defensin were absent in *Lepidopteran* species that were considered in this study even though Defensin-like proteins are reported in some *Lepidopteran* species like *Heliothis virescens* (Lamberty et al. 1999) and *Spodoptera frugiperda* (Destoumieux-Garzón et al. 2009). Similarly, for Transferrin 1 the studied *Hemiptera* species do not contribute to the corresponding orthogroup, although *Rhodinius prolixus*, *Riptortus clavatus* (Geiser & Winzerling 2012; Wu et al.

2018), *Nilaparvata lugens* have a corresponding ortholog for Transferrin 1 which has been tested for its role against insecticides (Wu et al. 2018). Besides this, Transferrin 2 and Transferrin 3 but not Transferrin 1, are known to be present in *Acyrthosiphon pisum* (Najera et al. 2021). These results could be explained by the divergent isoforms of immune genes present in insects, where one isoform is spread across insect species but the other remains restricted to a few, this can be understood by the example of Cathepsins. Due to the vast studied roles of Cathepsin L1 which includes reproductive diapause (Chen et al. 2022) and insect moulting (Liu et al. 2006) along with immunity, is present across insect species. Less is known about Cathepsin B beyond a few insect species and hence the exact reason behind its absence in *Hymenoptera* remains unexplainable. It would be beyond the scope of the study to capture the dynamics of all the isoforms present for each of the immune genes. A key observation though is the extent of duplication events in PGRP-SA where barring *Aphidius gifuensis* and *Acyrthosiphon pisum* all insects possessed the corresponding orthologs. The absence in the latter (Gerardo et al. 2010) is already known but that for *Aphidius gifuensis* is a novel report. The lack of orthologs in certain immune-related orthogroups in the non-*Dipteran* species further emphasizes the contribution of gene loss and innovation to the divergent evolution of immune defenses.

The crucial role of immune gene turnover in shaping immune diversity in insects can be further understood by the gene family expansion and contraction analysis. Species such as *Drosophila melanogaster* and *Apis mellifera* displayed notable expansion events, while others such as *Anopheles gambiae* and *Drosophila erecta* exhibited pronounced contractions. The pathogen-recognition genes and most of the effector (anit-microbial peptides) genes have been previously reported to have undergone gene family expansion in *Drosophila melanogaster* (Sackton et al. 2007). Contrary to our findings, fewer gene expansions were reported for *Apis mellifera* concerning immune genes like Thioester-containing protein and others in comparison to *Drosophila* and *Anopheles* (Evans et al. 2006), variations in the results can be attributed to the availability of better-annotated protein data now. Evidence on gene family contractions is scarce on *Drosophila erecta*, however, *Anopheles* have been associated with expansions (Christophides et al. 2002) although contractions in other gene families have not been reported yet. Most of the events inferred from the gene family expansion/contraction results are individual insect-specific and hence it becomes difficult to conclude coherent order-specific findings.

The reconstruction of immune-specific phylogeny using one-to-one orthologs for immune genes revealed noticeable deviations, though the topology overall remained in agreement with the whole-genome-based species tree. Rewiring of immunity during insect evolution can be noted at specific species level, in opposition to the branching noticed in the species tree. Such dissimilarities explain that immune pathway evolution is modelled by differential environmental selective pressures and functional constraints. This can also be explained through the differential rooting methods applied on both the trees.

Our study reports selection throughout the 27 insect species, signatures of positive selection were identified in Diptericin A and an elevated ω was noted for PGRP-SA. Previously, most of the immune genes evolving under purifying selection are reported in *Anopheles* (Lehmann et al. 2009). Immune effectors and recognition proteins as stated above are known to be rapidly expanding in response to the changing microbial community they encounter. Adaptive modifications have been reported in the pathogen-recognition proteins as a result of positive selection (Shultz & Sackton 2019). The co-existence of nonsynonymous mutation in Diptericin A in the natural populations of *Drosophila* along with positive selection, suggests adaptive evolution in the natural population. Insects depend heavily on innate immunity to combat pathogens and therefore have more flexible PGRP repertoire. The pathogen recognition depends on the accessibility of bacterial epitopes which themselves are highly conserved and functionally constraint in comparison to antimicrobial peptides that are directly involved in the killing of microbes.

A key limitation of the study lies in the extensive usage of protein data of non-model insect species. Presence of unannotated data has been an issue in insect species barring the model organism *Drosophila melanogaster* (McCartney et al. 2024) and finding orthologs from incomplete data could impact orthology detection rates. In a study performed with 15 insect species showed that on an average only 54% of genes in each species share orthology with *Drosophila melanogaster* (Asma et al. 2024). Incomplete protein data for insects could increase the risk of false positive results be it in determining orthology or selection. A study based on 12 *Drosophila* genomes found alarming above 45% false positive error rates in the estimation of site-specific positive selection (Markova-Raina & Petrov 2011). However, in our study we have ensured that at least three insects represent a particular insect order and have used latest protein data available in Ensembl. In spite of that the study is limited by the available data.

Our results emphasize that innate immunity in insects is a result of strong purifying selection that enforces the conservation of innate defence mechanisms and episodic positive selection facilitating adaptation to pathogenic pressures. The gene family expansion/contraction further reinforces that immune competence across diverse lineages of insects has been solidified due to adaptive gene duplications. These findings provide foundation for future studies on pest management and insect conservation. Immune adaptation has been linked to development of various pesticide resistance mechanisms as shown in *Plutella xylostella* against Cry1Ac toxin (Wu et al. 2023). Such adaptations are modulated through components of immune signaling pathways and enzymes involved both in immunity and metabolisms. In such cases, pest management strategies can be employed by targeting immune pathway modulators. Immune repertoires can help in identifying insect species that lack key immune pathways like *Acyrthosiphon pisum* does not have key genes of Imd pathway (Viljakainen 2015) which makes them susceptible to novel infections. This can aid in prioritizing conservation of vulnerable insect species.

4 Conclusion

This study provides a comprehensive analysis of immune gene evolution across a taxonomically broad and phylogenetically distant insect species. The combination of orthogroup analysis, gene family expansion and contraction profiling and positive selection test have demonstrated mosaic patterns of conservation and diversification in insect immune systems, where most of the genes evolve under purifying selection with instances of adaptive modification in some. The impact of pathogenic and ecological pressures in shaping the immune repertoires was strongly emphasized by the differential gene duplication and loss among insect species. Unique evolutionary trajectories captured through phylogenetic comparisons reveal lineage-specific adaptations. These findings have several practical implications as they may be useful in strategizing pest management and conservation of insects. Future studies correlating the comparative genomic analysis with the post-pathogen interaction transcriptome of varied insects would enhance the understanding of the mechanistic basis for adaptive changes. Further studies deep-diving into each isoform for the immune genes would capture an even more complex design in the evolution of immunity, which was out of the scope of the current study. It should also focus on mitigating annotation biases and reducing false positive rates in orthology determination.

5 Methods

5.1 Data Acquisition

Immune-related genes for *Drosophila melanogaster* belonging to the toll and imd signalling pathway, cellular immunity and antimicrobial peptide synthesis (Table 1) were filtered from Flybase (Gramates et al. 2022). The protein fasta and CDS fasta for the selected species (*Acromyrmex echinatior, Acyrthosiphon pisum, Aedes aegypti, Aethina tumida, Amblyteles armatorius, Anopheles gambiae, Anoplophora glabripennis, Aphidius gifuensis, Apis mellifera, Bactrocera latifrons, Bemisia tabaci, Bombus huntii, Bombyx mori, Cimex lectularius, Danaus plexippus, Drosophila ananassae, Drosophila erecta, Drosophila melanogaster, Drosophila pseudoobscura, Drosophila simulans, Drosophila virilis, Galleria mellonella, Glossina fuscipes, Helicoverpa armigera, Hermetia illucens, Leptinotarsa decemlineata, Tribolium castaneum, Musca domestica*) from Ensembl Genomes (Martin et al. 2022).

5.2 Identification of Orthogroups Among Immune Genes

Orthogroups were identified using OrthoFinder (version_2.5.5) (Emms & Kelly 2019), which finds orthogroups containing orthologs and paralogs in related insect species and infers rooted gene trees for the identified orthogroups. The tool was run with local alignment program diamond on 16 threads (Lu et al. 2023). A rooted species tree for the analyzed species is also provided by the tool. Comprehensive statistics for comparative analyses between the different species were deduced through the results of OrthoFinder. Initially, all the possible orthogroups between the 27 species were obtained and then orthogroups containing immune-related genes of *Drosophila melanogaster* were filtered from the Orthogroups.tsv file generated by the tool.

5.3 Expansion and Contraction Analysis of Immune Genes

The expansion and contraction analysis were performed using CAFE (version 5) (Mendes et al. 2021). The input provided was the Orthogroups.tsv file and species-rooted tree generated from OrthoFinder. The tool calculates lamba that is the most likely rate of change across the entire tree using maximum likelihood calculations. The analysis was performed without any among family rate variation along with other default parameters and the generated results were studied for identifying families with contraction and expansion in copy number. To identify the specific gene gain and loss in each species for a particular gene family, the results were further interpreted using CafePlotter (v0.2.0) (*GitHub - moshi4/CafePlotter: A Tool for Plotting CAFE5 Gene Family Expansion/contraction Result* n.d.).

5.4 One-to-One Orthologs and Gene Tree Prediction

The orthogroups corresponding to immune proteins had many-to-many orthologs which makes the downstream selection analysis complicated. Hence, sequences that are 1:1 orthologous had to be retrieved from the orthogroups. The sequences for each protein in the orthogroups were aligned using MAFFT (*GitHub - GSLBiotech/mafft: Align Multiple Amino Acid or Nucleotide Sequences* n.d.). The resultant alignment was then used to produce an orthogroup-specific tree using IQtree (version 2.3.0) (Minh et al. 2020). The aligned file along with the orthogroup tree is supplied to the tool Phylopypruner (*GitHub - fethalen/phylopypruner: Tree…*), a phylogenetic-tree-based orthology inference tool that produces 1:1 orthologous sequence. These 1:1 orthologs were identified by the largest subtree method (-prune LS) available in phylopypruner and the min-taxa was set to 4 which would discard output alignments with OTUs lesser than 4. The selected ortholog sequences were again aligned using MAFFT and gene trees were made by IQtree using the pruned aligned sequences. The alignments of all the orthogroups containing pruned orthologs were then concatenated into a single phyml file using catfasta2phyml (*GitHub - nylander/catfasta2phyml: Concatenates FASTA Formatted Files to One "Phyml" (PHYLIP) Formatted File,* n.d.). The concatenated phyml file was used to produce an exclusive species tree based on only the immune gene orthogroups selected for the study by IQtree with –ufboot 10000. All the tools were used with default parameters and the best-fit model for the trees produced was predicted by IQtree itself.

5.5 Selection Analysis

The new amino acid alignments for all the orthogroups were translated to corresponding nucleotide sequences using pal2nal (Suyama et al. 2006). Further, to investigate the protein evolution driven by positive selection, CODEML, a PAML (Yang 2007) package was used. The synonymous substitution (dS), and nonsynonymous (dN) substitution rates, along with protein substitution rates (dN/dS, ω) were estimated using the site model in CODEML. The pair of site models used were M7 (NSsites = 7 and Model = 0) vs M8 (NSsites = 8 and Model = 0) and CodonFreq = 7 (mutation-selection model). The log-likelihood difference between the model M7 and M8 was then calculated and the chi-square test was performed. The orthogroups where the LRT (log-likelihood ratio test) favours M8 which is positive selection, were then scrutinized to find specific sites in the genes under positive selection. This was inferred from the Bayes empirical Bayes (BEB) analysis reported in the CODEML output. The sites noted to be under positive selection were then examined in the DrosEU (*DrosEU – European Drosophila*

Population Genomics Consortium n.d.) variant file to identify variants present at the site in natural *Drosophila* populations.

Table 1. 27 Immune proteins and their functions

Immune Protein	Function and pathway	References
Immune Signaling Pathways		
cactus	NF-κB inhibitor, activates dif and regulates the toll pathway	Bangham et al. (2006)
Dorsal immunity factor	NF-κB proteins that activate antimicrobial peptides	Bangham et al. (2006)
pelle	Phosphorylates cactus leading to activation of NF-κB	Mahanta et al. (2023)
spatzle	Binds to toll and activates its pathway	Lima et al. (2021)
toll	Transmembrane receptor for responses against fungi and Gram-positive bacteria	Lima et al. (2021)
Imd	Adapter protein that gets activated post binding of peptidoglycan of Gram-positive bacteria to the receptors and further transduces the signal	Kleino & Silverman (2014)
relish	Transcription factor that regulates the production of antimicrobial peptides	Sanda et al. (2019)
straw	Activates the c-Jun N-terminal kinase (JNK) pathway that coordinates with Imd to amplify production of AMPs	Yu et al. (2022)
Pattern recognition receptors		
GNBP1	Pathogen-recognition receptor that binds Gram-positive bacteria and fungi	Bangham et al. (2006)
PGRP-SA	Pathogen-recognition receptor that binds Gram-positive bacteria and fungi	Bangham et al. (2006)
eater	Transmembrane receptor involved in phagocytosis	Bangham et al. (2006)
nimrod	Binds bacteria and activates phagocytosis in hematocytes	Estévez-Lao & Hillyer (2014)
Scavenger receptor class C	Membrane-bound receptors involved in phagocytosis	Kim et al. (2017)

(*continued*)

Table 1. (*continued*)

Immune Protein	Function and pathway	References
peste	Recognition and uptake of *Mycobacterium*	Philips et al. (2005)
Antimicrobial peptides		
Cecropin A1 and C	Antibacterial and antifungal activity	Hanson & Lemaitre (2020)
defensin	Antibacterial and antifungal activity	Hanson & Lemaitre (2020)
Diptericin A	Antibacterial activity	Hanson et al. (2019a, 2019b)
drosocin	Antibacterial activity	Hanson et al. (2019a, 2019b)
drosomycin	Antifungal activity	Hanson et al. (2019a, 2019b)
Cellular immunity		
Cathepsin B	Cellular proteases	Saikhedkar et al. (2015)
Cathepsin L1	Cellular proteases	Saikhedkar et al. (2015)
Serpin 88Eb	Serine protease inhibitor involved in melanisation	Zou et al. (2009)
peroxidases	Cellular response to oxidative stress and immunity	Zhao et al. (2001)
prophenoloxidase	Involved in melanin synthesis and encapsulates pathogens	A. Lu et al. (2014)
Superoxide dismutase	Detoxifies reactive oxygen species	Kobayashi et al. (2019)
thioester-containing protein	Opsonisation of pathogens	Williams & Baxter (2014)
Transferrin 1	Iron-binding protein	Iatsenko et al. (2020)

References

Alekseev, V., Napreenko, M., Napreenko-Dorokhova, T.: Ecological groups of Coleoptera (Insecta) as indicators of habitat transformation on drained and rewetted peatlands: a baseline study from a carbon supersite, Kaliningrad, Russia. Insects **15**(5) (2024). https://doi.org/10.3390/insects15050356

Asma, H., et al.: Regulatory genome annotation of 33 insect species. eLife **13** (2024). https://doi.org/10.7554/eLife.96738

Bangham, J., Jiggins, F., Lemaitre, B.: Insect immunity: the post-genomic era. Immunity **25**(1), 1–5 (2006)

Bradley, T.J., et al.: Episodes in insect evolution. Integr. Comp. Biol. **49**(5), 590–606 (2009)

Branstetter, M.G., et al.: Phylogenomic insights into the evolution of stinging wasps and the origins of ants and bees. Curr. Biol. CB **27**(7), 1019–1025 (2017)

Breeschoten, T., van der Linden, C.F.H., Ros, V.I.D., Schranz, M.E., Simon, S.: Expanding the menu: are polyphagy and gene family expansions linked across lepidoptera? Genome Biol. Evol. **14**(1) (2022). https://doi.org/10.1093/gbe/evab283

Bulmer, M.S., Crozier, R.H.: Variation in positive selection in termite GNBPs and Relish. Mol. Biol. Evol. **23**(2), 317–326 (2006)

Cabral, S., et al.: (Diptera: Culicidae) immune responses with different feeding regimes following infection by the entomopathogenic fungus. Insects **11**(2) (2020). https://doi.org/10.3390/insects11020095

Chen, J., et al.: Contributes to reproductive diapause by regulating lipid storage and survival of (Linnaeus). Int. J. Mol. Sci. **24**(1) (2022). https://doi.org/10.3390/ijms24010611

Chen, S., Zhou, Y., Chen, Y., Gu, J.: Fastp: an ultra-fast all-in-one FASTQ preprocessor. Bioinformatics **34**(17), i884–i890 (2018)

Christophides, G.K., et al.: Immunity-related genes and gene families in Anopheles gambiae. Science **298**(5591), 159–165 (2002)

Condamine, F.L., Clapham, M.E., Kergoat, G.J.: Global patterns of insect diversification: towards a reconciliation of fossil and molecular evidence? Sci. Rep. **6**, 19208 (2016)

De Gregorio, E., Spellman, P.T., Tzou, P., Rubin, G.M., Lemaitre, B.: The Toll and Imd pathways are the major regulators of the immune response in Drosophila. EMBO J. **21**(11), 2568–2579 (2002)

Destoumieux-Garzón, D., et al.: Spodoptera frugiperda X-tox protein, an immune-related defensin rosary, has lost the function of ancestral defensins. PLoS ONE **4**(8), e6795 (2009)

Dobin, A., et al.: STAR: ultrafast universal RNA-seq aligner. Bioinformatics **29**(1), 15–21 (2013)

DrosEU – European Drosophila Population Genomics Consortium (n.d.). https://droseu.net/. Accessed 7 Feb 2025

Eleftherianos, I.: Editorial: "evolutionary genetics of insect innate immunity". Genes **12**(5) (2021). https://doi.org/10.3390/genes12050725

Eleftherianos, I., et al.: Diversity of insect antimicrobial peptides and proteins - a functional perspective: a review. Int. J. Biol. Macromol. **191**, 277–287 (2021)

Emms, D.M., Kelly, S.: OrthoFinder: phylogenetic orthology inference for comparative genomics. Genome Biol. **20**(1), 238 (2019)

Estévez-Lao, T.Y., Hillyer, J.F.: Involvement of the Anopheles gambiae Nimrod gene family in mosquito immune responses. Insect Biochem. Mol. Biol. **44**, 12–22 (2014)

Evans, J.D., et al.: Immune pathways and defence mechanisms in honey bees Apis mellifera. Insect Mol. Biol. **15**(5), 645–656 (2006)

Geiser, D.L., Winzerling, J.J.: Insect transferrins: multifunctional proteins. Biochem. Biophys. Acta. **1820**(3), 437–451 (2012)

Gerardo, N.M., et al.: Immunity and other defenses in pea aphids, Acyrthosiphon pisum. Genome Biol. **11**(2), R21 (2010)

GitHub - fethalen/phylopypruner: Tree-based orthology inference with functionality for reducing contamination (n.d.). GitHub. https://github.com/fethalen/phylopypruner. Accessed 10 Feb 2025

GitHub - GSLBiotech/mafft: Align multiple amino acid or nucleotide sequences (n.d.). GitHub. https://github.com/GSLBiotech/mafft. Accessed 6 Feb 2025

GitHub - moshi4/CafePlotter: A tool for plotting CAFE5 gene family expansion/contraction result (n.d.). GitHub.https://github.com/moshi4/CafePlotter. Accessed 6 Feb 2025

GitHub - nylander/catfasta2phyml: Concatenates FASTA formatted files to one "phyml" (PHYLIP) formatted file (n.d.) GitHub.https://github.com/nylander/catfasta2phyml. Accessed 6 Feb 2025

Gramates, L.S., et al.: FlyBase: a guided tour of highlighted features. Genetics **220**(4) (2022). https://doi.org/10.1093/genetics/iyac035

Hanson, M.A., Lemaitre, B.: New insights on Drosophila antimicrobial peptide function in host defense and beyond. Curr. Opin. Immunol. **62**, 22–30 (2020)

Hanson, M.A., Dostálová, A., Ceroni, C., Poidevin, M., Kondo, S., Lemaître, B.: Correction: synergy and remarkable specificity of antimicrobial peptides in vivo using a systematic knockout approach. eLife **8** (2019a). https://doi.org/10.7554/eLife.48778

Hanson, M.A., Lemaitre, B., Unckless, R.L.: Dynamic evolution of antimicrobial peptides underscores trade-offs between immunity and ecological fitness. Front. Immunol. **10**, 2620 (2019b)

Harrison, P.W., et al.: Ensembl 2024. Nucleic Acids Res. **52**(D1), D891–D899 (2024)

Hoffmann, J.A.: Innate immunity of insects. Curr. Opin. Immunol. **7**(1), 4–10 (1995)

Hou, H.-X., Huang, D.-W., Xin, Z.-Z., Xiao, J.-H.: Genome-wide analysis of gene families of pattern recognition receptors in fig wasps (Hymenoptera, Chalcidea). Genes (2021) **12**(12). https://doi.org/10.3390/genes12121952

Iatsenko, I., Marra, A., Boquete, J.-P., Peña, J., Lemaitre, B.: Iron sequestration by transferrin 1 mediates nutritional immunity in. Proc. Natl. Acad. Sci. U.S.A. **117**(13), 7317–7325 (2020)

Jouault, C., Nel, A., Perrichot, V., Legendre, F., Condamine, F.L.: Multiple drivers and lineage-specific insect extinctions during the Permo-Triassic. Nat. Commun. **13**(1), 7512 (2022)

Kim, S.G., et al.: TmSR-C, scavenger receptor class C, plays a pivotal role in antifungal and antibacterial immunity in the coleopteran insect Tenebrio molitor. Insect Biochem. Mol. Biol. **89**, 31–42 (2017)

Kleino, A., Silverman, N.: The Drosophila IMD pathway in the activation of the humoral immune response. Dev. Comp. Immunol. **42**(1), 25–35 (2014)

Kobayashi, Y., et al.: Comparative analysis of seven types of superoxide dismutases for their ability to respond to oxidative stress in Bombyx mori. Sci. Rep. **9**(1), 2170 (2019)

Kurucz, E., et al.: Nimrod, a putative phagocytosis receptor with EGF repeats in Drosophila plasmatocytes. Curr. Biol. CB **17**(7), 649–654 (2007)

Lamberty, M., et al.: Insect immunity. Isolation from the lepidopteran Heliothis virescens of a novel insect defensin with potent antifungal activity. J. Biol. Chem. **274**(14), 9320–9326 (1999)

Lehmann, T., et al.: Molecular evolution of immune genes in the malaria mosquito Anopheles gambiae. PLoS ONE **4**(2), e4549 (2009)

Letunic, I., Bork, P.: Interactive tree of life (iTOL) v6: recent updates to the phylogenetic tree display and annotation tool. Nucleic Acids Res. **52**(W1), W78–W82 (2024)

Lima, L.F., Torres, A.Q., Jardim, R., Mesquita, R.D., Schama, R.: Evolution of Toll, Spatzle and MyD88 in insects: the problem of the Diptera bias. BMC Genomics **22**(1), 562 (2021)

Lin, X., Feng, C., Lin, T., Harris, A.J., Li, Y., Kang, M.: Jackfruit genome and population genomics provide insights into fruit evolution and domestication history in China. Hortic. Res. **9**, uhac173 (2022)

Liu, J., Shi, G.-P., Zhang, W.-Q., Zhang, G.-R., Xu, W.-H.: Cathepsin L function in insect moulting: molecular cloning and functional analysis in cotton bollworm, Helicoverpa armigera. Insect Mol. Biol. **15**(6), 823–834 (2006)

Love, M.I., Huber, W., Anders, S.: Moderated estimation of fold change and dispersion for RNA-seq data with DESeq2. Genome Biol. **15**(12), 550 (2014)

Lu, A., et al.: Insect prophenoloxidase: the view beyond immunity. Front. Physiol. **5**, 252 (2014)

Lu, M., Cao, M., Yang, J., Swenson, N.G.: Comparative transcriptomics reveals divergence in pathogen response gene families amongst 20 forest tree species. G3 (Bethesda, Md.) **13**(12) (2023). https://doi.org/10.1093/g3journal/jkad233

Ma, J., Faqir, Y., Tan, C., Khaliq, G.: Terrestrial insects as a promising source of chitosan and recent developments in its application for various industries. Food Chem. **373**(Pt A), 131407 (2022)

Ma, Y., Wang, J., Zhong, Y., Geng, F., Cramer, G.R., Cheng, Z.-M.M.: Subfunctionalization of cation/proton antiporter 1 genes in grapevine in response to salt stress in different organs. Hortic. Res. **2**, 15031 (2015)

Mahanta, D.K., et al.: Insect-pathogen crosstalk and the cellular-molecular mechanisms of insect immunity: uncovering the underlying signaling pathways and immune regulatory function of non-coding RNAs. Front. Immunol. **14**, 1169152 (2023)

Markova-Raina, P., Petrov, D.: High sensitivity to aligner and high rate of false positives in the estimates of positive selection in the 12 Drosophila genomes. Genome Res. **21**(6), 863–874 (2011)

Martin, F.J., et al.: Ensembl 2023. Nucleic Acids Res. **51**(D1), D933–D941 (2022)

McCartney, N., Kondakath, G., Tai, A., Trimmer, B.A.: Functional annotation of insecta transcriptomes: a cautionary tale from Lepidoptera. Insect Biochem. Mol. Biol. **165**, 104038 (2024)

Mendes, F.K., Vanderpool, D., Fulton, B., Hahn, M.W.: CAFE 5 models variation in evolutionary rates among gene families. Bioinformatics **36**(22–23), 5516–5518 (2021)

Minh, B.Q., et al.: IQ-TREE 2: new models and efficient methods for phylogenetic inference in the genomic era. Mol. Biol. Evol. **37**(5), 1530–1534 (2020)

Misof, B., et al.: Phylogenomics resolves the timing and pattern of insect evolution. Science **346**(6210), 763–767 (2014)

Mondal, S., Somani, J., Roy, S., Babu, A., Pandey, A.K.: Insect microbial symbionts: ecology, interactions, and biological significance. Microorganisms **11**(11) (2023). https://doi.org/10.3390/microorganisms11112665

Najera, D.G., Dittmer, N.T., Weber, J.J., Kanost, M.R., Gorman, M.J.: Phylogenetic and sequence analyses of insect transferrins suggest that only transferrin 1 has a role in iron homeostasis. Insect Sci. **28**(2), 495–508 (2021)

Obbard, D.J., Jiggins, F.M., Halligan, D.L., Little, T.J.: Natural selection drives extremely rapid evolution in antiviral RNAi genes. Curr. Biol. CB **16**(6), 580–585 (2006)

Obbard, D.J., Welch, J.J., Kim, K.-W., Jiggins, F.M.: Quantifying adaptive evolution in the Drosophila immune system. PLoS Genet. **5**(10), e1000698 (2009)

Philips, J.A., Rubin, E.J., Perrimon, N.: Drosophila RNAi screen reveals CD36 family member required for mycobacterial infection. Science **309**(5738), 1251–1253 (2005)

Raguso, R.A.: Don't forget the flies: dipteran diversity and its consequences for floral ecology and evolution. Appl. Entomol. Zool. **55**(1), 1–7 (2020)

Roux, J., Privman, E., Moretti, S., Daub, J.T., Robinson-Rechavi, M., Keller, L.: Patterns of positive selection in seven ant genomes. Mol. Biol. Evol. **31**(7), 1661–1685 (2014)

Sackton, T.B., Lazzaro, B.P., Schlenke, T.A., Evans, J.D., Hultmark, D., Clark, A.G.: Dynamic evolution of the innate immune system in Drosophila. Nat. Genet. **39**(12), 1461–1468 (2007)

Saikhedkar, N., Summanwar, A., Joshi, R., Giri, A.: Cathepsins of lepidopteran insects: aspects and prospects. Insect Biochem. Mol. Biol. **64**, 51–59 (2015)

Sanda, N.B., Hou, B., Muhammad, A., Ali, H., Hou, Y.: Exploring the role of relish on antimicrobial peptide expressions (AMPs) upon nematode-bacteria complex challenge in the nipa palm hispid beetle, Maulik (Coleoptera: Chrysomelidae). Front. Microbiol. **10**, 2466 (2019)

Seto, Y., Tamura, K.: Extensive differences in antifungal immune response in two Drosophila species revealed by comparative transcriptome analysis. Int. J. Genomics **2013**, 542139 (2013)

Shultz, A.J., Sackton, T.B.: Immune genes are hotspots of shared positive selection across birds and mammals. eLife **8** (2019). https://doi.org/10.7554/eLife.41815

Stadler, T., Bokma, F.: Estimating speciation and extinction rates for phylogenies of higher taxa. Syst. Biol. **62**(2), 220–230 (2013)

Starr, S.M., Wallace, J.R.: Ecology and biology of aquatic insects. Insects **12**(1) (2021). https://doi.org/10.3390/insects12010051

Suyama, M., Torrents, D., Bork, P.: PAL2NAL: robust conversion of protein sequence alignments into the corresponding codon alignments. Nucleic Acids Res. **34**, W609–W612 (2006). Web Server issue

Tanji, T., Hu, X., Weber, A.N.R., Ip, Y.T.: Toll and IMD pathways synergistically activate an innate immune response in Drosophila melanogaster. Mol. Cell Biol. **27**(12), 4578–4588 (2007)

The effects of agrochemicals on Lepidoptera, with a focus on moths, and their pollination service in field margin habitats. Agric. Ecosyst. Environ. **207**, 153–162 (2015)

Viljakainen, L.: Evolutionary genetics of insect innate immunity. Brief. Funct. Genomics **14**(6), 407–412 (2015)

Wang, P., et al.: The genome evolution and domestication of tropical fruit mango. Genome Biol. **21**(1) 60 (2020)

Williams, M., Baxter, R.: The structure and function of thioester-containing proteins in arthropods. Biophys. Rev. **6**(3–4), 261–272 (2014)

Wu, S.-F., Li, J., Zhang, Y., Gao, C.-F.: Transferrin family genes in the brown Planthopper, Nilaparvata lugens (Hemiptera: Delphacidae) in response to three insecticides. J. Econ. Entomol. **111**(1), 375–381 (2018)

Wu, X., et al.: Genome-wide identification and immune response analysis of mitogen-activated protein kinase cascades in tea geometrid, Ectropis grisescens Warren (Geometridae, Lepidoptera). BMC Genomics **24**(1), 344 (2023)

Yang, Z.: PAML 4: phylogenetic analysis by maximum likelihood. Mol. Biol. Evol. **24**(8), 1586–1591 (2007)

Yin, C., et al.: The genomic features of parasitism, Polyembryony and immune evasion in the endoparasitic wasp Macrocentrus cingulum. BMC Genomics **19**(1), 420 (2018)

Yu, S., Luo, F., Xu, Y., Zhang, Y., Jin, L.H.: Innate immunity involves multiple signaling pathways and coordinated communication between different tissues. Front. Immunol. **13**, 905370 (2022)

Zhao, X., Smartt, C.T., Li, J., Christensen, B.M.: Aedes aegypti peroxidase gene characterization and developmental expression. Insect Biochem. Mol. Biol. **31**(4–5), 481–490 (2001)

Zou, Z., Picheng, Z., Weng, H., Mita, K., Jiang, H.: A comparative analysis of serpin genes in the silkworm genome. Genomics **93**(4), 367–375 (2009)

Analyzing Sequence Similarity Distributions in Salmonidae: A Branching Process Approach

Yue Zhang(✉)

Thompson Rivers University, Kamloops, British Colombia, Canada
yuezhang@tru.ca

Abstract. This study examines genome evolution in the Salmonidae, focusing on the implications of whole-genome duplications (WGDs) and subsequent gene fractionation. By leveraging sequence similarity to infer key evolutionary parameters, such as gene fractionation rates, the research addresses the complexities using the 'syntenic crumble' phenomenon, which complicates the analysis of older polyploidization events due to the gradual disintegration of synteny blocks over time. A series of statistical analyses, along with a branching process model to assess duplicated gene pairs, are employed to refine the methodology, incorporating the analysis of unpaired genes—a technique adapted from plant genomics. Results demonstrate distinct gene retention and loss patterns following polyploidization, with notable differences in the 'crumble constant' (c). Analysis of synteny blocks across various salmon datasets underlines the impact of block size criteria on genomic evolution, pointing to species-specific evolutionary trajectories.

Keywords: whole-genome duplication · fractionation · branching process · sequence divergence · Salmonidae · comparative genomics

1 Introduction

The diversification of teleost fish has been significantly shaped by multiple whole genome duplications (WGDs). Two initial rounds of WGD in the early vertebrate lineage laid the groundwork for vertebrate diversification. Subsequently, the teleost lineage underwent an additional WGD, the teleost-specific third WGD (Ts3R), which is estimated to have occurred around 300 million years ago (Mya) [6,10,18,19]. A more recent fourth WGD, the Ss4R (salmonid-specific autotetraploidization), shaped the salmonid genomes, which occurred at approximately 80 Mya [14,15,17]. Recent genomic analyses such as that of chum salmon (*Oncorhynchus keta*), highlight patterns of gene retention, loss, and functional divergence [9]. Additionally, a homology guide for Pacific salmon has improved the resolution of orthologous and paralogous relationships, enhancing comparative genomic analyses [11]. Beyond salmonids, the catostomid-specific WGD

(Cat-4R) in the Chinese sucker (*Myxocyprinus asiaticus*), estimated at 25.2 Mya, reveals strong conserved synteny and genome stability, contrasting with the extensive rearrangements seen in other polyploid fish lineages [12].

Understanding the evolutionary consequences of WGD extends beyond identifying duplicated genes to examine the mechanisms driving their retention, divergence, and loss. Yu and Sankoff (2022) [22] developed a framework for analyzing syntenic block evolution, emphasizing the roles of gene fractionation, sequence divergence, and chromosomal dynamics in shaping genome structure. Their findings reveal that post-WGD gene loss, or fractionation, follows distinct evolutionary patterns influenced by genomic context and selective pressures. The fractionation process is crucial in influencing homolog similarity and shaping the genomic landscape [3,23].

This study focuses on paralogous genes derived from these WGD events, examining the distribution of coding-sequence similarities to provide evidence for ancient polyploidization events and aid in reconstructing their evolutionary history. To analyze these distributions, we employ a discrete-time branching process model to track gene pairs and examine their similarity distribution. This distribution is shaped primarily by two biological processes: 1) the duplication of genes through WGD, leading to gene pairs that may diverge within the same genome, and 2) the fractionation process, where one gene from each duplicated pair is preferentially lost over time due to biological processes.

The key aspect of our analysis is to decompose the mixture of paralog similarity distributions, identifying distinct peaks or local maxima that signify major evolutionary events, such as WGD or speciation. These peaks correspond to separate evolutionary events and provide insight into the temporal dynamics of genome evolution.

2 The Branching Process in WGD Context

Gene proliferation and loss within genomes can be modelled as a population of genes evolving stochastically (see Eq. 2). Discrete-time branching processes provide a good model for WGDs and gene loss.

Consistent with previous research methodologies [3,23], the survival or deletion of each gene in subsequent generations is treated as a multinomial event, with the assumption that at least one gene remains. This framework allows for the estimation of $u_j^{(i)}$ that is a probability distribution over the number of surviving progeny j at time t_{i+1}, given that each gene at one WGD time t_i produces r_i progeny. For times $t_1 < ... < t_{n-1}$ (where $n \geq 1$), each existing gene is replaced by $r_1, ..., r_{n-1}$ progeny, respectively, where each $r_i \geq 2$. In this context, t_{n-1} represents the occurrence of the final WGD event. This means:

$$t_1 \xrightarrow{u_j^{(1)}} t_2 \xrightarrow{u_j^{(2)}} \cdots \xrightarrow{u_j^{(i)}} t_{i+1} \xrightarrow{u_j^{(i+1)}} \cdots \xrightarrow{u_j^{(n-1)}} t_n$$

and

$$\sum_{j=1}^{r_i} u_j^{(i)} = 1 \qquad (1)$$

The observations are made at time t_n, reflecting the similarity measures (e.g., coding sequence similarity) among all gene pairs in the population, with the original m_1 genes considered unrelated or distantly related. The model initiates with $m_1 \geq 1$ genes at time t_1.

If m_i denotes the number of genes existing at time t_i, and at each time t_i, each gene produces r_i progeny, where $r_i \geq 1$. Let $a_j^{(i)}$ denote the number of genes that have survived in exactly j copies between time t_i and t_{i+1}. If the evolutionary process consists of a series of n 2r-ploidization events, where each duplication event produces $r_i \geq 2$ progeny, the total number of scenarios with the same probability is given by:

$$\prod_{i=1}^{n-1} \binom{m_i}{a_1^{(i)}, \ldots, a_{r_i}^{(i)}}, \tag{2}$$

Extending this to the probability distribution of evolutionary histories, represented by the sequence of duplication events $\mathbf{r} = (r_1, \ldots, r_{n-1})$ and survival variables $\mathbf{a} = \{a_j^{(i)}\}_{j=1,\ldots,r_i}^{i=1,\ldots,n-1}$, we obtain:

$$P(\mathbf{r}; \mathbf{a}) = \prod_{i=1}^{n-1} \binom{m_i}{a_1^{(i)}, \ldots, a_{r_i}^{(i)}} \prod_{j=1}^{r_i} u_j^{(i) a_j^{(i)}}, \tag{3}$$

Let $P^{(j,k)}(\mathbf{r}; \mathbf{a})$ is a conditional probability, representing the probability of evolutionary histories $(\mathbf{r}; \mathbf{a})$ given time t_j and t_k. A distinctive feature of our branching process model is its focus on tracking 'sibling' gene pairs. By focusing on expected values rather than individual gene pair trajectories, the model circumvents the complexity introduced by fractionation dependencies, where the loss of multiple gene pairs within synteny blocks may not be statistically independent. The expected number of genes at time t_j and t_k is:

$$E^{(j,k)}(m_k) = \sum_{\mathbf{r}; \mathbf{a}} P^{(j,k)}(\mathbf{r}; \mathbf{a}) m_j \tag{4}$$

Using expected values smooths out these effects, making calculations easier and keeping the model useful. This feature is pivotal for a comprehensive analysis of gene pair similarity distributions. Furthermore, the expected number of genes at time t_{i+1}, considering all possible retained copies, follows the recurrence relation:

$$E(m_{i+1}) = E(m_i) \cdot \left(\sum_{j=1}^{r_i} j \cdot u_j^{(i)} \right). \tag{5}$$

Upon establishing the method to calculate $E^{(j,k)}(m_k)$, and recognizing the independence of trajectories from any two sibling genes at time t_i, as well as their independence from the trajectory spanning time t_1 to t_i, we can now determine $E(N_i)$, the expected number of gene pairs at time t_n that originated at time t_i.

In our research, we delve into the evolution of gene pairs surviving from WGD events. A key aspect of our analysis is the independent evolution of each gene within a pair, characterized by a gradual decline in similarity. This divergence is primarily due to the accumulation of random single nucleotide mutations, which affect both the gene and amino acid sequences. Importantly, in the context of WGD, the simultaneous loss of both copies of a functional gene within a pair could have deleterious effects on the organism. Hence, our model explicitly excludes the possibility of concurrent gene deletion from a pair. By integrating this principle with the expected number of gene pairs emerging from a branching process, we can calculate the expected number of descendant gene pairs, $d(i, n)$, that share a common ancestor at a specific time point i, as expressed in the following Eq. (6):

$$E(d^{(i,n)}) = E(d^{(i,i+1)})[E^{(i+1,n)}(m_n)]^2 \tag{6}$$

where $E(d^{(i,i+1)})$ represents the expected number of gene pairs formed at generation i and surviving to generation $i + 1$, and $E^{(i+1,n)}(m_n)$ is the expected number of genes at time n starting from time t_{i+1}.

While the inclusion of multinomial coefficients and the potential for high-degree polynomials may seem computationally challenging, the model remains feasible in practice. This is due to the typical values of r_i being 2 or 3, which ensures computational tractability and allows for efficient analysis. For instance, consider a scenario with a single gene at time t_1, assuming all $r_i = 2$. Let $u(i), i = 1, ..., n - 1$ denote the probability that both progeny of a gene at time t_i survive to time t_{i+1} [6].

The expected number N_i of duplicate pairs originating at time t_i and observed at t_n is

$$E(N_1) = m_1 u(1) \prod_{j=2}^{n-1}(1 + u(j))^2, \tag{7}$$

$$E(N_i) = m_1 u(i)(\prod_{j=1}^{i-1}(1 + u(j)))(\prod_{j=i+1}^{n-1}(1 + u(j))^2), \tag{8}$$

$$E(N_{n-1}) = m_1 u(n - 1) \prod_{j=1}^{n-2}(1 + u(j)). \tag{9}$$

Equation (7) describes the expected number of duplicate gene pairs originating at the earliest stage t_1, incorporating the cumulative effect of all subsequent WGD events. Equation (8) generalizes this relationship for an intermediate duplication event, capturing both preceding and subsequent influences on gene survival. Equation (9) describes the final WGD event, where retention is determined solely by prior events.

Within the mathematical model, the model comprises $n-1$ vector parameters $u(1), \ldots u(n - 1)$, corresponding to an equal number of equations as outlined in Eqs. (7)–(9). The inclusion of an additional variable, m_1 (the total number of genes at time t_1), adds complexity to the system. As a result, solving the system by equating the observed number of gene pairs to the expected values derived from the model yields only the relative values of $u(j)$, rather than absolute ones. Despite this limitation, the model can predict a specific observable metric that cannot be directly inferred from the distribution of gene pair similarities.

The analysis centres on synteny blocks to explore the genomic evolution of the Salmonidae. By examining these conserved genomic regions, we endeavour to infer the fractionation rate after whole-genome duplications specific to Salmonidae.

2.1 Synteny Block

Two primary categories of gene pairs can be considered: individual pairs, and pairs within synteny blocks, where genes reside in conserved regions that retain their relative order across species. For these syntenic gene pairs, similarity is calculated as the average similarity of all gene pairs within the corresponding block, with the resulting value weighted by the block's length. This weighting ensures that longer synteny blocks, which reflect greater conservation, have a stronger influence on the overall similarity measure. Synteny blocks, compared to individual gene pairs, display distinct trends in similarity distributions.

In parallel, singleton genes—those not part of any synteny block—are identified (see Fig. 1). These singleton genes are compared to genes within synteny blocks to determine which blocks contain such genes. This step helps identify genes that may have undergone unique evolutionary processes. The identified singleton genes are documented for each synteny block for further analysis.

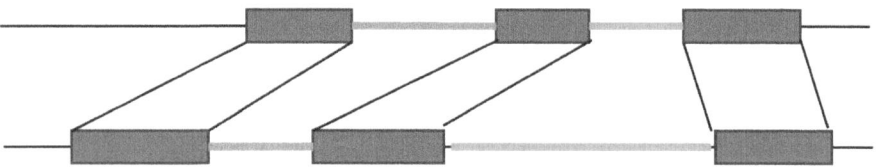

Fig. 1. Pairs and singletons: This figure illustrates gene pairs and singletons within a synteny block. The rectangles represent conserved gene pairs that remain in syntenic regions after WGD. The two rows correspond to homologous chromosomal regions in different species or paralogous regions within the same genome. The black diagonal lines indicate gene pairs that are still retained in both regions, preserving synteny. The horizontal green and purple lines represent singletons. (Color figure online)

To estimate the number of singleton genes, we subtract the count of genes in t_i pairs from the genome's total. This approach, however, requires a nuanced analysis due to the genes to appear in multiple pairs across various t_i events. Directly relying on the genome's total gene count can lead to inaccuracies due to gene expansions and duplications that occur after t_{n-1}. Our focus is on singletons within synteny blocks to accurately infer retention rates, identifying the t_i at which singletons were generated. This method offers a more precise alternative to simple subtraction, providing essential data for parameter estimation.

The model assumes a uniform probability of fractionation across gene pairs, with one pair selected for fractionation at each step. The impact of fractionation on multiple pairs within a block suggests a complex interaction, challenging the assumption of independence among adjacent gene pairs. But our analysis currently emphasizes expected values.

2.2 A Model for the Erosion of Synteny Blocks over Time

The presence of singletons within synteny blocks plays a crucial role in our analysis of gene complement sizes and the computation of "crumble" coefficients.

The process of fractionation gradually erodes synteny blocks, influenced by biological phenomena such as chromosomal rearrangement and gene pair divergence, as well as by technical challenges in detecting these blocks. This includes setting thresholds to distinguish significant collinearity from background noise. Termed 'block crumble,' this process affects our ability to infer retention rates during fractionation, often resulting in an underestimation of gene pairs and singletons within blocks for earlier evolutionary events. Consequently, this can bias retention rate estimations upwards.

To address this issue, our research uses the concept of a 'syntenic cohort' and employs a set of 'crumble coefficients' $c_1, \ldots, c_{n-1}$. These coefficients adjust the m_i values, allowing a more accurate approximation of gene complements and retention probabilities.

By incorporating these adjustments, we refine our understanding of synteny block erosion across evolutionary intervals. This is particularly relevant for genomes that have experienced multiple rounds of replication [23].

$$E(N_1) = m_1 c_1 u(1) \prod_{j=2}^{n-1} (1 + u(j))^2 \tag{10}$$

$$E(N_i) = m_1 c_i u(i) \prod_{j=1}^{i-1} (1 + u(j)) \prod_{j=i+1}^{n-1} (1 + u(j))^2 \tag{11}$$

$$E(N_n)) = m_1 u(n-1) \prod_{j=1}^{n-2} (1 + u(j)) \tag{12}$$

The expected values of singleton are as follows

$$E(S_1) = m_1 c_1 (1 - u(1)) \tag{13}$$

$$E(S_i) = m_1 c_i (1 - u(i)) \prod_{j=1}^{i-1} (1 + u(j)) \tag{14}$$

$$E(S_n) = m_1 (1 - u(n)) \prod_{j=1}^{i-1} (1 + u(j)) \tag{15}$$

Notes: the equations describing the models involved only the $u(\cdot)$, which are retention probabilities, not fractionation rates. Fractionation may simultaneously impact multiple duplicate gene pairs within a synteny block, resulting in a scenario where the loss or retention of gene copies among neighbouring pairs is not statistically independent. Our current model focuses on calculating expected values, and therefore is not directly influenced by this non-independence.

Finally, the number of gene pairs and singleton genes within the identified components is counted. These observed counts are substituted into evolutionary equations in Table 1, allowing for the estimation of evolutionary rates within the model.

Table 1. The expected number of pairs and singletons for two successive WGDs

event	observed	expected number
t_1	pairs	$c_1 m_1 u(1)(1+u(2))^2$
t_2	pairs	$m_1(1+u(1))u(2)$
t_1	singletons	$c_1 m_1(1-u(1))$
t_2	singletons	$m_1(1+u(1))(1-u(2))$

3 Methodology for Analyzing Genome Evolution: Synteny and Singleton Gene Identification

The analytical process begins with the extraction of relevant genomic sequences. In the current study, we adopted a systematic framework for analyzing synteny blocks and estimating evolutionary rates, as illustrated in Fig. 2.

The first step uses SYNMAP with default parameters to perform a self-comparison of the genome, identifying regions of conserved gene order and orientation, known as synteny blocks. The synteny blocks can be identified based on criteria such as the number of collinear genes, the distance between adjacent genes, and the level of sequence similarity. Additionally, parameters like the maximum 20 gene gap size allowed between genes and the minimum number of gene pairs (ranging from three to five) required to define a block can influence the identification process.

After identifying the synteny blocks, the analysis focuses on gene pairs within these regions. The goal is to calculate similarity scores or percentage identities for each gene pair to quantify their evolutionary conservation and divergence. A distribution plot of the similarity scores is generated to visualize the data. SYNMAP is used to obtain gene pair similarities (Fig. 3), and the observed values are substituted into a branching process model. To distinguish different evolutionary groups of gene pairs, a critical cutoff point is determined using the maximum likelihood method, as formulated in Eq. 16.

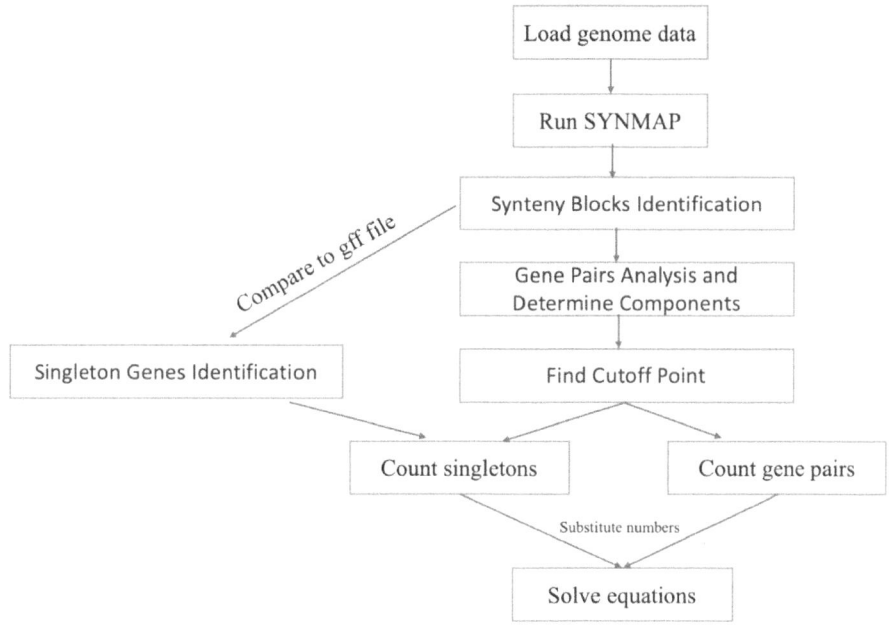

Fig. 2. Workflow for Counting t_i-Pairs Singletons through Successive WGDs

$$H = \arg\max_{h \in (0,1)} \prod_{x \leq h} \lambda_1 N(\mu_1, \sigma_1) \prod_{x > h} \lambda_2 N(\mu_2, \sigma_2) \qquad (16)$$

where $N(\mu, \sigma)$ represents the normal probability density function with mean μ and standard deviation σ, and h denotes the cutoff point to be estimated within the interval $(0, 1)$.

The critical cutoff point H is used to separate gene pairs into distinct similarity-based groups, reflecting different evolutionary origins or divergence levels. This cutoff helps determine whether a gene pair originates from the most recent WGD or from an earlier event. In practical terms, gene pairs with similarity scores above H are considered to have originated from a more recent duplication event, while those below H are likely derived from older duplication events and have diverged significantly.

The similarity distribution is further decomposed into component normal distributions using statistical tools such as *mixtools* [1], which helps estimate the means μ_i, standard deviations σ_i, and mixing proportions λ_i of each component in the Gaussian mixture model. Maximum likelihood estimation is used to identify the cutoff point that distinguishes the components within the distribution [20]. To simplify the optimization process, we apply the natural logarithm to the likelihood function, yielding the log-likelihood function:

$$\log(L) = \sum_{x \leq h} \log\left(\lambda_1 N(\mu_1, \sigma_1)\right) + \sum_{x > h} \log\left(\lambda_2 N(\mu_2, \sigma_2)\right) \tag{17}$$

The optimal cutoff point H, which maximizes the likelihood of correctly classifying observations into the respective distributions, is derived by setting the derivative of the log-likelihood with respect to h to zero. Solving the resulting quadratic equation yields the following analytical expression for the optimal threshold H:

$$H = \frac{2\left(\frac{\mu_1}{\sigma_1^2} - \frac{\mu_2}{\sigma_2^2}\right) \pm \sqrt{\left(2\left(\frac{\mu_1}{\sigma_1^2} - \frac{\mu_2}{\sigma_2^2}\right)\right)^2 - 4\left(\frac{1}{\sigma_1^2} - \frac{1}{\sigma_2^2}\right)\left(\frac{\mu_1^2}{\sigma_1^2} - \frac{\mu_2^2}{\sigma_2^2} - 2\log\left(\frac{\lambda_1}{\lambda_2}\right)\right)}}{2\left(\frac{1}{\sigma_1^2} - \frac{1}{\sigma_2^2}\right)} \tag{18}$$

The presence of the logarithmic ratio of mixing proportions $log(\frac{\lambda_1}{\lambda_2})$ adjusts the cutoff point based on the relative prevalence of each distribution, ensuring an optimal classification threshold. In cases where $\sigma_1 \approx \sigma_2$, the solution simplifies to the weighted average form:

$$H = \frac{\lambda_1 \frac{\mu_1}{\sigma_1^2} + \lambda_2 \frac{\mu_2}{\sigma_2^2}}{\lambda_1 \frac{1}{\sigma_1^2} + \lambda_2 \frac{1}{\sigma_2^2}} \tag{19}$$

These cutoff points are summarized in Table 2, 3 and 4.

Table 2. displays observed counts of gene pairs and singletons, cutoff values, and parameters for synteny blocks with 3 or more genes.

Type (block 3)	Singleton		Pair		cutoff	c_1	m_1	u(1)	u(2)
	t1	t2	t1	t2					
Altantic Salmon	7863	15668	8457	13384	0.852	0.544	21759	0.335	0.461
Lake trout	6727	9088	9627	11577	0.859	0.708	15081	0.37	0.56
Chum Salmon	3862	8596	5095	9504	0.852	0.431	14332.5	0.375	0.482
Chinook Salmon	5250	10206	8572	11960	0.853	0.563	15744	0.408	0.54
Rainbow trout	5202	13622	9283	14292	0.844	0.477	19407	0.438	0.512

4 Five Salmoninae Genomes

In this study, we utilized genomic data from the National Center for Biotechnology Information (NCBI) along with publicly available data on the CoGe (Comparative Genomics) platform [5], a suite of web-based tools designed for the visualization and comparison of genomic sequences. Specifically, we employed

Table 3. displays observed counts of gene pairs and singletons, cutoff values, and parameters for synteny blocks with 4 or more genes.

Type (block 4)	Singleton		Pair		cutoff	c_1	m_1	u(1)	u(2)
	t1	t2	t1	t2					
Atlantic Salmon	5033	15604	5114	12681	0.847	0.35	21326	0.326	0.448
Lake trout	4395	9218	5590	11163	0.853	0.445	15133	0.347	0.548
Chum Salmon	2341	8869	3050	9016	0.845	0.282	13098	0.365	0.504
Chinook Salmon	3775	10535	5605	11776	0.844	0.384	16065	0.389	0.528
Rainbow trout	3874	12605	6228	13920	0.838	0.348	18828	0.409	0.525

Table 4. displays observed counts of gene pairs and singletons, cutoff values, and parameters for synteny blocks with 5 or more genes.

Type (block 5)	Singleton		Pair		cutoff	c_1	m_1	u(1)	u(2)
	t1	t2	t1	t2					
Atlantic Salmon	3193	15382	3551	12235	0.843	0.239	20485	0.348	0.443
Lake trout	2982	9351	3838	10885	0.839	0.308	14963	0.352	0.538
Chum Salmon	1619	8982	2006	8717	0.836	0.193	13039	0.357	0.493
Chinook Salmon	2932	10671	4136	11638	0.841	0.291	16187	0.379	0.522
Rainbow trout	2856	12415	4622	13774	0.831	0.261	18573	0.41	0.526

DAGchainer and SYNMAP [7,8,16], integrated within CoGe, to identify gene pairs with sequence similarity and detect collinear gene pairs across genomes. DAGchainer parameters were configured with a maximum distance between matches (-D) of 20 genes and a minimum threshold of three to five aligned gene pairs (-A). To ensure data quality, the output was filtered to retain only the highest percent identity LASTZ hit for each query gene. The uniformity in pair similarity within a block, coupled with their collinearity, supports the parameter t_i, indicating that it retains the pre-replication gene order. This suggests that interspersed singletons are likely the result of gene pairs that were fractionated around the same time.

The genomic analysis encompassed several species within the salmonid family to provide a comprehensive overview of their evolutionary dynamics. The species analyzed were selected to represent a broad range of the salmonid family and included Atlantic salmon (*Salmo salar* [14]), lake trout (*Salvelinus namaycush* [21]), both of which had been uploaded and published. Additionally, three Pacific salmon species (downloaded from NCBI, uploaded privately): Chinook salmon (*Oncorhynchus tshawytscha* [4]), rainbow trout (*Oncorhynchus mykiss* [2]), and chum salmon (*Oncorhynchus keta* [13])—were included, with their genomic data privately downloaded from NCBI.

Atlantic salmon (*Salmo salar*, ID: 28938) has a genome size of 2.24 Gb, assembled into 29 chromosomes/contigs. The genome (NCBI version 2.0) is unmasked, with noncoding regions accounting for approximately 2.17 Gb. It was curated to remove unplaced chromosomes labeled as "NW_" in the GFF file for improved assembly quality.

Lake Trout (*Salvelinus namaycush*, ID 63989) has a genome size of 2.35 Gb, consisting of 4,121 contigs. The genome assembly (NCBI GCA_013841185.1) is unmasked and was released on May 24, 2022.

Chinook salmon (*Oncorhynchus tshawytscha*, ID 64176) with a genome size of 2.3 Gb and a total ungapped length of 2.3 Gb. The assembly consists of 34 chromosomes and approximately 9,982 contigs organized into 9,977 scaffolds, with 2,313 gaps between scaffolds. The Scaffold N50 and Contig N50 are both 2.9 Mb, indicating high assembly continuity.

Rainbow trout (*Oncorhynchus mykiss*, ID 63876) has a genome size of 2.3 Gb, with 32 chromosomes and 1,228 contigs within 938 scaffolds, containing 196 gaps. The Scaffold N50 is 39.2 Mb, and the Contig N50 is 15.6 Mb, indicating a high-quality assembly.

Chum salmon (*Oncorhynchus keta*, ID 64186) has a genome size of 2.6 Gb, assembled into 37 chromosomes and 17,497 contigs within 17,479 scaffolds, with 3,405 gaps. The Scaffold N50 and Contig N50 are both 2 Mb.

Table 5. Total gene pairs in synteny blocks for different species across block sizes 3, 4, and 5. The data compares the number of gene pairs in synteny blocks for Atlantic salmon, lake trout, chum salmon, chinook salmon, and rainbow trout.

Total pairs	block 3	block 4	block 5
Atlantic Salmon	21763	17770	15776
Lake trout	21143	16741	14721
Chum Salmon	14577	12063	10722
Chinook Salmon	20501	17377	15773
Rainbow trout	23548	20146	18395

Table 5 shows the total number of gene pairs in synteny blocks for different species. It is important to note that in some cases, the numbers may appear smaller than the summation of synteny block pairs from Table 2, 3 and 4. This discrepancy arises due to the potential double-counting of gene pairs when using the mean similarity for each block. Specifically, some gene pairs in one synteny block may have a higher similarity than the cutoff, while in another synteny block, the same gene pairs may have a lower similarity than the cutoff, leading to a reduction in the final count.

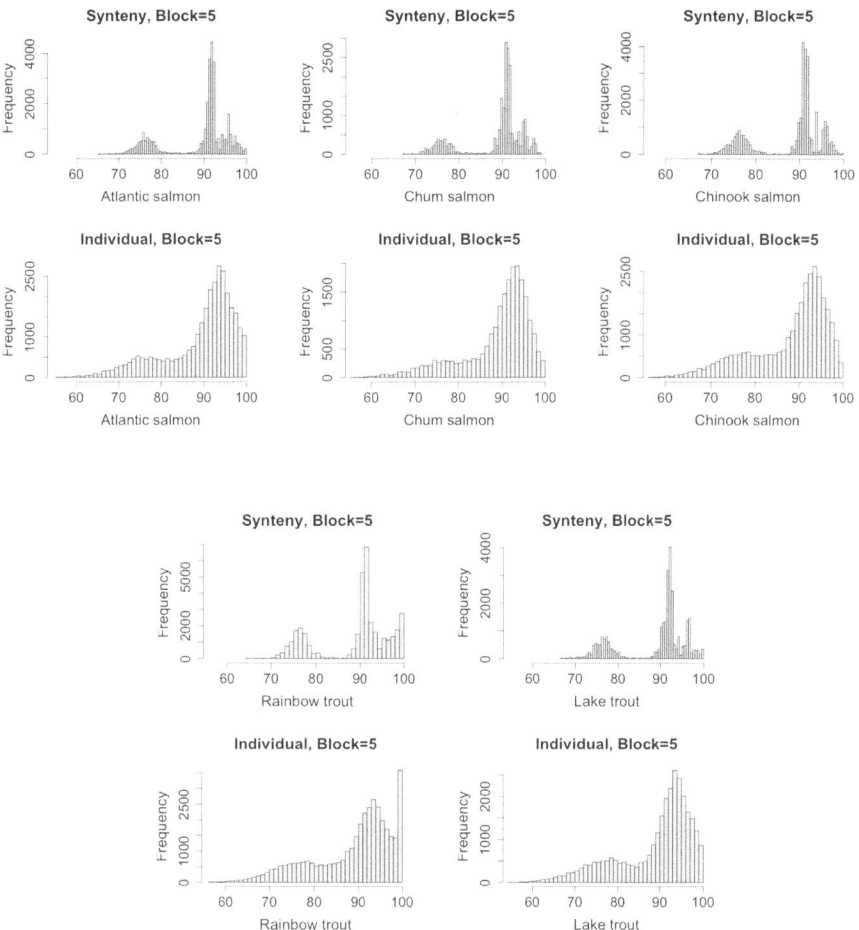

Fig. 3. Similarity Distributions of Individual Gene Pairs and Syntenic Block Pairs (Block = 5) Across Salmonid Species

Figure 3 illustrates the distribution of similarity scores for individual gene pairs and gene pairs within synteny blocks across the five salmonid species, providing valuable insights into WGD events. The local peaks observed in both distributions reflect the synchronicity of gene duplications associated with WGD. Distinct local peaks in these distributions reflect the synchronicity of gene duplications tied to WGD, corresponding to evolutionary periods marked by significant genomic changes. Analyzing these patterns allows inference of both the timing and ploidy levels of WGD events.

Notably, the two local peaks in the similarity distributions of synteny blocks, around 75% and 90%, underscore the differential conservation levels. This distribution pattern with that of individual gene pairs, where two distinct peaks similarly reflect varying degrees of conservation or divergence but may not capture the broader evolutionary context as effectively as synteny blocks. The difference in similarity distributions between individual pairs and those within synteny blocks enriches our understanding of genetic conservation and divergence.

The presence of an ambiguous peak around 95% in the distribution of synteny blocks points to a subset of highly conserved genomic segments. This observation likely suggests segmental duplications on the conservation of these genomic regions.

5 Results

The synteny block analysis across five datasets—Atlantic salmon, lake trout, chum salmon, chinook salmon, and rainbow trout—post-polyploidization reveals clear patterns in gene retention and loss dynamics. As expected, both singleton and gene pair counts increase when the block size criterion decreases from 5 to 3 due to the less stringent requirements for synteny block formation.

Despite variations in block sizes and species, the cutoff values remain relatively stable, ranging from 0.831 to 0.859. This consistency suggests that the criteria for defining synteny blocks—likely based on sequence similarity and gene collinearity—have been applied uniformly. The uniformity of these cutoff values is critical, as it ensures that comparisons across species and evolutionary time points are based on standardized parameters, enabling more accurate interpretations of genomic changes over time.

Interestingly, the parameter 'c', which may represent the proportion of conserved syntenic regions, exhibits a notable increase when moving from block size 5 to block size 3 (see Fig. 4). This upward shift suggests an accelerated rate of genomic rearrangement and gene loss in smaller synteny blocks, a phenomenon often referred to as 'block crumble'. The increase in 'c' highlights the potential for instability in regions with fewer conserved genes, where the structural integrity of syntenic blocks becomes more susceptible to evolutionary forces such as gene loss, duplication, and rearrangement.

The observed differences in the crumble constant c between five salmonid species and two plant orders—Poplar, Durian [6]—are notable. Salmonid species display approximately half the c value compared to these plant genomes, indicating a higher rate of genomic rearrangement and gene loss. In contrast, plants exhibit greater tolerance to polyploidization, likely retaining more duplicated genes.

The upward trend observed in the initial gene complement size, m_1, from block size 5 to block size 3 is intriguing but potentially misleading. Rather than indicating a direct increase in the number of genes within synteny blocks, this trend reflects the number of genes prior to WGDs. It may also be influenced by biases introduced through improved detection methods or by the historical erosion of synteny blocks. Such erosion can distort the perception of gene retention rates, potentially leading to an overestimation in more recent blocks due to an underestimation of the initial gene count.

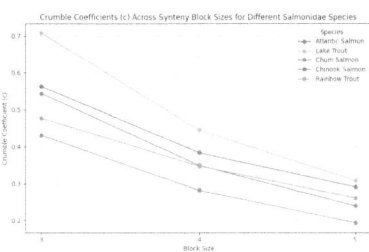 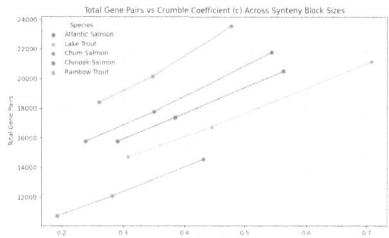

Fig. 4. Left plot showing the crumble coefficients (c) for different Salmonidae species across synteny block sizes (≥ 3, ≥ 4, and ≥ 5 genes). Right plot showing the total gene pairs for different Salmonidae species across synteny block sizes (≥ 3, ≥ 4, and ≥ 5 genes)

Further insights are provided by the survival probabilities, 'u' and 'v', which capture temporal dynamics of gene retention. Notably, u is consistently smaller than v across all datasets. This discrepancy likely reflects differences in evolutionary intervals, with u corresponding to a longer time span between the first and second points of measurement (approximately 200 Mya) and v representing a shorter interval between the second measurement and the present (approximately 80 Mya).

6 Conclusion and Future Work

By leveraging SYNMAP software on the CoGe platform, we assessed coding sequence similarities to evaluate fractionation rates and polyploidy events in salmonid genomes, utilizing a combination of algebraic and statistical methods. This approach emphasizes the role of singleton genes in the fractionation process, ensuring accurate identification and interpretation of retained and lost genes.

Future research should aim to standardize synteny block identification methods, enabling more accurate cross-species comparisons. Expanding the study to include a wider range of species will help clarify the evolutionary consequences of polyploidization. Furthermore, refining the models to incorporate processes such as gene conversion and subfunctionalization may improve the understanding of post-WGD genome evolution. Developing a robust computational pipeline will enhance the efficiency of fractionation analysis, and examining the functional roles of fractionated genes will provide further insights into genome diversification mechanisms.

Acknowledgements. This study was partial funded by the Natural Sciences and Engineering Research Council of Canada (NSERC) under Grant No. RGPIN-2024-06717

References

1. Benaglia, T., Chauveau, D., Hunter, D.R., Young, D.S.: mixtools: an R package for analyzing mixture models. J. Stat. Softw. **32**(6), 1–29 (2009). https://doi.org/10.18637/jss.v032.i06
2. Berthelot, C., et al.: The rainbow trout genome provides novel insights into evolution after whole-genome duplication in vertebrates. Nat. Commun. **5**, 3657 (2014). https://doi.org/10.1038/ncomms4657
3. Blanc, G., Wolfe, K.H.: Widespread paleopolyploidy in model plant species inferred from age distributions of duplicate genes. Plant Cell **16**(7), 1667–1678 (2004). https://doi.org/10.1105/tpc.021345, epub 18 June 2004
4. Christensen, K.A., et al.: Chinook salmon (Oncorhynchus tshawytscha) genome and transcriptome. PLoS ONE **13**(4), e0195461 (2018). https://doi.org/10.1371/journal.pone.0195461
5. Comparative Genomics (CoGe): Coge: Comparative genomics platform. https://genomevolution.org/coge/. Accessed 12 Oct 2025
6. Glasauer, S.M., Neuhauss, S.C.: Whole-genome duplication in teleost fishes and its evolutionary consequences. Mol. Genet. Genom. (MGG) **289**(6), 1045–1060 (2014). https://doi.org/10.1007/s00438-014-0889-2
7. Haas, B.J., Delcher, A.L., Wortman, J.R., Salzberg, S.L.: Dagchainer: a tool for mining segmental genome duplications and synteny. Bioinformatics **20**(18), 3643–3646 (2004). https://doi.org/10.1093/bioinformatics/bth397
8. Haug-Baltzell, A., Stephens, S.A., Davey, S., Scheidegger, C.E., Lyons, E.: Synmap2 and synmap3d: web-based whole-genome synteny browsers. Bioinformatics **33**(14), 2197–2198 (2017). https://doi.org/10.1093/bioinformatics/btx144
9. I., A., J., A.: Insights from chum salmon genome assembly reveal patterns of gene retention following whole-genome duplication. G3: Genes Genom. Genet. **13**(8), jkad127 (2023). https://doi.org/10.1093/g3journal/jkad127
10. Jaillon, O., et al.: Genome duplication in the teleost fish tetraodon nigroviridis reveals the early vertebrate proto-karyotype. Nature **431**(7011), 946–957 (2004). https://doi.org/10.1038/nature03025
11. Dimos, B., Phelps, M.: Developing a homology guide for pacific salmon to address challenges of multiple whole-genome duplications. Ecol. Evol. **13**(6), e9994 (2023). https://doi.org/10.1002/ece3.9994
12. Krabbenhoft, T.J., et al.: Chromosome-level genome assembly of Chinese sucker (*myxocyprinus asiaticus*) reveals strongly conserved synteny following a catostomid-specific whole-genome duplication. Genom. Biol. Evol. **13**(9), evab190 (2021). https://doi.org/10.1093/gbe/evab190
13. Lee, S.Y., Kim, Y.K.: Dataset for characterization of thrombospondin family in chum salmon (*Oncorhynchus keta*). Data Brief **22**, 866–870 (2019). https://doi.org/10.1016/j.dib.2019.01.008
14. Lien, S., et al.: The Atlantic salmon genome provides insights into rediploidization. Nature **533**(7602), 200–205 (2016). https://doi.org/10.1038/nature17164, epub 18 Apr 2016
15. Lien, S., Koop, B.F., Sandve, S.R., et al.: Genome-wide reconstruction of rediploidization following the salmonid whole-genome duplication. Genom. Biol. Evol. **13**, evab190 (2021). https://doi.org/10.1093/gbe/evab190
16. Lyons, E., et al.: Finding and comparing syntenic regions among arabidopsis and the outgroups papaya, poplar, and grape: coge with rosids. Plant Physiol. **148**(4), 1772–1781 (2008)

17. Macqueen, D.J., Johnston, I.A.: A well-constrained estimate for the timing of the salmonid whole genome duplication reveals major decoupling from species diversification. Proc. Roy. Soc. B: Biol. Sci. **281**, 20132881 (2014)
18. Near, T.J., et al.: Resolution of ray-finned fish phylogeny and timing of diversification. Proc. Natl. Acad. Sci. U. S. A. (PNAS) **109**(34), 13698–13703 (2012)
19. Robertson, F.M., Gundappa, M.K., Grammes, F., et al.: Lineage-specific rediploidization is a mechanism to explain time-lags between genome duplication and evolutionary diversification. Genom. Biol. **18**, 111 (2017). https://doi.org/10.1186/s13059-017-1241-z
20. Sankoff, D., Zheng, C., Zhang, Y., Meidanis, J., Lyons, E., Tang, H.: Models for similarity distributions of syntenic homologs and applications to phylogenomics. IEEE/ACM Trans. Comput. Biol. Bioinf. **16**(3), 727–737 (2019)
21. Smith, S.R., et al.: A chromosome-anchored genome assembly for lake trout (Salvelinus namaycush). Mol. Ecol. Resour. **22**(2), 679–694 (2022). https://doi.org/10.1111/1755-0998.13483, epub 14 Aug 2021
22. Yu, Z., Sankoff, D.: Syntenic dimensions of genomic evolution. In: Jin, L., Durand, D. (eds.) RECOMB-Comparative Genomics (RECOMB-CG) 2022. LNB, vol. 13234, pp. 21–30. Springer, Cham (2022). https://doi.org/10.1007/978-3-031-06220-9_2
23. Zhang, Y., Yu, Z., Zheng, C., Sankoff, D.: Integrated synteny- and similarity-based inference on the polyploidization-fractionation cycle. Interface Focus (2021)

Evolutionary Reconstruction of Hormone-bHLH Regulatory Networks in Solanaceae: Phylogenomics Insights from PSTVd-Tomato Interactions

Katia Aviña-Padilla[1], Octavio Zambada-Moreno[1], Manuel A. Barrios-Izás[2], Michelle Bustamante-Castillo[2], and Maribel Hernández-Rosales[1](✉)

[1] CINVESTAV-Irapuato, Lib.Norte Carretera Irapuato León Kilómetro 9.6, 36821 Irapuato, Guanajuato, México
maribel.hr@cinvestav.mx
[2] Universidad de San Carlos de Guatemala-Centro Universitario de Zacapa, entrada a Pueblo Modelo, 19001 Zacapa, Guatemala

Abstract. The evolutionary reconstruction of transcriptional regulatory networks plays a crucial role in plant adaptation to biotic stress. In Solanaceae, basic helix-loop-helix (bHLH) transcription factors (TFs) regulate hormone signaling pathways that balance growth and defense responses. In this study, we investigated the phylogenomic evolution of bHLH-mediated hormone networks in response to Potato Spindle Tuber Viroid (PSTVd) infection in tomato (*Solanum lycopersicum*) and its wild relatives. Among them, the jasmonic acid (JA)-associated bHLH *Solyc08g076930*, IA3/MYC2 exhibits strong evolutionary conservation across Solanaceae, whereas the auxin regulator *Solyc03g113560* miP-bHLH-ARF8 is a tomato-specific gene raising from an ancestral duplication, indicating neofunctionalization. Under PSTVd infection, we observe extensive regulatory rewiring, with bHLH hubs assuming dominant control over target genes typically regulated by multiple TFs in healthy conditions. The JA regulon undergoes the most pronounced shift, with *Solyc08g076930* directly regulating 91 genes, compared to shared regulation with over a thousand TFs in healthy plants. This rewiring reflects the adaptive plasticity of bHLH TFs in coordinating stress responses through hormonal crosstalk. These findings reveal key evolutionary pressures shaping Solanaceae stress responses, offering insights for improving viroid tolerance and crop resilience.

Keywords: Comparative genomics · Domestication · Solanaceae · Viroids · Regulatory networks · evolution · Genome-wide analysis

1 Introduction

Transcriptional regulatory networks play a fundamental role in plant adaptation to environmental stressors. Within these networks, the basic helix-loop-helix (bHLH) transcription factor family is one of the largest and most conserved, mediating a wide array of biological processes, including hormone signaling, development, and stress responses [1, 2]. The ability of bHLH TFs to bind specific DNA motifs, particularly E-box sequences, enables precise transcriptional control, which is essential for growth-defense trade-offs in plants. Recent studies have emphasized the importance of bHLH TFs in orchestrating responses to abiotic and biotic stress, such as drought, temperature fluctuations, and pathogen infections [3, 4].

Understanding how plants regulate their response to biotic stress is crucial for developing strategies to enhance disease resistance. In Solanaceae, hormone pathways regulated by bHLH TFs play a pivotal role in stress adaptation. Jasmonic acid (JA), auxin, and brassinosteroids are key hormones modulating plant defense responses, with bHLH TFs acting as central regulators in these pathways [5, 6]. PSTVd infection disrupts these hormonal networks, causing transcriptional reprogramming that can either enhance or suppress plant defense mechanisms [7]. Previous studies suggest that wild tomato species, such as *S. pennellii*, exhibit greater tolerance to PSTVd due to the retention of specific regulatory mechanisms that have been lost in domesticated varieties (*S. lycopersicum*) [8]. This evolutionary divergence raises the question of how bHLH-mediated transcriptional networks contribute to viroid resistance in wild species and whether these traits can be reintroduced into commercial crops.

The evolutionary history of bHLH TFs in plants suggests that gene duplication, neofunctionalization, and regulatory rewiring have played significant roles in shaping their functions [9]. Comparative genomic approaches provide insights into how bHLH TFs evolved to modulate plant stress responses. Studies in Arabidopsis and rice have demonstrated that bHLH genes underwent lineage-specific expansions associated with environmental adaptation [6]. However, little is known about the evolutionary dynamics of bHLH TFs in Solanaceae under pathogen-induced stress. By integrating phylogenomic and transcriptomic analyses, this study aims to elucidate the regulatory mechanisms governing bHLH-mediated hormone signaling in response to PSTVd infection.

This research seeks to bridge the gap in understanding how bHLH transcription factors have evolved to coordinate stress adaptation in Solanaceae. By identifying conserved and lineage-specific regulatory interactions, this study provides a foundation for future breeding strategies aimed at enhancing viroid tolerance in tomato and related crops.

2 Materials and Methods

A total of 161 bHLH transcription factors were identified in domesticated tomato (*S. lycopersicum*) and 172 in wild species (*S. pennellii*) using the PlantTFDB database (https://planttfdb.gao-lab.org). Seven Solanaceae species, including three wild and four domesticated varieties, were selected based on their viroid tolerance. Phylogenomic analyses were conducted using `REvolutionH-tl` to infer evolutionary trajectories of bHLH transcription factors across Solanaceae. Orthogroups, orthologs and paralogs, gene trees, species tree and reconciled trees with speciation, duplication and gene loss events were identified using the `REvolutionH-tl` framework [10]. The presence-absence patterns of bHLH genes were examined to determine lineage-specific expansions or contractions. To assess transcriptional regulatory shifts, microarray expression data from PSTVd-infected and healthy tomato plants were retrieved from public repositories. Gene regulatory networks (GRNs) were constructed using the `Corto` framework, identifying master transcriptional regulators (MTRs) and co-expression modules. Network rewiring analyses compared auxin, brassinosteroid, and jasmonic acid regulons between domesticated and wild tomato species, quantifying changes in connectivity and transcriptional influence of bHLH TFs.

3 Results

The regulatory network analysis of PSTVd-infected tomato revealed three key hormone-associated bHLH transcription factors. *Solyc08g076930* (IA3/MYC2) emerged as a highly conserved jasmonic acid regulator, playing a pivotal role in stress adaptation. *Solyc03g113560* (miP-bHLH-ARF8) was identified as a tomato-specific auxin modulator, likely undergoing neofunctionalization. Meanwhile, *Solyc05g007210* (miP-bHLH-BR) was conserved across *Solanum* but absent in *Capsicum*, indicating lineage-specific regulatory divergence. Figure 1 illustrates the complex transcriptional network associated with these bHLH TFs and their role in modulating hormone signaling during PSTVd infection.

Comparative network analysis revealed that the brassinosteroid regulatory network is conserved in *Solanum* but not in *Capsicum*, while the auxin regulatory hub is exclusive to domesticated tomato. The jasmonic acid regulatory network showed progressive divergence from *Capsicum* to domesticated tomato, demonstrating the impact of evolutionary pressures on transcriptional control. Figure 2 provides a visualization of the orthologous distribution of bHLH regulons across different Solanaceae species, highlighting the degree of conservation and divergence among the analyzed species. These findings indicate that bHLH TFs have undergone significant lineage-specific diversification to mediate stress responses.

A striking observation from the study is the extensive rewiring of bHLH-mediated regulatory networks under viroid infection. Figure 3 demonstrates how these networks shift from highly modular structures in healthy plants to a more centralized and adaptive regulatory configuration under stress. The jasmonic acid-associated bHLH (IA3/MYC2) shifted from a co-regulatory architecture, where its regulon was controlled by multiple transcription factors in healthy conditions, to a dominant regulatory role, directly governing 91 genes under infection. Similarly, the auxin and brassinosteroid regulons underwent extensive rewiring, with bHLH regulators assuming control over genes typically regulated by other transcription factor families in healthy plants [11].

4 Discussion

This study highlights the evolutionary conservation and regulatory divergence of bHLH transcription factors in Solanaceae, demonstrating the impact of domestication and pathogen-induced stress responses. The strong conservation of jasmonic acid-associated regulatory networks suggests their critical role in stress resilience, whereas the emergence of species-specific auxin and brassinosteroid regulators underscores the adaptive flexibility of these transcription factors. The evolutionary trajectory of bHLH regulatory networks illustrates how selective pressures have shaped stress adaptation in wild and domesticated tomato species [12]. The observed regulatory rewiring in response to

Evolutionary Reconstruction of Hormone-bHLH Regulatory Networks in Solanaceae 287

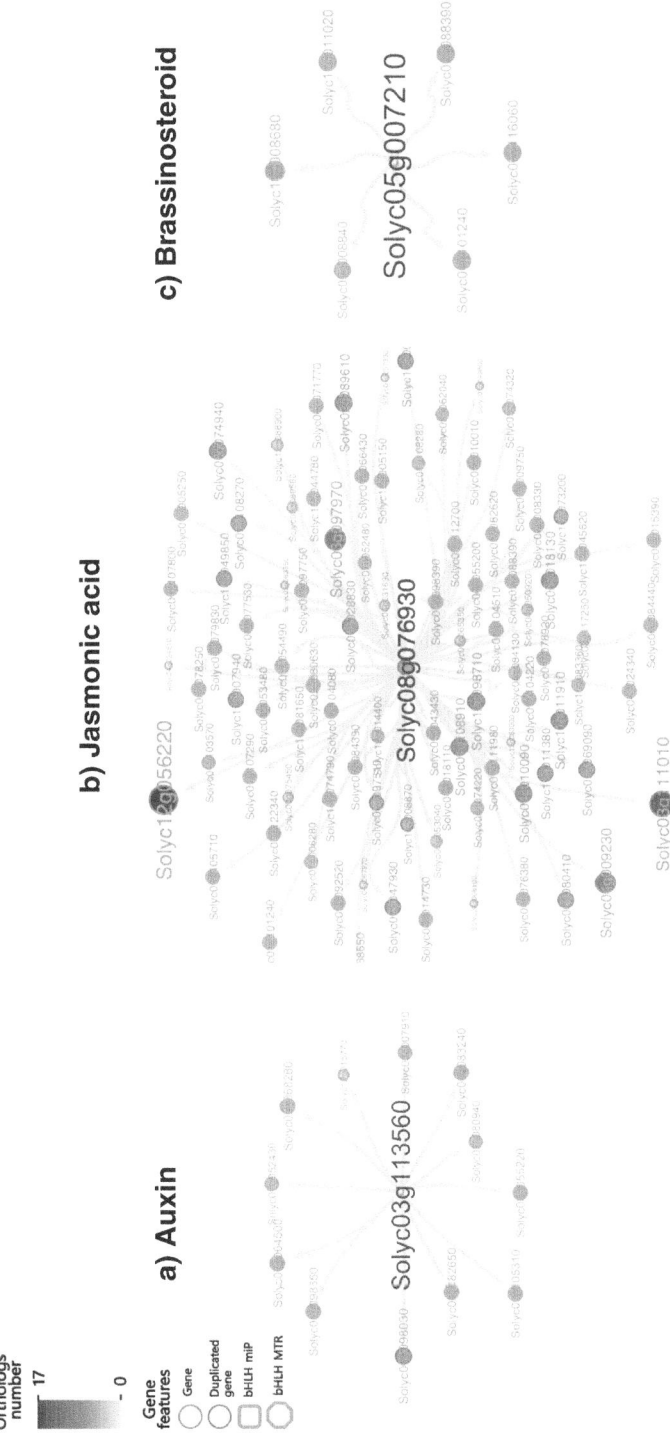

Fig. 1. PSTVd Regulatory Network and Orthologs in Tomato. The networks depict the regulatory modules associated with (a) auxin (*Solyc03g113560*), (b) jasmonic acid (*Solyc08g076930*), and (c) brassinosteroid (*Solyc05g007210*) hormone signaling pathways. Nodes represent genes, with duplicated genes outlined in pink. Gray squares indicate bHLH miPs, while hexagons denote bHLH master transcriptional regulators (MTR). The color gradient represents the number of orthologs identified across Solanaceae species, ranging from 0 (yellow) to 17 (purple). This analysis highlights key regulatory hubs modulating transcriptional responses under PSTVd infection. (Color figure online)

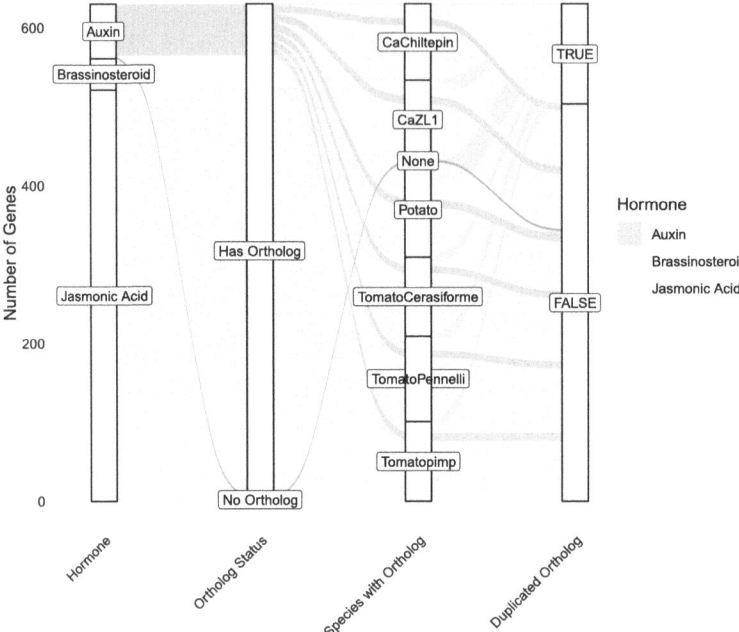

Fig. 2. Sankey diagram showing ortholog distribution for genes involved in bHLH-ARF8 (pink), bHLH-brassinosteroid (blue), and IA3/MYC2 (green) regulons across Solanaceae species. The plot displays the number of genes by hormone pathway (left), their ortholog status (middle-left), species with identified orthologs (middleright), and duplication status (right, TRUE if the gene has more than one co-ortholog in a species, FALSE if there are none). Genes related to JA signaling are the most prevalent and widely conserved, while auxin and brassinosteroid genes show more lineage-specific patterns, especially in Solanum species. (Color figure online)

PSTVd infection further emphasizes the functional plasticity of bHLH TFs, supporting their role in rapid transcriptional reprogramming under biotic stress conditions.

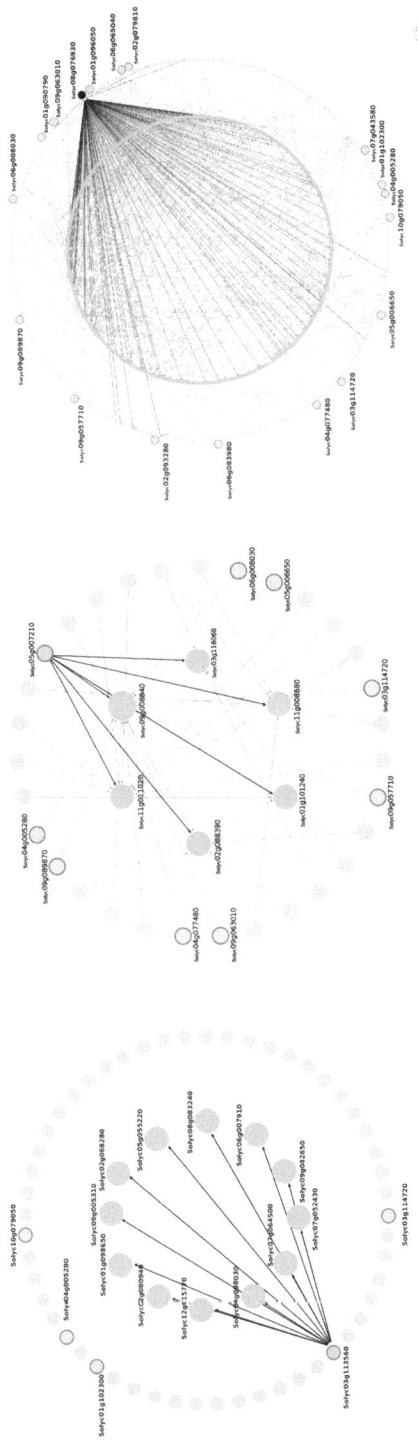

Fig. 3. Rewiring of regulatory networks comparing viroid disease and healthy stages. The inferred regulatory networks show extensive rewiring under PSTVd infection. The auxin-related bHLH regulon (left) gains regulatory control over genes usually governed by other TFs. The auxin network (middle) features a bHLH hub targeting genes normally regulated by multiple TFs. The jasmonic acid network (right) is dominated by a bHLH that regulates 91 genes, replacing over a thousand TFs under normal conditions.

References

1. Heim, M.A., Jakoby, M., Werber, M., Martin, C., Weisshaar, B., Bailey, P.C.: The basic helix-loop-helix transcription factor family in plants: a genome-wide study of protein structure and functional diversity. Plant Mol. Biol. **52**(4), 653–669 (2003)
2. Carretero-Paulet, L., Galstyan, A., Roig-Villanova, I., Martínez-García, J.F., Bilbao-Castro, J.R., Robertson, D.L.: Genome-wide classification and evolutionary analysis of the bHLH family. Plant Physiol. **153**(3), 1398–1412 (2010)
3. Pham, V.N., Kathare, P.K., Huq, E.: Phytochromes and phytochrome interacting factors. Plant Physiol. **176**(2), 1025–1038 (2018)
4. Meraj, T.A., et al.: Transcriptional factors regulate plant stress responses through mediated signaling. Genes **11**(4), 346 (2020)
5. Chini, A., Fonseca, S., Chico, J.M., Fernández-Calvo, P., Solano, R.: The JAZ family of repressors is the missing link in jasmonate signaling. Nature **448**(7154), 666–671 (2007)
6. Pires, N.D., Dolan, L.: Origin and diversification of basic-helix-loop-helix proteins in plants. Mol. Biol. Evol. **27**(4), 862–874 (2010)
7. Owens, R.A., Tech, K.B., Shao, J.Y., Sano, T., Baker, C.J.: Global analysis of tomato gene expression during potato spindle tuber viroid infection reveals a complex array of changes affecting hormone signaling. Mol. Plant Microbe Interact. **25**(4), 582–598 (2012)
8. Bergougnoux, V.: The history of tomato: from domestication to biopharming. Biotechnol. Adv. **32**(1), 170–189 (2014)
9. Hudson, C., Hudson, M.E.: A classification of basic helix-loop-helix transcription factors in the genomes of Arabidopsis thaliana and Oryza sativa. BMC Genomics **12**, 83 (2011)
10. Ramírez-Rafael, J.A., et al.: REvolutionH-tl: reconstruction of evolutionary histories tool. In: Scornavacca, C., Hernández-Rosales, M. (eds.) RECOMB-CG 2024. LNCS, vol, 14616, pp. 89–109. Springer, Cham (2024). https://doi.org/10.1007/978-3-031-58072-7_5

11. Aviña-Padilla, K., Zambada-Moreno, O., Jimenez-Limas, M.A., Hammond, R.W., Hernández-Rosales, M.: Dynamic co-expression modular network analysis of bHLH transcription factors regulation in the potato spindle tuber viroid-tomato pathosystem (2023). bioRxiv. https://doi.org/10.1101/2023.11.10.566618
12. Deneweth, J., Peer, Y., Vermeirssen, V.: Nearby transposable elements impact plant stress gene regulatory networks: a meta-analysis in A. Thaliana and S. lycopersicum. BMC Genomics **23**(1) (2022)

Author Index

A
Ali, Sarwan 202
Aviña-Padilla, Katia 284

B
Barrios-Izás, Manuel A. 284
Bayzid, Md. Shamsuzzoha 131, 150
Bustamante-Castillo, Michelle 284

C
Chan, Michael 231
Chen, Eric CH 231
Chen, Pin-Yu 202
Chen, Siyu 219
Chourasia, Prakash 202

F
Frith, Martin C. 197, 247

G
Gao, Meijun 181
Gupta, Ishaan 251
Gupta, Vanika 251

H
Hasan, Navid Bin 150
Hernández-Rosales, Maribel 284
Hu, Songdi 27
Huang, Muyao 247

J
Jassem, Agatha 231
Jones, Mark 107

K
Kalhor, Reza 51
Khan, Imdad Ullah 202
Kille, Bryce 87

L
Lafond, Manuel 51, 69
Lam, Anthea 231
Landry, Kaari 157
Liu, Kevin J. 181
Liu, Yushu 9
López Sánchez, Alitzel 69

M
Mansoor, Haris 202
Momin, Rabib Jahin Ibn 131
Murad, Taslim 202

N
Nakhleh, Luay 9, 87
Newman, Tara 231

O
Ogilvie, Huw A. 87

P
Patterson, Murray 202
Prystajecky, Natalie 231

R
Reinharz, Vladimir 27
Russell, Shannon 231

S
Sankoff, David 141, 219
Schestag, Jannik 107
Scholz, Guillaume E. 69
Scornavacca, Celine 51, 107
Shami-Schnitzer, Ofek 3
Shelke, Triveni 251
Sohaib, 150
Stadler, Peter F. 69

T
Treangen, Todd J. 87
Tremblay-Savard, Olivier 27, 157
Tuller, Tamir 3
Tyson, John 231

V
van Iersel, Leo 107

W
Weller, Mathias 107

Y
Yin, Yongze 87
Yu, Hao 137

Z
Zaman, Tanjeem Azwad 131
Zambada-Moreno, Octavio 284
Zhang, Louxin 137
Zhang, Yue 268
Zhou, Xintong 141
Zlosnik, James 231

The manufacturer's authorised representative in the EU is Springer Nature Customer Service Centre GmbH, Europaplatz 3, 69115 Heidelberg, Germany. If you have any concerns regarding our products, please contact ProductSafety@springernature.com

Printed and bound by CPI Group (UK) Ltd, Croydon, CR0 4YY

26/03/2026

02078994-0001